Probability, Random Variables, Statistics, and Random Processes

Probability, Random Variables, Statistics, and Random Processes

Fundamentals & Applications

Ali Grami

Registered Office
John Wiley & Sons, Inc., 111 River Street, Hoboken, NJ 07030, USA

Editorial Office
111 River Street, Hoboken, NJ 07030, USA

For details of our global editorial offices, customer services, and more information about Wiley products visit us at www.wiley.com.

Wiley also publishes its books in a variety of electronic formats and by print-on-demand. Some content that appears in standard print versions of this book may not be available in other formats.

Library of Congress Cataloging-in-Publication data applied for

Hardback: 9781119300816

Cover image: © Markus Gann/Shutterstock
Cover design: Wiley

Set in 10/12pt WarnockPro by SPi Global, Chennai, India

Printed in the United States of America

V10008519_030419

To my beloved mother,
the exceptional symbol of hard work, perseverance, and confidence
and
In loving memory of my father,
the epitome of integrity and generosity

Contents

Preface

I enjoy writing a book, but more importantly, I enjoy having written a book, as my hope is to make a difference by helping students study, learn, and apply subject matter. The purpose of this textbook is to present a modern introduction to fundamentals and applications of probability, statistics, and random processes for students in various disciplines of engineering and science.

The importance of the material discussed in this book lies in the fact that randomness and uncertainty, which always go hand in hand, exist in virtually every aspect of life, and their impacts are expanding exponentially. Think of the thousands of hourly weather forecasts, the millions of insurance policies issued each day, the billions of stock exchange transactions per week, and the trillions of wireless phone calls, text messages, e-mails, and web searches each month, they are all based on probability models and statistical analyses.

This book provides a simple and intuitive approach, while maintaining a reasonable level of mathematical rigor and subtleties. The style of writing is less formal than most textbooks, so students can build more interest in reading the book and having it as a self-study reference. Most of the text does not require any previous knowledge apart from first year calculus and linear algebra, the exception to this is the analysis and processing of random processes.

I firmly believe that before statistics and random processes can be understood, students must have a good knowledge of probability and random variables. As over time, discrete random variables and discrete-time random processes have rapidly grown in importance, the mathematics of discrete random variables is introduced separately from the mathematics of continuous random variables. An attempt has also been made to have chapters that are rather independent, so a couple of topics are visited more than once. This allows the instructor to follow any sequence of chapters.

The pedagogy behind this book and its choice of contents evolved over many years. Most of the material in the book has been class-tested and proven very effective. There are about 200 examples to help understand the fundamental concepts and approximately 250 problems at the end of chapters to test understanding of the material. Moreover, there are also dozens of practical applications, along with the introduction of dozens of widely known, frequently encountered probability distributions.

This textbook contains 12 chapters. As the title of the book suggests, it consists of four parts of Probability (Chapters 1–3), Random Variables (Chapters 4–7), Statistics (Chapters 8–10), and Random Processes (Chapters 11 and 12). There is intentionally more in this book than can be taught in one semester (of roughly 36 lecture hours) so as to provide the instructor flexibility in the selection of topics. The coverage of the book, in terms of depth and breadth as well as tone and scope, supports a variety of syllabi for various courses. As a third-year course, the coverage may include Chapters 1, 2 (select sections), 4, 5 (select sections), 6, 7, and 8. As a fourth-year course, the coverage may include Chapters 1–8 inclusive as well as Chapter 11. As a graduate course taken by first year master's students, the entire book should be covered, beginning first with a complete review of Chapters 1–8, followed by an in-depth coverage of Chapters 9–12.

Upon request from the publisher, a Solutions Manual can be obtained only by instructors who use the book for adoption in a course. No book is flawless, and this book is certainly no different from others. Your comments and suggestions for improvements are always welcomed. I would greatly appreciate it if you could send your feedback to ali.grami@uoit.ca.

January 2019 *Ali Grami*

Acknowledgments

Acknowledgements are viewed by the reader as a total bore, but regarded by the author as a true source of incredible joy.

An expression of great respect and gratitude is due to all who have ever taught me. To my teachers at Shahrara Elementary School and Hadaf High School, as well as to my professors at the University of Manitoba, McGill University, and the University of Toronto, I owe an unrepayable debt of gratitude.

Writing a textbook is in some sense collaborative, as one is bound to lean on the bits and pieces of materials developed by others. I would therefore like to greatly thank the many authors whose writings helped me. I am heavily indebted to many individuals for their contributions to the development of this text, including Dr. I. Dincer, Dr. X. Fernando, Dr. Y. Jing, Dr. M. Nassehi, and Dr. H. Shahnasser, as well as the anonymous reviewers who provided helpful comments. I would also like to thank Nickrooz Grami who provided several examples and reviewed the book to improve coherence and clarity as well as Neeloufar Grami who prepared multiple figures and tables. My deep gratitude goes to Ahmad Manzar who carefully reviewed the entire manuscript, provided valuable comments, and helped improve many aspects of the book.

The financial support of Natural Sciences and Engineering Research Council (NSERC) of Canada was also crucial to this project. I am grateful to the staff of John Wiley & Sons for their support throughout various phases of this project, namely Tiina Wigley, Jon Gurstelle, Viniprammia Premkumar and all members of the production team.

Last, but certainly most important of all, no words can ever express my most heartfelt appreciation to my wife, Shirin, as well as my children, Nickrooz and Neeloufar, for their immeasurable love and support, without which this book would not have been possible.

About the Companion Website

This book is accompanied by a companion website:
www.wiley.com/go/grami/PRVSRP

The website includes: Solutions Manual and PowerPoint Slides with all Figures and Tables

Part I
Probability

1

Basic Concepts of Probability Theory

Randomness and uncertainty, which always go hand in hand, exist in virtually every aspect of life. To this effect, almost everyone has a basic understanding of the term *probability* through intuition or experience. The study of probability stems from the analysis of certain games of chance. Probability is the measure of chance that an event will occur, and as such finds applications in disciplines that involve uncertainty. Probability theory is extensively used in a host of areas in science, engineering, medicine, and business, to name just a few. As claimed by Pierre-Simon Laplace, a prominent French scholar, probability theory is nothing but common sense reduced to calculation. The basic concepts of probability theory are discussed in this chapter.

1.1 Statistical Regularity and Relative Frequency

An *experiment* is a measurement procedure or observation process. The *outcome* is the end result of an experiment, where if one outcome occurs, then no other outcome can occur at the same time. An *event* is a single outcome or a collection of outcomes of an experiment.

If the outcome of an experiment is certain, that is the outcome is always the same, it is then a *deterministic experiment*. In other words, a deterministic experiment always produces the same output from a given starting condition or initial state. The measurement of the temperature in a certain location at a given time using a thermometer is an example of a deterministic experiment.

In a *random experiment*, the outcome may unpredictably vary when the experiment is repeated, as the conditions under which it is performed cannot be predetermined with sufficient accuracy. In studying probability, we are concerned with experiments, real or conceptual, and their random outcomes. In a random experiment, the outcome is not uniquely determined by the causes and cannot be known in advance, because it is subject to chance. In a lottery game, as an example, the random experiment is the drawing, and the outcomes

Probability, Random Variables, Statistics, and Random Processes: Fundamentals & Applications, First Edition. Ali Grami.

are the lottery number sequences. In such a game, the outcomes are uncertain, not because of the inaccuracy in how the experiment may be performed, but because how it has been designed to produce uncertain results. As another example, in a dice game, such as craps or backgammon, the random experiment is rolling a pair of dice and the outcomes are positive integers in the range of one to six inclusive. In such a game, the outcomes are uncertain, because it has been designed to produce uncertain results as well as the fact that there exists some inaccuracy and inconsistency in how the experiment may be carried out, i.e. how a pair of dice may be rolled.

In random experiments, the making of each measurement or observation, i.e. each repetition of the experiment, is called a ***trial***. In ***independent trials***, the observable conditions are the same and the outcome of one trial has no bearing on the other. In other words, the outcome of a trial is independent of the outcomes of the preceding and subsequent trials. For instance, it is reasonable to assume that in coin tossing and dice rolling repeated trials are independent. The conditions under which a random experiment is performed influence the probabilities of the outcomes of an experiment. To account for uncertainties in a random experiment, a probabilistic model is required. A ***probability model***, as a simplified approximation to an actual random experiment, details enough to include all major aspects of the random phenomenon. Probability models are generally based on the fact that averages obtained in long sequences of independent trials of random experiments almost always give rise to the same value. This property, known as ***statistical regularity***, is an experimentally verifiable phenomenon in many cases.

The ratio that represents the number of times a particular event occurs over the number of times the trial has been repeated is defined as the ***relative frequency*** of the event. When the number of times the experiment being repeated approaches infinity, the relative frequency of the event, which approaches a limit because of statistical regularity, is called the ***relative-frequency definition of probability***. Note that this limit, which is based on an a posteriori approach, cannot truly exist, as the number of times a physical experiment can be repeated may be very large, but always finite.

Figure 1.1 shows a sequence of trials in a coin-tossing experiment, where the coin is fair (unbiased) and N_h represents the number of heads in a sequence of N independent trials. The relative frequency fluctuates widely when the number of independent trials is small, but eventually settles down near $\frac{1}{2}$ when the number of independent trials is increased. If an ideal (fair) coin is tossed infinitely many times, the probability of heads is then $\frac{1}{2}$.

If outcomes are equally likely, then the probability of an event is equal to the number of outcomes that the event can have divided by the total number of possible outcomes in a random experiment. This probability, known as the

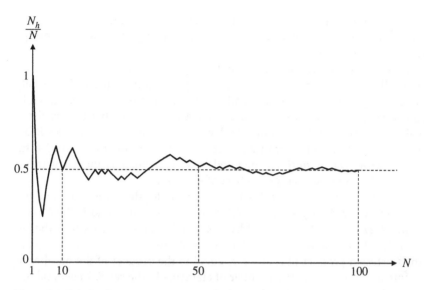

Figure 1.1 Relative frequency in a tossing of a fair coin.

classical definition of probability, is determined a priori without actually carrying out the random experiment. For instance, if an ideal (fair) die is rolled, the probability of getting a 4 is then $\frac{1}{6}$.

1.2   Set Theory and Its Applications to Probability

In probability models associated with random experiments, simple events can be combined using set operations to obtain complicated events. We thus briefly review the set theory as it is the mathematical basis for probability.

A *set* is a collection of objects or things, which are called *elements* or *members*. As shorthand, a set is generally represented by the symbol { }. It is customary to use capital letters to denote sets and lowercase letters to refer to set elements. If x is an element of the set A, we use the notation $x \in A$, and if x does not belong to the set A, we write $x \notin A$. It is essential to have a clear definition of a set either by using the *set roster method*, that is listing all its elements between curly brackets, such as {3, 6, 9}, or by using the *set builder notation*, that is describing some property held only by all members of the set, such as $\{x \mid x$ is a positive integer less than 10 that is a multiple of 3$\}$. Noting the order of elements presented in a set is immaterial, the number of distinct elements in a set A is called the *cardinality* of A, written as $|A|$. The cardinality of a set may be finite or infinite.

We use **Venn diagrams,** as shown in Figure 1.2, to pictorially illustrate an important collection of widely known sets and their logical relationships through geometric intuition.

The **universal set** U, also represented by the symbol Ω, is defined to include all possible elements in a given setting. The universal set U is usually represented pictorially as the set of all points within a rectangle, as shown in Figure 1.2a. The set B is a **subset** of the set A if every member of B is also a member of A. We use the symbol $\subset$ to denote subset, $B \subset A$ thus implies B is a subset of A, as shown in Figure 1.2b. Every set is thus a subset of the universal set. The **empty set** or **null set**, denoted by $\emptyset$, is defined as the set with no elements. The empty set is thus a subset of every set.

The set of all subsets of a set A, which also includes the empty set $\emptyset$ and the set A itself, is called the **power set** of A. Given two sets of A and B, the **Cartesian product** of A and B, denoted by $A \times B$ and read as A cross B, is the set of all ordered pairs (a, b), where $a \in A$ and $b \in B$. The number of ordered pairs in the Cartesian product of A and B is equal to the product of the number of elements in the set A and the number of elements in the set B. In general, we have $A \times B \neq B \times A$, unless $A = B$.

The sets A and B are considered to be **equal sets** if and only if they contain the same elements, as shown in Figure 1.2c. In other words, we have

$$A = B \Leftrightarrow A \subset B \text{ and } B \subset A \tag{1.1a}$$

The **union** of two sets A and B is the set of all elements that are in A or in B or in both, as shown in Figure 1.2d. The operation union corresponds to the logical "or" operation:

$$A \cup B \triangleq \{x \mid x \in A \text{ or } x \in B\} \tag{1.1b}$$

The **intersection** of two sets A and B is the set of all elements that contain in both A and B, as shown in Figure 1.2e. The operation intersection corresponds to the logical "and" operation:

$$A \cap B \triangleq \{x \mid x \in A \text{ and } x \in B\} \tag{1.1c}$$

The union and intersection operations can be repeated for an arbitrary number of sets. Thus, the union of n sets is the set of all elements that are in at least one of the n sets and the intersection of n sets is the set of all elements that are shared by all n sets. The intersection of n sets is a subset of each of the n sets, and in turn, each of the n sets is a subset of the union of n sets.

The **complement** of a set A, with respect to the universal set U, denoted by A^c, is the set of all elements that are not in A, as shown in Figure 1.2f. The operation complement corresponds to the logical "not" operation:

$$A^c \triangleq \{x \mid x \notin A\} \tag{1.1d}$$

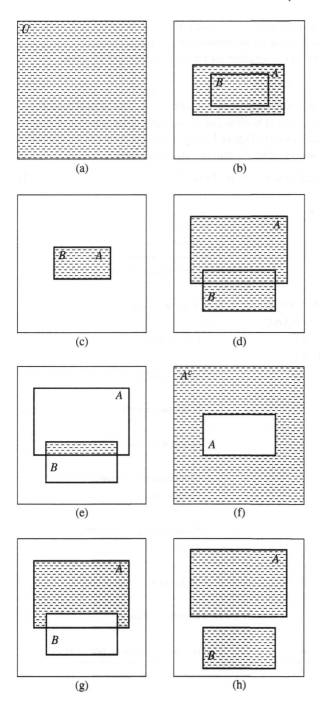

Figure 1.2 Sets and their relationships through geometric intuition: (a) universal set; (b) subset; (c) equal sets; (d) union of sets; (e) intersection of sets; (f) complement of a set; (g) difference of sets; and (h) mutually exclusive sets.

Note that the complement of the universal set is the null set, and vice versa. The *difference* of sets A and B is the set of elements in A that are not in B, as shown in Figure 1.2g. The difference operation can be represented as follows:

$$A - B \triangleq \{x \mid x \in A \text{ and } x \notin B\} \tag{1.1e}$$

Note that the set $A - B$ is different from the set $B - A$. The sets A and B are *mutually exclusive*, also known as *disjoint*, if and only if the sets A and B have no common elements, as shown in Figure 1.2h. The intersection of two mutually exclusive sets (disjoint sets) is thus an empty set, i.e. we have

$$A \text{ and } B \text{ are mutually exclusive} \Leftrightarrow A \cap B \triangleq \emptyset \tag{1.1f}$$

Table 1.1 Set identities.

Identity	Name
$A \cup B = B \cup A$	Commutative laws
$A \cap B = B \cap A$	
$(A \cup B) \cup C = A \cup (B \cup C) = A \cup B \cup C$	Associative laws
$(A \cap B) \cap C = A \cap (B \cap C) = A \cap B \cap C$	
$A \cup (B \cap C) = (A \cup B) \cap (A \cup C)$	Distributive laws
$A \cap (B \cup C) = (A \cap B) \cup (A \cap C)$	
$A \cup \emptyset = A$	Identity laws
$A \cap U = A$	
$A \cup U = U$	Domination laws
$A \cap \emptyset = \emptyset$	
$A \cup A = A$	Idempotent law
$A \cap A = A$	
$(A^c)^c = A$	Complementation law
$A \cup A^c = U$	Complement laws
$A \cap A^c = \emptyset$	
$A - B = A \cap B^c$	Relative complement law
$A \cup (A \cap B) = A$	Absorption laws
$A \cap (A \cup B) = A$	
$(A \cup B)^c = A^c \cap B^c$	De Morgan's laws
$(A \cap B)^c = A^c \cup B^c$	
$A \subseteq B \text{ iff } A \cup B = B$	Consistency laws
$A \subseteq B \text{ iff } A \cap B = A$	
$|A \cup B| = |A| + |B| - |A \cap B|$	Inclusion–exclusion principle

It is important to note that a set A is a collection of n mutually exclusive subsets of A, i.e. $A_1, A_2, \ldots, A_n$, whose union equals A, where n is a positive integer. The cardinality of a union of two finite sets A and B can be found using the **principle of inclusion–exclusion**, i.e. we have

$$|A \cup B| \triangleq |A| + |B| - |A \cap B| \tag{1.1g}$$

Note that $|A| + |B|$ counts each element that is in set A, but not in set B once, and in set B, but not in set A once, and each element that is in both sets A and B exactly twice. The number of elements that is in both A and B, i.e. $|A \cap B|$, is subtracted from $|A| + |B|$ so as to count the elements in the intersection $A \cap B$ only once. Note that if A and B are mutually exclusive, then we have $|A \cup B| \triangleq |A| + |B|$.

The principle of inclusion–exclusion can be easily extended to more than two sets. In general, for n sets, where n is a positive integer, the principle of inclusion–exclusion has a maximum of $2^n - 1$ terms. However, some of these terms may be zero, as it is possible that some of the n sets are mutually exclusive.

Some basic set operations can be combined to form other sets, as shown in Table 1.1. Due to the **principle of duality** in set theory, any identity involving sets is also true if we replace unions by intersections, intersections by unions, and sets by their complements. This principle is used when the complements of events are easier to define and specify than the events themselves. For instance, **De Morgan's laws** as defined below use the principle of duality:

$$(A \cup B)^c = A^c \cap B^c$$
$$(A \cap B)^c = A^c \cup B^c \tag{1.1h}$$

Example 1.1

Consider the universal set U consisting of all positive integers from 1 to 10 inclusive. We define the event A consisting of all positive integers from 1 to 10 inclusive that are multiples of 2 and the event B consisting of all positive integers from 1 to 10 inclusive that are multiples of 3. Verify the following set of identities: $(A \cup B)^c = A^c \cap B^c$, $(A \cap B)^c = A^c \cup B^c$, $A - B = A \cap B^c$, and $|A \cup B| = |A| + |B| - |A \cap B|$.

Solution

Since we have

$$\begin{cases} A = \{2, 4, 6, 8, 10\} \\ B = \{3, 6, 9\} \end{cases} \rightarrow \begin{cases} A^c = \{1, 3, 5, 7, 9\} \\ B^c = \{1, 2, 4, 5, 7, 8, 10\} \end{cases}$$

$$\rightarrow \begin{cases} A^c \cap B^c = \{1, 5, 7\} \\ A^c \cup B^c = \{1, 2, 3, 4, 5, 7, 8, 9, 10\} \end{cases}$$

and

$$\begin{cases} A = \{2,4,6,8,10\} \\ B = \{3,6,9\} \end{cases} \rightarrow \begin{cases} A \cup B = \{2,3,4,6,8,9,10\} \\ A \cap B = \{6\} \end{cases}$$

$$\rightarrow \begin{cases} (A \cup B)^c = \{1,5,7\} \\ (A \cap B)^c = \{1,2,3,4,5,7,8,9,10\} \end{cases}$$

we obtain the following:

$$(A \cup B)^c = A^c \cap B^c = \{1,5,7\}$$
$$(A \cap B)^c = A^c \cup B^c = \{1,2,3,4,5,7,8,9,10\}$$
$$A - B = A \cap B^c = \{2,4,8,10\}$$

Since we have

$$|A| = 5, \quad |B| = 3, \quad |A \cup B| = 7, \quad |A \cap B| = 1$$

we obtain the following:

$$|A \cup B| = |A| + |B| - |A \cap B| = 5 + 3 - 1 = 7 \qquad \blacksquare$$

The **sample space** S of a random experiment is defined as the set of all possible outcomes of an experiment. In a random experiment, the outcomes, also known as **sample points**, are mutually exclusive, i.e. they cannot occur simultaneously. When no single outcome is any more likely than any other, we then have **equally likely outcomes**. An **event** is a subset of the sample space of an experiment, i.e. a set of one or more sample points, and the symbol { } is used as shorthand for representing an event.

Two **mutually exclusive events**, also known as **disjoint events**, have no common outcomes, i.e. the occurrence of one precludes the occurrence of the other. The **union** of two events is the set of all outcomes that are in either one or both of the two events. The **intersection** of two events, also known as the **joint event**, is the set of all outcomes that are in both events. A **certain (sure) event** consists of all outcomes, and thus always occurs. A **null (impossible) event** contains no outcomes, and thus never occurs. The **complement** of an event contains all outcomes not included in the event. A sure event and an impossible event are thus complements of one another.

Example 1.2

Consider a random experiment that constitutes rolling a fair six-sided cube-shaped die and coming to rest on a flat surface, where the face of the die that is uppermost yields the outcome. Provide specific examples to highlight the above definitions associated with the sample space and events in a die-rolling experiment.

Solution

The sample space S includes six sample points 1, 2, 3, 4, 5, and 6. Since the die is fair (unbiased), all six outcomes are equally likely. Some different events can be defined as follows: one with even outcomes (i.e. 2, 4, and 6), one with odd outcomes (i.e. 1, 3, and 5), and one whose outcomes are divisible by 3 (i.e. 3 and 6). Two mutually exclusive events may be an event with even outcomes and an event with odd outcomes. The union of two events, where one is with odd outcomes and the other is with outcomes that are divisible by 3, consists of 1, 3, 5, and 6, and their intersection consists of only 3. A certain (sure) event consists of outcomes that are integers between 1 and 6 inclusive. A null (impossible) event consists of outcomes that are less than 1 or greater than 6. The complement of the event whose outcomes are divisible by 3 is an event that contains 1, 2, 4, and 5. ∎

When the sample space S is countable, it is known as a ***discrete sample space***. In a discrete sample space, the probability law for a random experiment can be specified by giving the probabilities of all possible outcomes. With a finite nonempty sample space of equally likely outcomes, the probability of an event that is a subset of the sample space is the ratio of the number of outcomes in the event to the number of outcomes in the sample space.

Example 1.3

Suppose a pair of fair (ideal) dice is rolled. Determine the probability when the sum of the outcomes is three.

Solution

For mere clarity, it is assumed that one of the two dice is red and the other is green. There are six possible outcomes for the red die and six possible outcomes for the green die. The outcome of one die is independent of the outcome of the other die. In other words, for a specific outcome for the red die, there can be six different outcomes for the green die, and for a specific outcome for the green die, there can be 6 different outcomes for the red die. We therefore have 36 (=6 × 6) possible outcomes in rolling a pair of dice, and as the dice are fair, we have a total of 36 equally likely outcomes. It is thus a discrete sample space. To have a sum of three, the outcomes must then be a 1 and a 2. But to get a sum of 3, there are two different possible scenarios that include the red die is a 1 and the green die is a 2 or the red die is a 2 and the green die is a 1. Hence, the probability that a sum of three comes up is $\frac{2}{36} = \frac{1}{18}$. ∎

When the sample space S is uncountably infinite, it is known as a ***continuous sample space***. In a continuous sample space, the probability law for a random experiment specifies a rule for assigning numbers to intervals of the real line. In the case of continuous sample space, an outcome that occurs only once in an infinite number of trials has a zero relative frequency that is the probability that the outcome takes on a specific value is zero. However, this does not imply that it cannot occur, but rather that its occurrence is extremely infrequent, hence zero relative frequency.

Example 1.4
In a game of darts, the dartboard has a radius of 22.5 cm. Consider the sample space is the entire dartboard, it is thus a continuous sample space. Assuming a dart is thrown at random and lands on the board, determine the probability that it lands in a region within 7.5 cm from the center of the board, and the probability that the dart lands exactly at the center of the board.

Solution
The total area of the dartboard is $(22.5)^2 \pi$ cm^2. The event of interest, which represents the area of a region within 7.5 cm from the center, is $(7.5)^2 \pi$ cm^2. Hence, the probability of interest is $\frac{(7.5)^2 \pi}{(22.5)^2 \pi} = \left(\frac{7.5}{22.5}\right)^2 = \left(\frac{1}{3}\right)^2 = \frac{1}{9}$. Although it is possible that the dart lands exactly at the center of the board, the probability of its occurrence is zero, as the center of the board is one of the infinitely many points that the dart can hit. ■

1.3 The Axioms and Corollaries of Probability

Axioms are self-evidently true statements that are unproven, but accepted. Axiomatic probability theory was developed by Andrey N. Kolmogorov, a prominent Russian mathematician, who was the founder of modern probability theory and believed the theory of probability as a mathematical discipline can and should be developed from axioms in exactly the same way as geometry and algebra. In the *axiomatic definition of probability*, the probability of the event A, denoted by $P(A)$, in the sample space S of a random experiment is a real number assigned to A that satisfies the following *axioms of probability*:

Axiom I: $P(A) \geq 0$

Axiom II: $P(S) = 1$

Axiom III: If $A_1, A_2, \dots$ is a countable sequence of events such that
$A_i \cap A_j = \emptyset$ for all $i \neq j$ where $\emptyset$ is the null event,
that is they are pairwise disjoint (mutually exclusive)
events, then $P(A_1 \cup A_2 \cup \cdots) = P(A_1) + P(A_2) + \cdots$ (1.2)

Note that these results do not indicate the method of assigning probabilities to the outcomes of an experiment, they merely restrict the way it can be done. These axioms satisfy the intuitive notion of probability. *Axiom I of probability*

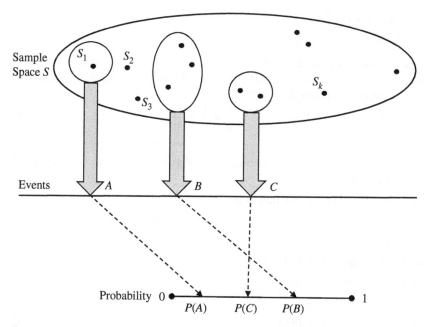

Figure 1.3 Relationship among sample space, events, and probability.

highlights that the probability of an event is nonnegative, i.e. chances are always at least zero. ***Axiom II of probability*** emphasizes that the probability of all possible outcomes is 1, i.e. the chance that something happens is always 100%. ***Axiom III of probability*** underscores that the total probability of a number of disjoint (mutually exclusive) events, i.e. nonoverlapping events, is the sum of the individual probabilities. Figure 1.3 shows the relationship among sample space, events, and probability.

Example 1.5
A box contains 20 balls, numbered from 1 to 20 inclusive. Determine the probability that the number on a ball chosen at random is either a prime number or a multiple of 4.

Solution
Let A_1 be the event that includes prime numbers, and A_2 be the event that includes numbers that are multiples of 4, that is we have

$$A_1 = \{2, 3, 5, 7, 11, 13, 17, 19\} \rightarrow P(A_1) = \frac{8}{20} = \frac{2}{5}$$

$$A_2 = \{4, 8, 12, 16, 20\} \rightarrow P(A_2) = \frac{5}{20} = \frac{1}{4}$$

We also have the following:

$$A_1 \cap A_2 = \emptyset \rightarrow P(A_1 \cap A_2) = 0$$

As the events A_1 and A_2 are mutually exclusive, we can use Axiom III to determine the probability that the number on a ball chosen at random is either a prime number or a multiple of 4. We thus have

$$P(A_1 \cup A_2) = P(A_1) + P(A_2) = \frac{2}{5} + \frac{1}{4} = \frac{13}{20} \qquad \blacksquare$$

From the three axioms of probability, the following important **corollaries of probability** can be derived:

(i) $P(\emptyset) = 0$, where $\emptyset$ is a null event

(ii) $P(A) + P(A^c) = 1$, where event A^c is the complement of event A

(iii) $P(A) \leq 1$

(iv) $P(A) = P(A \cap B) + P(A \cap B^c)$

(v) $A \subset B \rightarrow P(A) \leq P(B)$

(vi) $P(A \cup B) = P(A) + P(B) - P(A \cap B)$

(vii) $P(A \cup B) \leq P(A) + P(B)$ \hfill (1.3)

The abovementioned corollaries provide the following insights:

(i) The impossible event has probability zero, and it provides a symmetry to Axiom II.

(ii) If the sample space is partitioned into two mutually exclusive events, the sum of the probabilities of these two events is then one. Quite often, one of these two probabilities is much easier than the other to compute.

(iii) The probability of an event is always less than or equal to one, and provides a sanity check on whether the results are correct, as the probabilities can never be negative or greater than one.

(iv) The probability of an event is the sum of the probabilities of two mutually exclusive events, and this divide-and-conquer approach can often help to easily compute the probability of interest.

(v) Probability is a nondecreasing function of the number of outcomes in an event, that is with more outcomes, the probability may either increase or remain the same.

(vi) It provides the generalization of Axiom III, when the two events are not mutually exclusive.

(vii) The sum of two individual probabilities is lower bounded by the probability of the union of two events. Clearly, equality holds if and only if the events are mutually exclusive. This property can be viewed as a special case of the union bound, as the **union bound**, also known as **Boole's inequality**, states that the probability of a finite union of events is upper bounded by the sum of the probabilities of the constituent events.

Example 1.6
A sequence of 20 bits is randomly generated. Noting that the probability of a bit being 0 is equal to that being 1, determine the probability that at least one of these bits is 0.

Solution
There are many possibilities where at least 1 of the 20 bits is a zero. For instance, there may only be one 0 in the sequence, and this constitutes 20 different possibilities, such as the first one is a 0 or the second one is a 0, so on and so forth. There may be only two zeros or three zeros or four zeros in the sequence or even more, where the locations of the zeros could be anywhere in the sequence of 20 bits. In fact, there are quite many possibilities, in each of which there is at least one 0 in the sequence of 20 bits, and the sum of the probabilities of all these possibilities would then determine the probability that at least one of these bits is 0.

There is an alternative strategy for finding the probability of an event when a direct approach is long and difficult. Instead of determining the probability of an event A, the probability of its complement A^c can be first found. The complement of the event that has at least one bit being 0 is the event that all bits are 1s. There are $2^{20} = 1\,048\,576$ different sequences of 20 bits, i.e. there are $1\,048\,576$ sample points in the sample space. The probability of a sequence with all 1s is thus $\frac{1}{1\,048\,576}$. Using $P(A) = 1 - P(A^c)$, the probability that at least one of these bits is 0 is then as follows: $1 - \frac{1}{1\,048\,576} = \frac{1\,048\,575}{1\,048\,576}$. ∎

Example 1.7
A box contains 10 balls, numbered from 2 to 11 inclusive. A ball is drawn at random. Determine the probability that the drawn ball has an odd number, a prime number, both a prime number and an odd number, a prime number or an odd number or both, and neither an odd number nor a prime number.

Solution
Let A be the event with balls having odd numbers and B be the event with balls having prime numbers, we thus have the following:

$$A = \{3,5,7,9,11\} \rightarrow P(A) = \frac{5}{10} = \frac{1}{2}$$
$$B = \{2,3,5,7,11\} \rightarrow P(B) = \frac{5}{10} = \frac{1}{2}$$
$$A \cap B = \{3,5,7,11\} \rightarrow P(A \cap B) = \frac{4}{10} = \frac{2}{5}$$
$$A \cup B = \{2,3,5,7,9,11\} \rightarrow P(A \cup B) = \frac{6}{10} = \frac{3}{5}$$
$$A^c \cap B^c = (A \cup B)^c = \{4,6,8,10\} \rightarrow P(A^c \cap B^c) = \frac{4}{10} = \frac{2}{5}$$ ∎

Example 1.8
Of all bit sequences of length 8, an 8-bit sequence is selected at random. Assuming that the probability of a bit being 0 is equal to that being 1, determine the probability that the selected bit sequence starts with a 1 or ends with the two bits 00.

Solution
The sample space S includes 256 equally likely outcomes, as we can construct a bit sequence of length eight in $2^8 = 256$ ways, i.e. $|S| = 256$. The event A includes bit sequences of length 8 that all begin with a 1, we can also construct such sequences in $2^{8-1} = 2^7 = 128$ ways, i.e. $|A| = 128$. The event B includes bit sequences of length eight that all end with the two bits 00, we can also construct such sequences in $2^{8-2} = 2^6 = 64$ ways, i.e. $|B| = 64$. Some of the ways to construct a bit sequence of length 8 starting with a 1 are the same as the ways to construct a bit sequence of length 8 that ends with the two bits 00, there are $2^{8-3} = 2^5 = 32$ ways to construct such a sequence, i.e. $|A \cap B| = 32$.

Using the principle of inclusion–exclusion, the number of bit sequences starting with the bit 1 or ending with the two bits 00 is thus as follows:

$$|A \cup B| = |A| + |B| - |A \cap B| = 128 + 64 - 32 = 160$$

The probability of selecting the random sequence of interest is then $\frac{160}{256} = \frac{5}{8}$. ∎

Example 1.9
A pair of fair (ideal) cube-shaped dice is rolled. Determine the probability that the sum on the dice is n, where n is an integer ranging between 2 and 12 inclusive, and also the probability that at least one of the two dice is m, where m is an integer between 1 and 6 inclusive.

Solution
Since the dice are fair, all 36 (=6×6) outcomes are equally likely. In the following, we present the values of the first and second dice by an ordered pair. Let A_n denote the event that the sum on the dice is n, we thus have the following:

$$n = 2 \rightarrow A_2 = \{(1,1)\} \rightarrow P(A_2) = \frac{1}{36}$$

$$n = 3 \rightarrow A_3 = \{(1,2),(2,1)\} \rightarrow P(A_3) = \frac{2}{36}$$

$$n = 4 \rightarrow A_4 = \{(1,3),(2,2),(3,1)\} \rightarrow P(A_4) = \frac{3}{36}$$

$$n = 5 \rightarrow A_5 = \{(1,4),(2,3),(3,2),(4,1)\} \rightarrow P(A_5) = \frac{4}{36}$$

$$n = 6 \rightarrow A_6 = \{(1,5),(2,4),(3,3),(4,2),(5,1)\} \rightarrow P(A_6) = \frac{5}{36}$$

$$n = 7 \rightarrow A_7 = \{(1,6),(2,5),(3,4),(4,3),(5,2),(6,1)\} \rightarrow P(A_7) = \frac{6}{36}$$

$$n = 8 \rightarrow A_8 = \{(6,2),(5,3),(4,4),(3,5),(2,6)\} \rightarrow P(A_8) = \frac{5}{36}$$

$$n = 9 \rightarrow A_9 = \{(6,3),(5,4),(4,5),(3,6)\} \rightarrow P(A_9) = \frac{4}{36}$$

$$n = 10 \rightarrow A_{10} = \{(6,4),(5,5),(4,6)\} \rightarrow P(A_{10}) = \frac{3}{36}$$

$$n = 11 \rightarrow A_{11} = \{(6,5),(5,6)\} \rightarrow P(A_{11}) = \frac{2}{36}$$

$$n = 12 \rightarrow A_{12} = \{(6,6)\} \rightarrow P(A_{12}) = \frac{1}{36}$$

Suppose i is the number appearing on the first die, where $i = 1, 2, ..., 6$, and j is the number appearing on the second die, where $j = 1, 2, ..., 6$, and $a_{i,j}$ denotes such an outcome. Let B_m denote the event that at least one of the two dice is m, where $m = 1, 2, ..., 6$. Noting that in rolling a pair of dice, the outcome of one die is independent of the outcome of the other die, we have

$$P(B_m) = P(a_{1,m} \cup \cdots \cup a_{6,m} \cup a_{m,1} \cup \cdots \cup a_{m,6}) - P(a_{m,m})$$
$$= P(a_{1,m}) + \cdots + P(a_{6,m}) + P(a_{m,1}) + \cdots + P(a_{m,6})$$
$$- P(a_{m,m}) \quad m = 1, ..., 6$$

Noting we have $P(a_{i,m}) = \frac{1}{36}, i = 1, 2, ..., 6$, $P(a_{m,j}) = \frac{1}{36}, j = 1, 2, ..., 6$, and $P(a_{m,m}) = \frac{1}{36}$, we therefore have

$$P(B_1) = P(B_2) = P(B_3) = P(B_4) = P(B_5) = P(B_6) = \frac{12}{36} - \frac{1}{36} = \frac{11}{36} \quad \blacksquare$$

Example 1.10
Suppose for the events A and B, we have $P(A) = 0.8$ and $P(B) = 0.6$. Show that $P(A \cap B) \geq 0.4$.

Solution
We have

$$P(A \cup B) = P(A) + P(B) - P(A \cap B) \leq 1 \rightarrow 0.8 + 0.6$$
$$-P(A \cap B) \leq 1 \rightarrow P(A \cap B) \geq 0.4 \quad \blacksquare$$

The probability of an impossible (null) event is zero; however, the converse is not true, that is the probability of an event can be zero, but that event may not be an impossible event. For instance, the probability of randomly selecting an integer that is both odd and even is zero, simply because such an integer does not exist, as it is an impossible event. However, the probability of randomly selecting an integer out of infinitely many possible integers is zero, as we have $\lim_{n \to \infty} \frac{1}{n} = 0$, yet such an event can occur, for it is not an impossible event.

1.4 Joint Probability and Conditional Probability

The probability of the occurrence of a single event, such as $P(A)$ or $P(B)$, which takes a specific value irrespective of the values of other events, i.e. unconditioned on any other events, is called a *marginal probability*. For instance, in rolling a die, the probability of getting a two represents a marginal probability.

The probability that both events A and B simultaneously occur is known as the *joint probability* of events A and B and is denoted by $P(A \cap B)$ or $P(A, B)$, and read as probability of A and B. From the joint probability of two events, all relevant probabilities involving either or both events can be obtained. For instance, in rolling a pair of dice, the probability of getting a two on one die and a five on the other die represents a joint probability.

The following case should provide some insight into the utility of conditional probability. Consider two individuals, a man and a woman, while noting that one individual has four children and the other has none. Suppose we need to guess which of the two individuals is a parent. If no information is given, we say that the probability that the man is the parent is 0.5. If, however, we have the information that the man is 60 years old and the woman is 20 years old, then we say that the man is most likely the parent. Of course, this could be wrong, but it is likely to be true.

If we assume the probability of event B is influenced by the outcome of event A and we also know that event A has occurred, then the probability that event B will occur may be different from $P(B)$. The probability of event B when it is known that event A has occurred is defined as the *conditional probability*, denoted by $P(B \mid A)$ and read as probability of B given A. The conditional probability $P(B \mid A)$ captures the information that the occurrence of event A provides about event B.

In Figure 1.4, assuming the probabilities are represented by the areas of the events, the occurrence of the event A brings about two effects: (i) the original sample space S (the large square) becomes the event A (the smaller square)

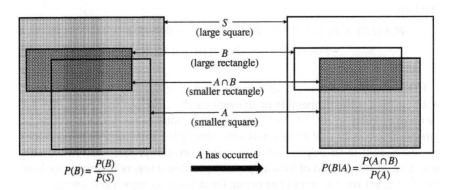

Figure 1.4 Occurrence of event A, as probabilities represented by areas.

and (ii) the event B (the large rectangle) becomes the event $A \cap B$ (the smaller rectangle). In essence, due to the occurrence of the event A, rather than dealing with a subset B of a sample space S, we are now concerned with a subset $A \cap B$ of the restricted sample space A. The conditional probability $P(B|A)$ and the conditional probability $P(A|B)$ are, respectively, defined as follows:

$$P(B \mid A) = \frac{P(A \cap B)}{P(A)} \quad P(A) > 0$$

$$P(A \mid B) = \frac{P(A \cap B)}{P(B)} \quad P(B) > 0 \tag{1.4}$$

In conditional probabilities $P(B|A)$ and $P(A|B)$, we must have $P(A) \neq 0$ and $P(B) \neq 0$, respectively. The reason lies in the fact that if we have $P(A) = 0$ (i.e. A never occurs) or $P(B) = 0$ (i.e. B never occurs), then the corresponding conditional probability, i.e. the probability of B, given that A occurs or the probability of A, given that B occurs, becomes totally meaningless. Note that $P(A)$ or $P(B)$ in the denominator normalizes the probability of the joint event $P(A \cap B)$. The normalization process in the calculation of conditional probability, i.e. dividing the joint probability by the marginal probability, ensures that the sum of all conditional probabilities in a sample space is one. Note that all three axioms of probability and the resulting important properties hold also for conditional probabilities. In addition, the conditional probability does not commute, that is we have $P(B|A) \neq P(A|B)$, unless $P(A) = P(B)$.

Example 1.11
Determine the conditional probability that a family with two children has two girls, given they have at least one girl. Assume the probability of having a girl is the same as the probability of having a boy.

Solution
The four possibilities GG, GB, BG, and BB, each reflecting the order of birth, are equally likely, where G represents a girl and B represents a boy. Let A be the event that a family with two children has two girls, and let C be the event that a family with two children has at least one girl. We therefore have the following events and probabilities:

$$A = \{GG\} \rightarrow P(A) = \frac{1}{4}$$

$$C = \{GG, GB, BG\} \rightarrow P(C) = \frac{3}{4}$$

$$A \cap C = \{GG\} \rightarrow P(A \cap C) = \frac{1}{4}$$

We can thus conclude

$$P(A \mid C) = \frac{P(A \cap C)}{P(C)} = \frac{\frac{1}{4}}{\frac{3}{4}} = \frac{1}{3} \quad \blacksquare$$

Based on the definition of the conditional probability, the joint probability can be obtained in the following two ways:

$$P(A, B) = P(A \cap B) = P(B \mid A)P(A)$$

$$P(A, B) = P(A \cap B) = P(A \mid B)P(B) \tag{1.5}$$

Example 1.12
There are 10 000 smart phones in a warehouse, out of which 10 are defective. Two phones are selected at random and tested one by one. Determine the probability that both phones are defective.

Solution
Let A denote the event that the first phone tested is defective, we therefore have

$$P(A) = \frac{10}{10000} = \frac{1}{1000}$$

Let B denote the event that the second phone tested is defective. Therefore, the event $B \mid A$ represents that the second phone tested is also defective, and we have

$$P(B \mid A) = \frac{9}{9999} = \frac{1}{1111}$$

We can thus obtain

$$P(A \cap B) = P(B \mid A)P(A) = \left(\frac{1}{1111}\right)\left(\frac{1}{1000}\right) \cong 0.9 \times 10^{-6} \qquad \blacksquare$$

The rule on joint events can be generalized for multiple events $A_1, A_2, \ldots, A_n$ via iterations. Assuming $A_1 \cap A_2 \cap \cdots \cap A_j \neq \emptyset, j = 1, 2, \ldots, n$, we then have the following **multiplication rule of probability**, also known as the **chain rule of probability**:

$$
\begin{aligned}
P(A_1 \cap \cdots \cap A_n) &= \frac{P(A_1 \cap \cdots \cap A_n)}{P(A_1 \cap \cdots \cap A_{n-1})} \times \frac{P(A_1 \cap \cdots \cap A_{n-1})}{P(A_1 \cap \cdots \cap A_{n-2})} \\
&\quad \times \cdots \times \frac{P(A_1 \cap A_2 \cap A_3)}{P(A_1 \cap A_2)} \\
&\quad \times \frac{P(A_1 \cap A_2)}{P(A_1)} \times P(A_1) \\
&= P(A_n \mid A_1 \cap \cdots \cap A_{n-1}) \times P(A_{n-1} \mid A_1 \cap \cdots \cap A_{n-2}) \\
&\quad \times \cdots \times P(A_3 \mid A_1 \cap A_2) \\
&\quad \times P(A_2 \mid A_1) \times P(A_1)
\end{aligned} \tag{1.6}
$$

Example 1.13

Suppose there are 50 female students and 50 male students in a classroom. The students are leaving the classroom at random one by one. Determine the probability of leaving the classroom for the first five students, if their order of leaving is female, male, female, female, and male.

Solution

We have the following events and corresponding probabilities:

$$A_1 = \{\text{the first student is a female}\} \rightarrow P(A_1) = \frac{50}{100}$$

$$A_2 = \{\text{the second student is a male}\} \rightarrow P(A_2 \mid A_1) = \frac{50}{99}$$

$$A_3 = \{\text{the third student is a female}\} \rightarrow P(A_3 \mid A_1 \cap A_2) = \frac{49}{98}$$

$$A_4 = \{\text{the fourth student is a female}\}$$

$$\rightarrow P(A_4 \mid A_1 \cap A_2 \cap A_3) = \frac{48}{97}$$

$$A_5 = \{\text{the fifth student is a male}\}$$

$$\rightarrow P(A_5 \mid A_1 \cap A_2 \cap A_3 \cap A_4) = \frac{49}{96}$$

We then obtain the joint probability of interest as follows:

$$P(A_1 \cap A_2 \cap A_3 \cap A_4 \cap A_5) = \frac{50}{100} \times \frac{50}{99} \times \frac{49}{98} \times \frac{48}{97} \times \frac{49}{96} \cong 0.032 \qquad \blacksquare$$

1.5 Statistically Independent Events and Mutually Exclusive Events

If the occurrence of event A has some bearing on the occurrence of event B, then the conditional probability of event B given event A, vis-à-vis the marginal probability of event B, may result in a larger probability (even one) or in a smaller probability (even zero) or may even yield no change in probability. In other words, the knowledge that event A has occurred may increase or reduce the probability that event B has occurred or may even leave it unchanged.

Example 1.14

In rolling a fair die, event B represents the outcomes that are multiples of 3. Determine the probability of B and the conditional probability $P(B \mid A)$ for the following distinct cases:

(a) Event A represents the outcomes that are multiples of 3.
(b) Event A represents the outcomes that are greater than 4.

(c) Event A represents the outcomes that are greater than 3.

(d) Event A represents the outcomes that are less than 5.

(e) Event A represents the outcomes that are not multiples of 3.

Solution

Since it is a fair die, we have the event B and its probability as follows:

$$B = \{3,6\} \rightarrow P(B) = \frac{2}{6} = \frac{1}{3}$$

The conditional probabilities of interest can be thus obtained as follows:

(a) $A = \{3,6\} \rightarrow A \cap B = \{3,6\} \rightarrow P(B \mid A) = \dfrac{P(A \cap B)}{P(A)} = \dfrac{\frac{2}{6}}{\frac{2}{6}} = 1 \rightarrow$

$P(B \mid A) > P(B)$

(b) $A = \{5,6\} \rightarrow A \cap B = \{6\} \rightarrow P(B \mid A) = \dfrac{P(A \cap B)}{P(A)} = \dfrac{\frac{1}{6}}{\frac{2}{6}} = \dfrac{1}{2} \rightarrow$

$P(B \mid A) > P(B)$

(c) $A = \{4,5,6\} \rightarrow A \cap B = \{6\} \rightarrow P(B \mid A) = \dfrac{P(A \cap B)}{P(A)} = \dfrac{\frac{1}{6}}{\frac{3}{6}} = \dfrac{1}{3} \rightarrow$

$P(B \mid A) = P(B)$

(d) $A = \{1,2,3,4\} \rightarrow A \cap B = \{3\} \rightarrow P(B \mid A) = \dfrac{P(A \cap B)}{P(A)} = \dfrac{\frac{1}{6}}{\frac{4}{6}} = \dfrac{1}{4} \rightarrow$

$P(B \mid A) < P(B)$

(e) $A = \{1,2,4,5\} \rightarrow A \cap B = \{\emptyset\} \rightarrow P(B \mid A) = \dfrac{P(A \cap B)}{P(A)} = \dfrac{0}{\frac{4}{6}} = 0 \rightarrow$

$P(B \mid A) < P(B)$ ∎

If the occurrence of event A has no statistical impact on the occurrence of event B, we then say events A and B are statistically independent. Independence is an important, but simple, concept. Statistical independence often arises from the physical independence of events and experiments. In random experiments, it is common to assume that the events of separate trials are independent. For instance, in a series of coin tossing experiments where a coin is tossed many times, the outcomes are assumed to be independent because the outcome of a toss is not affected by the outcomes of the preceding tosses and does not affect the outcomes of the subsequent tosses.

If the knowledge of event A does not change the probability of the occurrence of event B, then we have $P(B \mid A) = P(B)$. If the knowledge of event B does not change the probability of the occurrence of event A, then we have $P(A \mid B) = P(A)$. Therefore, in both of these cases, the joint probability of events A and B is equal to the product of the marginal probability of event A and the marginal probability of event B, i.e. $P(A \cap B) = P(A)P(B)$, and these events are then said to be ***statistically independent***. If two events are statistically independent, then it must be proven mathematically that $P(A \cap B) = P(A)P(B)$, rather than explained intuitively.

It is important to emphasize that statistical independence between two events does not mean one event does not affect another event. It merely means the probability of the joint event is equal to the product of the probabilities of individual events.

Example 1.15
In rolling a fair die, we define the event A when the outcome is a three or a four, and the event B when the outcome is an odd number. Are the events A and B statistically independent?

Solution
We have the following events and probabilities:

$$A = \{3, 4\} \rightarrow P(A) = \frac{2}{6} = \frac{1}{3}$$

$$B = \{1, 3, 5\} \rightarrow P(B) = \frac{3}{6} = \frac{1}{2}$$

$$A \cap B = \{3\} \rightarrow P(A \cap B) = \frac{1}{6}$$

We thus have

$$P(A \mid B) = \frac{P(A \cap B)}{P(B)} = \frac{\frac{1}{6}}{\frac{1}{2}} = \frac{1}{3}$$

As a result, we have

$$P(A \mid B) = P(A)$$

Hence, events A and B are statistically independent, yet, knowledge of event B occurring has affected the possible outcomes. The event $A \cap B = \{3\}$ has half as many outcomes as $A = \{3, 4\}$, but the sample space $B = \{1, 3, 5\}$ also has half as many outcomes as the original sample space $S = \{1, 2, 3, 4, 5, 6\}$. ∎

It can be shown that if the events A and B are independent, then the events A^c and B, the events A and B^c, as well as the events A^c and B^c are also statistically independent. Note that if $A \subset B$, then A and B cannot be independent, unless B is the entire sample space.

The notion of statistical independence can be exploited to compute probabilities of events that involve noninteracting, independent subevents, such as calculations of probabilities in link connections and counting with permutations and combinations.

If the joint probability of events A and B is zero, i.e. $P(A \cap B) = 0$, these two events are then said to be ***mutually exclusive*** or ***disjoint***. The following highlights the condition under which two events are statistically independent or that under which they are mutually exclusive:

$$\text{Independent events} \Leftrightarrow P(A \cap B) = P(A)P(B)$$
$$\Leftrightarrow P(A \cup B) = P(A) + P(B) - P(A)P(B)$$
$$\text{Mutually exclusive events} \Leftrightarrow P(A \cap B) = 0$$
$$\Leftrightarrow P(A \cup B) = P(A) + P(B) \tag{1.7}$$

To distinguish between the statistical independence and mutual exclusiveness (disjointness) of a collection of events $A_1, A_2, \ldots, A_n$, we have the following:

$$P\left(\bigcap_{j=i_1}^{i_k} A_j\right) = \prod_{j=i_1}^{i_k} P(A_j) \Leftrightarrow A_1, A_2, \ldots, A_n$$

are statistically independent events

$$k = 2, \ldots, n \text{ and } 1 \leq i_1 < \cdots < i_k \leq n$$

$$P\left(\bigcup_{i=1}^{n} A_i\right) = \sum_{i=1}^{n} P(A_i) \Leftrightarrow A_1, A_2, \ldots, A_n$$

are mutually exclusive events $\tag{1.8}$

In the case of the statistically independent events, the probability of the intersection of events is equal to the product of the probabilities of individual events and also a similar equality holds for every subset (sub-collection) of the n events. Note that pairwise independence is not sufficient. In fact, in order to have n statistically independent events, a total of $2^n - (n+1)$ equations are needed. In general, if there are n independent events, then any one of them is also independent of any event formed by unions, intersections, and complements of the other events.

Example 1.16
The probability that A hits a target is 90%, the probability that B hits the target is 80%, and the probability that C hits the target is 70%. Assuming all three shoot for the target, determine the probability that at least one of them hits the target.

Solution
The events A, B, and C are all statistically independent, and their probabilities are as follows:

$$P(A) = 0.9, \quad P(B) = 0.8, \quad P(C) = 0.7$$

By extending the principle of inclusion–exclusion from two events to three events, we thus have the following:

$$\begin{aligned}
P(A \cup B \cup C) &= P(A) + P(B) + P(C) - P(A \cap B) - P(A \cap C) \\
&\quad - P(B \cap C) + P(A \cap B \cap C) \\
&= P(A) + P(B) + P(C) - P(A)P(B) \\
&\quad - P(A)P(C) - P(B)P(C) \\
&\quad + P(A)P(B)P(C) \\
&= 0.9 + 0.8 + 0.7 - 0.9 \times 0.8 - 0.9 \times 0.7 \\
&\quad - 0.8 \times 0.7 + 0.9 \times 0.8 \times 0.7 \\
&= 0.994
\end{aligned}$$

There is another method to determine the probability of interest. Using the complement property and De Morgan's laws and noting that the events A, B, and C are all statistically independent, we have the following:

$$\begin{aligned}
P(A \cup B \cup C) &= 1 - P((A \cup B \cup C)^c) = 1 - P(A^c \cap B^c \cap C^c) = 1 \\
&\quad - P(A^c)P(B^c)P(C^c) \\
&= 1 - (1 - P(A))(1 - P(B))(1 - P(C)) = 1 \\
&\quad - (1 - 0.9)(1 - 0.8)(1 - 0.7) \\
&= 0.994
\end{aligned}$$ ∎

Example 1.17

In rolling a pair of fair dice, the event A denotes when the sum of the outcomes of the two dice is equal to three and the event B is when the outcome of at least one of the dice is three. Are the events A and B mutually exclusive? Are they statistically independent?

Solution

For the event B, the sum of the outcomes of the two dice is at least four. Therefore, the events A and B are mutually exclusive, because they cannot occur together, i.e. we have

$$P(A \cap B) = 0$$

The event A, with the sum of three, occurs when the outcome is either $(1, 2)$ or $(2, 1)$. The total number of possible outcomes is 36 ($= 6 \times 6$), as each die can have 6 equally likely outcomes. We thus have

$$P(A) = \frac{2}{36} = \frac{1}{18}$$

Since we have

$$P(B) = 1 - P(B^c) = 1 - \left(\frac{5}{6}\right)\left(\frac{5}{6}\right) = \frac{11}{36}$$

we have

$$\frac{1}{18} \times \frac{11}{36} \neq 0 \rightarrow P(A)P(B) \neq P(A \cap B)$$

This clearly points to the fact that the events A and B are not statistically independent. ∎

Example 1.18

Determine the probability that in a randomly generated sequence of n bits, where $n \geq 4$, the sequence begins with a 11 and/or ends with a 00, assuming the probability of a bit 1 is p and the probability of a bit 0 is $1 - p$,

Solution

Let E be the event that the bit sequence begins with a 11 and F be the event that the bit sequence ends with a 00. As the bits in a sequence are statistically independent, we have

$$P(E) = p^2$$

and

$$P(F) = (1 - p)^2$$

Moreover, the events E and F are statistically independent, we thus have

$$P(E \cap F) = P(E)P(F) = p^2(1 - p)^2$$

Therefore, the probability of interest is as follows:

$$P(E \cup F) = P(E) + P(F) - P(E \cap F) = p^2 + (1 - p)^2 - p^2(1 - p)^2$$
$$= 1 - 2p + p^2 + 2p^3 - p^4 \quad ∎$$

It is of great importance to make a clear distinction between the concept of statistically independent events and the concept of mutually exclusive events. Both concepts seem to imply separation and distinctness, but in fact two mutually exclusive events cannot generally be independent of one another. Note that if the two events A and B are mutually exclusive, then the events A and B cannot occur at the same time, i.e. the occurrence of one implies that the other has zero probability of happening. Hence, mutually exclusive events can be considered to be dependent events. If the two events A and B are both mutually exclusive and statistically independent, then it implies that as a consequence of the Kolmogorov axioms, at least one of the two events A and B has zero probability, i.e. we have $P(A \cap B) = P(A)P(B) = 0$.

Example 1.19
A box contains 10 balls, numbered from 11 to 20 inclusive. A ball is drawn from the box at random. We define the following five distinct events:

A : {Number on ball drawn is an odd number}
B : {Number on ball drawn is a prime number}
C : {Number on ball drawn is an even number}
D : {Number on ball drawn is a multiple of 5}
E : {Number on ball drawn is a multiple of 25}

For each of the following four cases, determine if the two events are mutually exclusive and also if they are statistically independent: the events A and B, the events A and C, the events A and D, and the events A and E. Comment on the results.

Solution
The probabilities of the five events are as follows:

$$A = \{11, 13, 15, 17, 19\} \rightarrow P(A) = \frac{5}{10}$$
$$B = \{11, 13, 17, 19\} \rightarrow P(B) = \frac{4}{10}$$
$$C = \{12, 14, 16, 18, 20\} \rightarrow P(C) = \frac{5}{10}$$
$$D = \{15, 20\} \rightarrow P(D) = \frac{2}{10}$$
$$E = \{\emptyset\} \rightarrow P(E) = 0$$

We then identify the four joint events of interest and their corresponding probabilities:

$$A \cap B = \{11, 13, 17, 19\} \rightarrow P(A \cap B) = \frac{4}{10}$$
$$A \cap C = \{\emptyset\} \rightarrow P(A \cap C) = 0$$
$$A \cap D = \{15\} \rightarrow P(A \cap D) = \frac{1}{10}$$
$$A \cap E = \{\emptyset\} \rightarrow P(A \cap E) = 0$$

The results can then be summarized as follows:

Events	Statistically independent?	Mutually exclusive?
$A \& B$	$P(A \cap B) \neq P(A)P(B) \rightarrow$ No	$P(A \cap B) \neq 0 \rightarrow$ No
$A \& C$	$P(A \cap C) \neq P(A)P(C) \rightarrow$ No	$P(A \cap C) = 0 \rightarrow$ Yes
$A \& D$	$P(A \cap D) = P(A)P(D) \rightarrow$ Yes	$P(A \cap D) \neq 0 \rightarrow$ No
$A \& E$	$P(A \cap E) = P(A)P(E) \rightarrow$ Yes	$P(A \cap E) = 0 \rightarrow$ Yes

We can thus conclude that any two events, no matter how they are defined, always fall into one of these four distinct cases, i.e. whether being statistically independent or not as well as whether being mutually exclusive or not. ∎

1.6 Law of Total Probability and Bayes' Theorem

In order to determine the probability of event A, it is sometimes best to separate all possible causes leading to event A. If the events $B_1, B_2, \ldots, B_n$ are all mutually exclusive events whose union forms the entire sample space S, that is we have $S = B_1 \cup B_2 \cup \cdots \cup B_n$, we then refer to these sets as a ***partition*** of S. As shown in Figure 1.5, we can then use the set of joint probabilities to obtain $P(A)$, i.e. we can have the following:

$$P(A) = P(A \cap B_1) + P(A \cap B_2) + \ldots + P(A \cap B_n) \tag{1.9a}$$

Using the corresponding conditional and marginal probabilities for the above joint probabilities, we can obtain the ***law of total probability***, also known as the ***theorem on total probability***, as follows:

$$P(A) = P(A \mid B_1) P(B_1) + P(A \mid B_2) P(B_2) + \ldots + P(A \mid B_n) P(B_n) \tag{1.9b}$$

This divide-and-conquer approach is a practical tool used to determine the probability of event A. This is simply due to the fact that the probability of

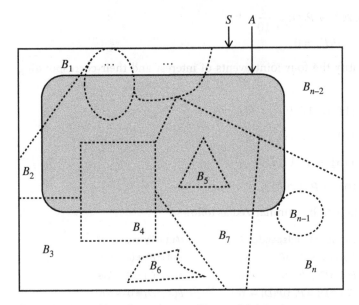

Figure 1.5 A partition of sample space S into n disjoint sets.

event A can be expressed as a combination of the probabilities of the mutually exclusive events $B_1, B_2, \ldots, B_n$ that form the partition of the sample space S. The law of total probability thus highlights the probabilities of effects given causes when causes do not deterministically select effects.

Example 1.20
Suppose the following information regarding the upcoming election in a country is available: 25% of the population who can vote is under 25 years old, of which 40% is likely to vote, 60% of the population who can vote is between 25 and 65 years old, of which 75% is likely to vote, and the remaining 15% of the population who can vote is over 65 years old, of which 60% is likely to vote. Determine the probability that a person votes in the election.

Solution
Let A denote the event that a person votes, B_1 denote the event that a person is under 25 years old, B_2 denote the event that a person is between 25 and 65 years, and B_3 denote the event that a person is over 65 years old. Noting that we have the following probabilities:

$$P(B_1) = 0.25, \quad P(B_2) = 0.6, \quad P(B_3) = 0.15$$

and

$$P(A \mid B_1) = 0.4, \quad P(A \mid B_2) = 0.75, \quad P(A \mid B_3) = 0.6$$

the probability of interest is then as follows:

$$P(A) = 0.25 \times 0.4 + 0.6 \times 0.75 + 0.15 \times 0.6 = 0.64 \qquad ∎$$

Example 1.21
Suppose the following information regarding the incidence of rabies in an ecosystem is available:

Animals	Percent of population	Probability of infection
Rabbits	30	0.1
Foxes	10	0.4
Wolves	20	0.2
Raccoons	20	0.6
Skunks	10	0.5
Coyotes	10	0.3

Determine the probability that rabies will appear within the population of animals.

Solution
Let event A denote the population of animals with rabies, event B_1 denote the population of rabbits, event B_2 denote the population of foxes, event B_3 denote the population of wolves, event B_4 denote the population of raccoons, event B_5 denote the population of skunks, and event B_6 denote the population of coyotes. Noting that we have the following probabilities:

$$P(B_1) = 0.3, \quad P(B_2) = 0.1, \quad P(B_3) = 0.2,$$
$$P(B_4) = 0.2, \quad P(B_5) = 0.1, \quad P(B_6) = 0.1$$

and

$$P(A \mid B_1) = 0.1, P(A \mid B_2) = 0.4, P(A \mid B_3) = 0.2,$$
$$P(A \mid B_4) = 0.6, P(A \mid B_5) = 0.5, P(A \mid B_6) = 0.3$$

the probability of interest is then as follows:

$$P(A) = 0.3 \times 0.1 + 0.1 \times 0.4 + 0.2 \times 0.2$$
$$+ 0.2 \times 0.6 + 0.1 \times 0.5 + 0.1 \times 0.3 = 0.31 \qquad \blacksquare$$

When one conditional probability is given, but the reversed conditional probability is required, the following relation, known as **Bayes' theorem** or **Bayes' rule**, which is based on the law of total probability, can be used:

$$P(B_1 \mid A) = \frac{P(A \cap B_1)}{P(A)}$$
$$= \frac{P(A \mid B_1) P(B_1)}{P(A \mid B_1) P(B_1) + P(A \mid B_2) P(B_2) + \dots + P(A \mid B_n) P(B_n)} \qquad (1.10)$$

where the events $B_1, B_2, \dots, B_n$ are all mutually exclusive events whose union makes the entire sample space S. Bayes' theorem highlights inference from an observed effect, and is used to estimate probabilities based on partial evidence. It can be used to synthesize decision rules that attempt to determine the most probable cause in light of an observation. In short, the law of total probability is about effects from causes and Bayes' theorem is about causes from effects.

It is also important to note that $P(B_i)$, the probability of an event B_i before the experiment is performed, is referred to as *a priori probability*, and $P(B_i \mid A)$, the probability of an event B_i after the experiment has been performed and the event A has occurred, is called *a posteriori probability*. Bayes' theorem connects the a posteriori probability with the a priori probability.

Example 1.22
The probability that a student passes an exam is 0.95, given that he studied. The probability that he passes the exam without studying is 0.15. As he is a lazy student, we know that the probability that the student studies for an exam is

just 50%. Given the student passed the exam, determine the probability that he studied.

Solution

Let P be the event that the student passes the exam and S be the event the student studies for the exam. We thus have the following probabilities:

$$P(P|S) = 0.95 \quad \text{and} \quad P(P|S^c) = 0.15$$

and

$$P(S) = 0.5 \rightarrow P(S^c) = 0.5$$

Using Bayes' rule, we have

$$P(S \mid P) = \frac{P(P \mid S)P(S)}{P(P)}$$

$$= \frac{P(P \mid S)P(S)}{P(P \mid S)P(S) + P(P \mid S^c)P(S^c)} = \frac{0.95 \times 0.5}{0.95 \times 0.5 + 0.15 \times 0.5} \cong 0.86$$

Before we had any extra information, we assumed that the probability that he studied was 0.5. However, with the extra information that he passed the exam, this probability was expectedly increased to about 0.86. ∎

Example 1.23

There are three boxes. The first box contains ten black balls and five white balls, the second box contains three black balls and three white balls, and the third box contains two black balls and six white balls. The probability of selecting any one box is one third. If a ball selected at random from one of the boxes is white, determine the probability that it was drawn from the first box.

Solution

There are three boxes A_1, A_2, and A_3, and the probabilities of selecting them are as follows: $P(A_1) = P(A_2) = P(A_3) = \frac{1}{3}$. Let W represent the event that the ball is white. We thus have

$$P(W \mid A_1) = \frac{5}{15} = \frac{1}{3}, \quad P(W \mid A_2) = \frac{3}{6} = \frac{1}{2}, \quad P(W \mid A_3) = \frac{6}{8} = \frac{3}{4}$$

Using Bayes' rule, the probability of interest is then as follows:

$$P(A_1 \mid W) = \frac{P(W \mid A_1)P(A_1)}{P(W)}$$

$$= \frac{P(W \mid A_1)P(A_1)}{P(W \mid A_1)P(A_1) + P(W \mid A_2)P(A_2) + P(W \mid A_3)P(A_3)}$$

$$= \frac{\dfrac{1}{3} \times \dfrac{1}{3}}{\dfrac{1}{3} \times \dfrac{1}{3} + \dfrac{1}{2} \times \dfrac{1}{3} + \dfrac{3}{4} \times \dfrac{1}{3}} = \frac{4}{19} \cong 21\%$$

The probability that a ball selected from the first box reduced from one-third, when no extra information was available, to about 21%, once we knew that the ball selected was white. ∎

1.7 Summary

In this chapter, we first developed the basics of probability, events, and random experiments by building on a basic foundation of set theory, and then presented the axioms of probability and their properties. The focus then turned toward the joint and conditional probabilities, as well as the concepts of statistically independent events and mutually exclusive events. Lastly, the law of total probability, which is about effects from causes, and Bayes' theorem, which is about causes from effects, were introduced.

Problems

1.1 Given that $P(A) = 0.9$, $P(B) = 0.8$, and $P(A \cap B) = 0.7$, determine $P(A^c \cap B^c)$.

1.2 Police report that among drivers stopped on suspicion of impaired driving 70% took test A, 30% test B, and 20% both tests A and B. Determine the probability that a randomly selected suspect is given (a) test A or test B or both, (b) test A or test B, but not both, and (c) neither test A nor test B.

1.3 The probabilities that a husband and wife will be alive in 30 years from now are given by 0.7 and 0.8, respectively. Assuming that the events that the husband and wife will be alive in 30 years from now are independent, find the probability that in 30 years (a) both, (b) neither, and (c) at least one, will be alive.

1.4 An experiment consists of rolling three fair dice at the same time. Event G is when the sum of the dots on the faces of the three dice that are uppermost is equal to 16. Find the probability of event G.

1.5 A string whose length is 1 meter is cut into three strings. What is the probability that these three pieces can form a triangle?

1.6 The event A is defined when at least one 6 in 4 tosses of a fair die is obtained, and the event B is defined when at least one double 6 in 24 tosses of a pair of fair dice is obtained. Which one of these two events is more likely?

1.7 An integer between 1 and 30 is selected at random. Find the probability of selecting a prime number if an integer between 1 and 15 is twice as likely to occur.

1.8 A fair die is tossed twice. Determine the probability of getting a 4 or a 5 or a 6 on the first toss and a 1 or a 2 or a 3 or a 4 on the second toss.

1.9 There are $n \geq 1$ identical bags. In each bag, there are 10 balls, numbered from 1 to 10 inclusive. We randomly pick a ball from each bag. Determine the probability that the product of the n numbers on the n balls is a multiple of 3.

1.10 Assuming all dice are fair, which one of the following three cases is more likely when the sum in each case is six: (i) rolling two dice, (ii) rolling three dice, and (iii) rolling four dice?

1.11 You are given the opportunity to guess the color on the opposite side of a card placed on a table. The card can be any one of a set of three. One card is red on both sides, one is black on both sides, and the third is red on one side and black on the other. A card is placed on the table, the side that is showing is red. What is the probability that the other side of the card is black?

1.12 There are two events A and B such that $P(A \cup B) = 0.8$, $P(A \cap B) = 0.1$, and $P(A \,|\, B) = 0.25$. Determine $P(B \,|\, A)$.

1.13 A bag contains 12 red marbles, 8 green marbles, and 4 blue marbles. A marble is drawn from the bag and it happens not to be a red marble. What is the probability that it is a blue marble?

1.14 A coin is tossed twice. Note that the probability of getting a head is p and the probability of getting a tail is thus $1 - p$. Given that the first toss resulted in a head, determine the probability that both tosses result in heads.

1.15 Show that the conditional $P(A \,|\, B)$, where $P(B) > 0$, satisfies the three axioms of probability.

1.16 In a well-shuffled deck of cards, we observe the bottom card is a red card, what is the probability that the bottom card is the king of hearts?

1.17 There is a bag containing six red balls, four white balls, and five blue balls. Three balls are drawn at random successively from the bag. Determine the probability that they are drawn in the order red, white and blue, if each ball is (a) replaced, and (b) not replaced.

1.18 A box contains 25 cookies of which x of them are bad. Two cookies are eaten one by one. Assuming the probability that both cookies are good is 0.57, determine the value of x.

1.19 A and B play a game in which they alternately toss a pair of fair dice. The one who is first to get a sum of seven wins the game. Noting that A tosses first, which one is more likely to win first?

1.20 There are two bags of balls, the first bag contains five red balls and four blue balls and the second bag contains three red balls and six blue balls. One ball is taken from the first bag and put in the second bag, without seeing what the color of the ball is. Determine the probability that a ball now drawn from the second bag is blue.

1.21 At a university, 60% of students live on campus, 70% have a meal plan, and 40% do both. Are living on campus and having a meal plan statistically independent?

1.22 Consider rolling a single fair die, and define two events $H = \{1, 2, 3, 4\}$ and $J = \{3, 4, 5\}$. Are events H and J mutually exclusive? Are they statistically independent?

1.23 Two numbers x and y are selected at random each between 0 and 2. Let events A, B, and C be defined as follows: $A = \{1 < x < 2\}$, $B = \{1 < y < 2\}$, and $C = \{x > y\}$. Are there any two events that are statistically independent?

1.24 Suppose A is the event that a randomly generated bit sequence of length four begins with a 1 and B is the event that the four-bit sequence contains an even number of 1s. Assuming all 16 bit sequences of length four are equally likely, are the events A and B statistically independent?

1.25 The suppliers A, B, and C have records of providing microchips with 10%, 5%, and 20% defective rates, respectively. Suppose 20%, 35 %, and 45% of the current supply come from the suppliers A, B, and C, respectively. If a randomly selected microchip is defective, what is the probability that it comes from the supplier B?

1.26 Suppose we have two coins, one fair that is $P(H) = P(T) = 0.5$, and one is unfair with $P(H) = 0.51$ and $P(T) = 0.49$, where $P(H)$ represents the probability of getting a head in a coin toss and $P(T)$ represents the probability of a tail. In case 1, we pick one of the two coins at random, i.e. with the probability of 0.5, toss it once, put it back, pick one of the two coins at random again, and toss it. In case 2, we pick one of the two coins at random and toss it twice. In each of these two cases, determine the probability that both tosses are heads. In the second case, determine which one of the two coins is more likely to have been picked.

1.27 In assessing the strength of evidence in a legal investigation, a police detective always approaches his two informers to get information. The detective gets his information 80% of the time from the informer A who tells a lie 75% of the time and 20% of the time from the informer B who tells a lie 60% of the time. Suppose the information the detective has received is truthful, what is the probability that the information was received from the informer B?

1.28 Suppose in a city, there are five car brands A, B, C, D, and E, that share 1%, 4%, 15%, 30%, and 50% of the market share, respectively. The probabilities that a car will need major repair during their first five years of purchase for brands A, B, C, D, and E are 1%, 4%, 15%, 30%, and 50%, respectively. (a) Determine the probability that a car in this city will need major repair during its first five years of purchase. (b) Suppose a car needs major repair during its first five years of purchase, determine the probability that it is made by the manufacturer A or B or C or D or E.

1.29 In a certain city, 20% of teenage drivers text while driving. The research record indicates that 40% of those who text have a car accident and 1% of those who do not text have a car accident. If a teenage driver has an accident, what is the probability that he was texting?

1.30 An exam consists of three types of questions: 20% essay-type questions, 50% short-answer questions, and 30% multiple-choice questions. A student knows how to answer correctly 80% of the essay-type questions, 50% of the short-answer questions, and 70% of the multiple-choice questions. What is the probability that her answer to a question is correct?

1.31 Prove the following, which is known as Boole's inequality or the union bound:

$$P\left(\bigcup_{i=1}^{n} A_i\right) \leq \sum_{i=1}^{n} P(A_i)$$

2

Applications in Probability

Noting there are numerous applications in probability, our focus here is on just a few applications. In this chapter, some major probability applications in science and engineering, such as systems reliability, medical diagnostic testing, and Bayesian spam filtering, as well as some interesting probability applications in life, such as odds and risk, gambler's ruin problem, and how to make the best choice problem, are briefly highlighted.

2.1 Odds and Risk

In the world of gambling, games of chance, and sports, probabilities are often expressed by odds. Odds are also involved in the statistical analysis of frequency data, such as in medicine and business. Odds measures are calculated when the study is either retrospective (backward looking), such as the impacts of various treatments in medicine, or prospective (forward looking), such as the prospects of teams winning in sports events.

The ***odds for*** (in favor of) an event A, which are the ratio of the probability of A to the probability of A^c, reflect the likelihood that the event A will take place. The ***odds against*** (not in favor of) an event A, which are the ratio of the probability of A^c to the probability of A, highlight the likelihood that the event A will not take place. We thus have the following:

$$\text{Odds for } A = \frac{P(A)}{P(A^c)} = \frac{P(A)}{1 - P(A)} \rightarrow P(A) = \frac{\text{Odds for } A}{1 + \text{Odds for } A}$$

$$\text{Odds against } A = \frac{P(A^c)}{P(A)} = \frac{1 - P(A)}{P(A)} \rightarrow P(A) = \frac{1}{1 + \text{Odds against } A}$$

$$(2.1)$$

For a given odds, the probability of the event A can be thus easily determined. Note that the odds against an event A are the same as the odds in favor of the event A^c and the odds in favor of the event A are the same as the odds against the event A^c. When the odds for an event is larger than one, the event is more likely

Probability, Random Variables, Statistics, and Random Processes: Fundamentals & Applications,
First Edition. Ali Grami.
© 2020 John Wiley & Sons, Inc. Published 2020 by John Wiley & Sons, Inc.
Companion Website: www.wiley.com/go/grami/PRVSRP

to occur than its complement, when it is one, the event and its complement have the same probability of occurrence, and when it is less than one, the event is less likely to occur than its complement. If an event is more likely not to occur than to occur, it is customary to give odds that it will not occur rather than the odds that it will occur. Oftentimes, odds for A are written as $P(A) : P(A^c)$ and read as $P(A)$ to $P(A^c)$, and odds against A are written as $P(A^c) : P(A)$ and read as $P(A^c)$ to $P(A)$.

Example 2.1

(a) The probability of rain tomorrow is 65%, determine the odds against rain tomorrow.
(b) A card is randomly drawn from a well-shuffled standard deck of cards, determine the odds for the ace of hearts.
(c) A fair coin is tossed, determine the odds in favor of heads.
(d) A fair die is rolled, determine the odds of getting a 5.

Solution
The odds of interest are as follows:

(a) $P(A) = 65\% \rightarrow$ odds against $A = \frac{1-0.65}{0.65} = \frac{7}{13}$ or $7 : 13$ (odds are 7 to 13 against).

(b) $P(A) = \frac{1}{52} \rightarrow$ odds for $A = \frac{\frac{1}{52}}{1-\frac{1}{52}} = \frac{1}{51}$ or $1 : 51$ (odds are 1 to 51 for).

(c) $P(A) = \frac{1}{2} \rightarrow$ odds in favor of $A = \frac{\frac{1}{2}}{1-\frac{1}{2}} = 1$ or $1 : 1$ (odds are 1 to 1 for).

(d) $P(A) = \frac{1}{6} \rightarrow$ odds for $A = \frac{\frac{1}{6}}{1-\frac{1}{6}} = \frac{1}{5}$ or $1 : 5$ (odds are 1 to 5 for). ∎

An **odds ratio** is a relative measure of effect and is used to compare the odds for two groups. An odds ratio is calculated by dividing the odds in one group by the odds in the other group. Suppose the odds in one group are $\frac{a}{c}$ and those in the other are $\frac{b}{d}$, we therefore have the following:

$$\text{Odds Ratio} = \frac{\frac{a}{c}}{\frac{b}{d}} = \frac{a \times d}{c \times b} \tag{2.2}$$

An odds ratio of about one indicates that the condition or event under study is almost equally likely to occur in both groups. An odds ratio greater than one indicates that the condition or event is more likely to occur in the first group. An odds ratio less than one indicates that the condition or event is less likely to occur in the first group. The easily calculated odds and odds ratio can provide valuable insights into tabled data for decision-making.

Example 2.2

In a country, data regarding employment have been gathered for both male and female populations as well as for both high school and university graduates. The following two tables show employment status by gender and level of education:

	Employed	Unemployed		Employed	Unemployed
Male	8 000 000	1 000 000	High school	4 000 000	2 000 000
Female	6 000 000	3 000 000	University	10 000 000	2 000 000

Using odds ratios, interpret the data.

Solution

The data regarding employment and gender thus results in the following odds ratio:

$$\text{Odds ratio} = \frac{8\,000\,000 \times 3\,000\,000}{1\,000\,000 \times 6\,000\,000} = 4$$

The odds in favor of being employed for males are significantly more than that for females. The data regarding employment and education, however, results in the following odds ratio:

$$\text{Odds ratio} = \frac{4\,000\,000 \times 2\,000\,000}{2\,000\,000 \times 10\,000\,000} = 0.4$$

The odds in favor of being employed for the high school graduates are much less than that for the university graduates. ∎

Risk or *absolute risk* mixes up the chance that an event will occur with the impact of the occurrence of the event. Risk measures are calculated when the study is generally prospective (forward looking). The probability of a risk occurring can range anywhere from just above 0% (almost never) to just below 100% (almost certain). Note that it cannot be exactly 100%, as it would then be a certainty not a risk, and it cannot be exactly 0%, as it would not be a risk. Risk is restricted to adverse outcomes, i.e. negative impacts, the resulting impact could be anything from insignificant to catastrophic. There is generally a probability-impact matrix associated with a risk, which lists the relative probability of a risk vis-à-vis the relative impact of the risk.

Relative risk or *risk ratio* is a measure of the risk of a certain event occurring in an exposed group compared to the risk of the same event occurring in an unexposed group. Relative risk is used frequently in the statistical analysis of binary outcomes where the outcome of interest has relatively low probability. For instance, in medicine, it can be used to compare the risk of developing a disease in people not receiving a new medical treatment as compared to the people who are receiving a well-established treatment.

Table 2.1 Statistics about developing disease versus receiving treatment.

Receiving treatment	Developing disease	Not developing disease	Total number in the group
Yes	a	b	$a+b$
No	c	d	$c+d$

The relative risk (of the disease for an exposure) is thus defined as the ratio of the incidence rate (of the disease) in the exposed group to the incidence rate (of the disease) in the unexposed group. Based on Table 2.1, which provides statistics about developing disease versus receiving treatment, we have the following:

$$\text{Relative risk} = \frac{\dfrac{a}{a+b}}{\dfrac{c}{c+d}} = \frac{a \times (c+d)}{(a+b) \times c} \tag{2.3}$$

A relative risk of almost one means there is almost no difference between two groups in terms of their risk of the disease, based on whether or not they were exposed to a certain substance or factor, or how they responded to two treatments being compared. A relative risk of greater than one or less than one usually means that being exposed to a certain substance or factor either increases (relative risk greater than one) or decreases (relative risk less than one) the risk of disease.

Example 2.3

Suppose the probability of developing lung cancer among smokers was 28% and among nonsmokers 2%. Determine the relative risk of cancer associated with smoking.

Solution

Using Table 2.1, we have $a = 28$, $b = 100 - 28 = 72$, $c = 2$, and $d = 100 - 2 = 98$. We thus have

$$R = \frac{\dfrac{28}{100}}{\dfrac{2}{100}} = 14$$

Therefore, smokers would be 14 times as likely as nonsmokers to develop lung cancer. ∎

The relative risk is different from the odds ratio, although the relative risk asymptotically approaches the odds ratio for small probabilities. For instance, with $a \ll b$ and $c \ll d$, the relative risk $\frac{a \times (c+d)}{(a+b) \times c}$ approaches the odds ratio

$\frac{a \times d}{c \times b}$. Relative risk is, however, a more intuitive measure of effectiveness, and the distinction between these two measures is important in cases of high probabilities. For instance, if the risk in one group is 99.99% and that in another group is 99%, then their relative risk is just 1.01 $\left(= \frac{99.99\%}{99\%} \right)$, while their odds ratio is 101 $\left(= \frac{\frac{99.99\%}{0.01\%}}{\frac{99\%}{1\%}} \right)$.

In the context of conditional probability, the ratio $\frac{P(B|A)}{P(B|A^c)}$ is known as the *relative risk of B given A*. Note that if the events A and B are independent, then the relative risk is 1. If they are dependent, the relative risk is not 1, and the more dependent they are, the further the relative risk will be from 1.

Example 2.4
Suppose two doctors, A and B, test all patients coming into a clinic for a certain disease. Suppose doctor A diagnoses 10% of all patients as positive, doctor B diagnosis 17% of all patients as positive, and both doctors diagnose 8% of all patients as positive. Determine the relative risk of doctor B diagnoses positive given doctor A diagnoses positive, i.e. $\frac{P(B^+|A^+)}{P(B^+|A^-)}$.

Solution
We have the following:

$$\frac{P(B^+ \mid A^+)}{P(B^+ \mid A^-)} = \frac{\frac{P(B^+, A^+)}{P(A^+)}}{\frac{P(B^+, A^-)}{P(A^-)}} = \frac{\frac{P(B^+, A^+)}{P(A^+)}}{\frac{P(B^+) - P(B^+, A^+)}{1 - P(A^+)}} = \frac{\frac{0.08}{0.10}}{\frac{0.17 - 0.08}{1 - 0.10}} = 8$$

Since we have $P(B^+|A^+) = 8\, P(B^+|A^-)$, we can conclude that doctor B is eight times as likely to diagnose a patient as positive when doctor A diagnoses the patient as positive more than when doctor A diagnoses the patient as negative. ∎

2.2 Gambler's Ruin Problem

Gambling is to play a game of chance in which the player can win or lose money. Gambling has become exceedingly more popular than before, and there are now numerous types of games played for money in casinos, online, and other places. Although many religions and jurisdictions either totally ban gambling or heavily control it, it is a very major national and international commercial activity. The way they principally do it is to make the payout on the bet slightly less than the actual odds of winning.

Some gamblers believe that a particular outcome is more likely, because it has not happened recently, or conversely because a particular outcome has recently occurred, it will be less likely in the immediate future. They mistakenly believe past streaks must eventually even out over the next short while. The *law of averages*, sometimes referred to as the *gambler's fallacy*, is a layman's term for a false belief that the statistical distribution of outcomes in a small sample size must reflect the distribution of outcomes across the population as a whole. The law of averages has absolutely no base in probability theory. It is imperative to note that the long period during which one vehemently holds on to a belief and the fact that a significant number of people also have the very same belief are no proof at all that the belief is true. All gamblers who believe in the law of averages are wrong.

Example 2.5

Two players A and B start to gamble until one of them is bankrupt, A starts with a capital of $\$a$ and B starts with a capital of $\$b$. The loser pays $\$1$ to the winner in each game, and the games are independent. Let p be the probability of winning each game for the player A and q for the player B, where $p + q = 1$. Determine the probability of ruin for each player, if no limit is set for the number of games they can play.

Solution

Let $P(k)$ denote the probability of ultimate ruin for the player A when he has k dollars, where $0 \leq k \leq a + b$. Note that we have $P(0) = 1$, that is the player A has no money left to play and is thus ruined, and $P(a + b) = 0$, that is the player A has $a + b$ dollars and the player B has no money left to play and is thus ruined.

The ruin of the player A can occur in one of the two mutually exclusive ways. Either the player A wins the next game with probability p and his money is thus increased to $k + 1$ dollars and the probability of being ruined ultimately then equals $P(k + 1)$ or the player A loses the next game with probability q and his money is thus decreased to $k - 1$ dollars and the probability of being ruined ultimately then equals $P(k - 1)$. By using the law of total probability, we thus have

$$P(k) = p\,P(k + 1) + q\,P(k - 1)$$

The solution to the above difference equation with conditions $P(0) = 1$ and $P(a + b) = 0$ can be shown to be as follows:

$$P(k) = \begin{cases} \dfrac{1 - \left(\dfrac{p}{q}\right)^{a+b-k}}{1 - \left(\dfrac{p}{q}\right)^{a+b}} & p \neq q \\[4ex] 1 - \dfrac{k}{a + b} & p = q \end{cases}$$

Similarly we can get $Q(k)$, the probability of ultimate ruin for the player B when she has k dollars:

$$
Q(k) = \begin{cases} \dfrac{1 - \left(\dfrac{q}{p}\right)^{a+b-k}}{1 - \left(\dfrac{q}{p}\right)^{a+b}} & p \neq q \\[2em] 1 - \dfrac{k}{a+b} & p = q \end{cases}
$$

By substituting $k = a$ in $P(k)$ and $k = b$ in $Q(k)$, we get the following probabilities when the game is going to start:

$$
P(a) = \begin{cases} \dfrac{1 - \left(\dfrac{p}{q}\right)^{b}}{1 - \left(\dfrac{p}{q}\right)^{a+b}} & p \neq q \\[2em] \dfrac{b}{a+b} & p = q \end{cases}
\qquad
Q(b) = \begin{cases} \dfrac{1 - \left(\dfrac{q}{p}\right)^{a}}{1 - \left(\dfrac{q}{p}\right)^{a+b}} & p \neq q \\[2em] \dfrac{a}{a+b} & p = q \end{cases}
$$

We observe that we have $P(a) + Q(b) = 1$, which means that the series of games will not go on forever and either A or B will eventually win. Note that when both players are of equal skills, i.e. $p = q = \frac{1}{2}$, their probabilities of ruin are not equal, and in fact they are inversely proportional to the money the players start with. With equal probability, the player who has a great deal less money than the other player will certainly go bankrupt. In other words, if $a \gg b$, then $P(a) \to 0$ and $Q(b) \to 1$, and if $b \gg a$, then $Q(b) \to 0$ and $P(a) \to 1$. It is critically important to highlight that the player who is more skillful and has more money will never be ruined in the course of the games and the other player's ruin is certain in the long run. All casinos work on this principle that is why the house always wins. ∎

2.3 Systems Reliability

Reliability is an important aspect of analysis, design, and operation of systems. Since it is possible that some of the subsystems fail, it is essential that the system continues operation in case of some subsystem failures. With redundant components in the system, probability of system failure can be minimized.

To assess the reliability of a system with a number of components, we assume that the components can fail independently, and the probability of failure in the ith component is p_i, where $0 \leq p_i \leq 1$, that is its probability of success or functioning is $1 - p_i$. It is possible that a system consists of components in series or in parallel, examples of such configurations are shown in Figure 2.1. Note that

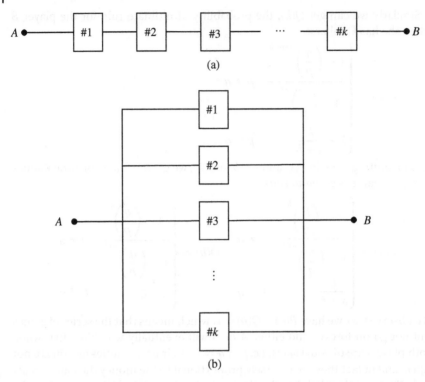

Figure 2.1 Link configuration with k components: (a) series and (b) parallel.

when it is difficult to calculate directly the probability of interest, we can often calculate the probability of the complementary event and then subtract it from one to determine the probability of interest.

Consider a system that consists of k components in series. Such a system functions if all k components are functioning or it fails if any one component fails. Note that the probability that the system functions is lower than the functioning probability of the weakest component. The probabilities that a system with k components in series functions or fails are thus, respectively, as follows:

$$\text{Probability of Functioning} = \prod_{i=1}^{k}(1 - p_i)$$

$$\text{Probability of Failure} = 1 - \prod_{i=1}^{k}(1 - p_i) \tag{2.4}$$

In a system with components in series, it is always much easier to determine the probability of functioning first, and then the probability of failure.

Consider a system that consists of k components in parallel. Such a system functions if at least a component is functioning or it fails if all k components fail. Note that the probability that the system functions is higher than the functioning probability of the strongest component. The probabilities that a system with k components in parallel fails or functions are thus, respectively, as follows:

$$\text{Probability of Failure} = \prod_{i=1}^{k} p_i$$

$$\text{Probability of Functioning} = 1 - \prod_{i=1}^{k} p_i \qquad (2.5)$$

In a system with components in parallel, it is always much easier to determine the probability of failure first, and then the probability of functioning.

Sometimes, there are some components in a system due to which the system does not consist of only series and parallel components. In such a complex system, we can take conditional probability approach through which mutually exclusive conditions are considered. We first assume some components have failed; the system can then boil down to a system consisting of just series and parallel components and the conditional probability of interest is found. We then assume that the same components can never fail; the system can then boil down to a system consisting of just series and parallel components and the conditional probability of interest is found. Using the law of total probability, the probability of interest is then determined.

Example 2.6

As shown in Figure 2.2a, a network connecting the end points A and D has five identical links. The links are assumed to fail independently, and the probability of failure in a link is p. Determine the probability of failure between the end points A and D in terms of the probability p.

Solution

We first define $P(F_i)$ as the probability that link #i fails, for $i = 1, 2, 3, 4, 5$, and $P(F)$ as the probability that the entire network fails. One way to find the probability of network failure is to include all 32 ($=2^5$) possible events that involve in the failure or nonfailure of the five links. However, the law of total probability offers a simpler approach.

Link #5 (the link between B and C) makes this problem complex, because it is causing the network not to be consisting of just series and parallel links. We therefore find the conditional probability of network failure under two mutually exclusive conditions.

In one condition, we assume that link #5 has failed and can thus be viewed as nonexistent. This in a way is similar to the concept of open circuit in the

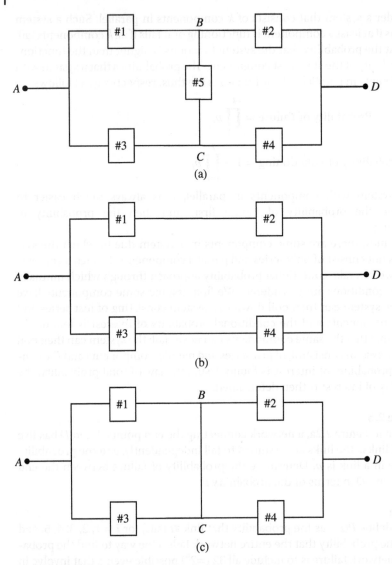

Figure 2.2 (a) Link configuration. (b) Configuration with permanently failed link #5. (c) Configuration with never-failing link #5.

context of circuit theory. As shown in Figure 2.2b, we then have two branches in parallel, and in each branch, there are two links in series. The conditional probability that the network fails given that link #5 has failed is thus as follows:

$$P(F \mid F_5) = P((F_1 \cup F_2) \cap (F_3 \cup F_4)) = (1 - (1 - p)^2)^2 = p^2(2 - p)^2$$

In the other condition, we assume that link #5 is a permanent link with no possibility of failure. This in a way is similar to the concept of short circuit in the context of circuit theory. As shown in Figure 2.2c, we then have two parts in series and in each part, there are two links in parallel. The conditional probability that the network fails given that link #5 never fails is thus as follows:

$$P(F \mid \overline{F_5}) = P((F_1 \cap F_3) \cup (F_2 \cap F_4)) = 1 - (1 - p^2)^2 = p^2(2 - p^2)$$

The law of total probability thus determines the probability of system failure:

$$
\begin{aligned}
P(F) &= P(F \mid F_5) P(F_5) + P(F \mid \overline{F_5})P(\overline{F_5}) \\
&= (p^2(2 - p)^2)p + (p^2(2 - p^2))(1 - p) \\
&= p^2(2p^3 - 5p^2 + 2p + 2)
\end{aligned}
$$

Note that $P(F)$ is not a linear function of p. As expected, for $p = 0$, we have $P(F) = 0$, for $p = 1$, we have $P(F) = 1$, and for $p = \frac{1}{2}$, we have $P(F) = \frac{1}{2}$. Since in practice we have $0 < p \ll 1$, we have $P(F) \cong 2p^2$. ∎

2.4 Medical Diagnostic Testing

In this section, we highlight an interesting application of Bayes' rule to medicine. The results obtained are quite surprising, and often perceived as counter-intuitive. Suppose that there is a particular rare disease that is independently and identically distributed throughout the general population. We further assume that genetics and environmental factors do not play any role, a randomly selected individual can thus have it.

As no medical test is perfect, the diagnostic test result can be negative for a person who has the disease, known as *false negative* or can be positive for a person who does not have the disease, known as *false positive*. A successful result is either a true positive or a true negative. Assuming A is defined as the event that a person selected at random has the disease, and B is defined as the event that the test result is positive, the following set of information is assumed to be known:

– $P(B^c \mid A) = \alpha$, which is the probability of having a negative test result given the person has the disease. This is known as the *false negative probability*, as the test misses something that does exist. We thus have $P(B \mid A) = 1 - \alpha$, which is the probability of having a positive test result given the person has the disease, also known as the *true positive probability*.
– $P(B \mid A^c) = \beta$, which is the probability of having a positive test result given the person does not have the disease. This is known as the *false positive probability*, as the test detects something that does not exist. We thus have $P(B^c \mid A^c) = 1 - \beta$, which is the probability of having a negative test result

given the person does not have the disease, also known as the ***true negative probability***.

- $P(A) = \rho$, which is the probability that a person selected at random has the disease. We thus have $P(A^c) = 1 - \rho$, which is the probability that a person in general population does not have the disease.

Based on the above information, we are interested to find the following a posteriori probabilities of interest, in terms of α, β, and ρ:

- $P(A \mid B)$, which is the probability that a person who tests positive for the disease has the disease.
- $P(A^c \mid B^c)$, which is the probability that a person who tests negative for the disease does not have the disease.
- $P(A^c \mid B) = 1 - P(A \mid B)$, which is the probability that a person who tests positive for the disease does not have the disease.
- $P(A \mid B^c) = 1 - P(A^c \mid B^c)$, which is the probability that a person who tests negative for the disease has the disease.

Note that for a perfect test, the a posteriori probabilities $P(A \mid B)$ and $P(A^c \mid B^c)$ must be one or equivalently, the a posteriori probabilities $P(A^c \mid B)$ and $P(A \mid B^c)$ must be zero. However, even for the most accurate medical diagnostic test, these ideal a posteriori probabilities cannot be achieved, as tests are always flawed. Using Bayes' rule, we can get the following results:

$$
\begin{cases}
P(A \mid B) = \dfrac{(1 - \alpha)\rho}{(1 - \alpha)\rho + \beta(1 - \rho)} \\[2mm]
P(A^c \mid B) = \dfrac{\beta(1 - \rho)}{(1 - \alpha)\rho + \beta(1 - \rho)}
\end{cases}
$$

$$
\begin{cases}
P(A^c \mid B^c) = \dfrac{(1 - \beta)(1 - \rho)}{(1 - \beta)(1 - \rho) + \alpha\rho} \\[2mm]
P(A \mid B^c) = \dfrac{\alpha\rho}{(1 - \beta)(1 - \rho) + \alpha\rho}
\end{cases}
\tag{2.6}
$$

Example 2.7
Suppose that 1% of women have breast cancer. The research record indicates that 80% of mammograms detect breast cancer when it is there, and 10% of mammograms detect breast cancer when it is not there. Determine the probability that a woman who tests negative for the cancer does not have cancer, and the probability that a woman who tests positive for the cancer has cancer. Comment on the results.

Solution
Assuming A is defined as the event that a woman has cancer, and B is defined as the event that the mammogram test result is positive, we have $P(A) = \rho = 1\%$, $P(B \mid A) = 1 - \alpha = 80\%$, and $P(B \mid A^c) = \beta = 10\%$. It is thus important to note

that 99% of women do not have cancer, as we have $P(A^c) = 1 - \rho = 99\%$, 20% of mammograms miss cancer, as we have $P(B^c \mid A) = \alpha = 20\%$, and 90% correctly return a negative test result, as we have $P(B^c \mid A^c) = 1 - \beta = 90\%$. The following table captures the available information:

	With cancer (1%)	No cancer (99%)
Test positive	80%	10%
Test negative	20%	90%

For a positive test result, we are in the top row of the table, it could then be either a true positive or a false positive, and for a negative test result, we are in the bottom row of the table, it could then be either a true negative or a false negative, but only 1% of the time, we are in the first column and 99% of the time, we are in the second column. The following table provides all four possible cases with their corresponding probabilities:

	With cancer	No cancer
Test positive →	80% × 1% = 0.8% (true positive)	10% × 99% = 9.9% (false positive)
Test negative →	20% × 1% = 0.2% (false negative)	90% × 99% = 89.1% (true negative)

Note that, as expected, the sum of all four probabilities in the above table is 1 (=0.8 + 0.2 + 9.9 + 89.1). We can thus get the following probabilities:

$$P(A^c \mid B^c) = \frac{(1 - 0.1)(1 - 0.01)}{(1 - 0.1)(1 - 0.01) + 0.2 \times 0.01} \cong 99.8\%$$

$$P(A \mid B) = \frac{(1 - 0.2) \times 0.01}{(1 - 0.2) \times 0.01 + 0.1 \times (1 - 0.01)} \cong 7.5\%$$

As the above result may seem counterintuitive, we need to provide some insight into it. A positive mammogram only means a woman has about a 7.5% chance of having cancer. This makes sense simply because the test gives a false positive 10% of the time, and there are quite many false positives in any given population, as most people do not have cancer. In fact, there are so many positives that most of them will be wrong. By examining the above tables, we can simply say that if we take, for instance, 1000 women in general population, only 10 of them will have cancer and 8 of which will be true positive. Of the 990 remaining women who do not have cancer, 10% will test positive, so there will be 99 false positives. We therefore have a total of 107(= 8 + 99) positives, of which only 8 are true positive, so there is about $\frac{8}{107} \cong 7.5\%$ chance of having cancer given a positive test, as confirmed earlier.

In view of the fact that no medical diagnostic test is ideal, the probability that a person who tests positive for the disease has the disease is always less than 100%. This in turn may be a good enough reason to seek out a second opinion. ∎

2.5 Bayesian Spam Filtering

Spam e-mails, also known as junk e-mails, are unwanted, irrelevant e-mails sent on the Internet to a large number of recipients, and may contain disguised links, malware, and fraudulent messages. *Bayesian spam filtering* is a popular spam-filtering technique that relies on word probabilities and can tailor itself to the e-mail needs of individual users. For instance, we may say if an e-mail contains certain words or group of words – such as urgent, bank, no obligation, no catch, act now, risk free, no gimmick, account, money, special, million, deal, virus, update – that are often used in spam, and some words – such as food, appointment, car, see, dinner, kids, book, hope, time, good, glasses, sun screen – that are hardly used in spam, it is then likely to be spam. Bayesian spam filters look for occurrences of particular words. Such words have particular probabilities of occurring in a spam e-mail vis-à-vis in a nonspam e-mail.

Let S be the event that the e-mail is spam and S^c be the event that the e-mail is not spam. Suppose we have received a new e-mail with a particular word, say X_1. Let W_1 be the event that the e-mail contains the word X_1. We also have the empirical probability that a spam e-mail contains the word X_1 is $P(W_1 \mid S) = a(X_1)$ and the empirical probability that an e-mail that is not spam and contains the word X_1 is $P(W_1 \mid S^c) = b(X_1)$. A Bayesian spam detection software generally makes the assumption that there is no a priori reason for any incoming e-mail to be spam rather than nonspam, and considers both cases to have equal probabilities, i.e. $P(S) = P(S^c) = 0.5$. Using Bayes' theorem, p_1 the probability that the e-mail is spam, given that it contains X_1, is as follows:

$$P(S \mid W_1) = \frac{P(W_1 \mid S)P(S)}{P(W_1 \mid S)P(S) + P(W_1 \mid S^c)P(S^c)} \rightarrow p_1 = \frac{a(X_1)}{a(X_1) + b(X_1)}$$

(2.7)

If p_1 is greater than a certain threshold, then the software classifies it as spam. Using a single word for the detection of spam may not be very effective, as it can lead to excessive false positives and false negatives. If we use k words, say $X_1, X_2,$ ..., X_k, we can then increase significantly the probability of detecting spam correctly. Assuming that $W_1, W_2, ..., W_k$ are the events that the e-mail contains the words $X_1, X_2, ..., X_k$, the number of incoming spam e-mails is approximately the same as the number of incoming e-mails that are not spam, and the events $W_1 \mid S, W_2 \mid S, ..., W_k \mid S$ are independent, by using Bayes' theorem, p_k

the probability that an e-mail containing all the words X_1, X_2, ..., X_k is spam can thus be obtained as follows:

$$
\begin{aligned}
p_k &= P(S \mid (W_1 \cap W_2 \cap \cdots \cap W_k)) \\
&= \frac{(P(W_1 \mid S) \times \cdots \times P(W_k \mid S))}{(P(W_1 \mid S) \times \cdots \times P(W_k \mid S)) + (P(W_1 \mid S^c) \times \cdots \times P(W_k \mid S^c))} \\
&= \frac{(a(X_1) \times \cdots \times a(X_k))}{(a(X_1) \times \cdots \times a(X_k)) + (b(X_1) \times \cdots \times b(X_k))}
\end{aligned}
\tag{2.8}
$$

Example 2.8

Suppose a Bayesian spam filter is trained on a set of 10 000 spam e-mails and 5000 e-mails that are not spam. The word *information* appears in 1500 spam e-mails and 20 e-mails that are not spam, the word *account* appears in 800 spam e-mails and 200 e-mails that are not spam, and the word *time* appears in 200 spam e-mails and 500 e-mails that are not spam. Estimate the probability that a received e-mail containing all three words of "information," "account," and "time" is spam. Will the e-mail be rejected as spam if the threshold for rejecting spam is set at 0.9?

Solution

Assuming X_1, X_2, and X_3 refer to the words "information," "account," and "time," respectively, we have the following probabilities:

$$a(X_1) = \frac{1500}{10000} = 0.15 \quad \text{and} \quad b(X_1) = \frac{20}{5000} = 0.004$$

$$a(X_2) = \frac{800}{10000} = 0.08 \quad \text{and} \quad b(X_2) = \frac{200}{5000} = 0.04$$

$$a(X_3) = \frac{200}{10000} = 0.02 \quad \text{and} \quad b(X_3) = \frac{500}{5000} = 0.1$$

We can thus obtain:

$$p_3 = \frac{0.15 \times 0.08 \times 0.02}{0.15 \times 0.08 \times 0.02 + 0.004 \times 0.04 \times 0.1} = 0.9375$$

As $p_3 > 0.9$, an incoming e-mail containing all these three words will be rejected. ∎

2.6 Monty Hall Problem

The Monty Hall problem was named after Monty Hall, the host of the original TV game show Let's Make a Deal. This problem is controversial, as most people find the solution counterintuitive.

The **Monty Hall problem** is as follows: You are a participant on a game show where given the choice of three doors, you will win what is behind your chosen door. Behind one door is a valuable prize, such as a new car, and behind the

Table 2.2 Results of staying or switching in a Monty Hall problem.

The door originally picked	A door not picked	Another door not picked	Staying at original door	Switching to the other closed door
Valuable prize	Worthless prize	Worthless prize	*Valuable prize*	Worthless prize
Worthless prize	*Valuable prize*	Worthless prize	Worthless prize	*Valuable prize*
Worthless prize	Worthless prize	*Valuable prize*	Worthless prize	*Valuable prize*

other two are worthless prizes, such as a donkey or a goat. All prizes are placed randomly behind the doors before the show starts.

The rules of the game show are as follows: After you have chosen a door, the door remains closed for the time being. The game show host, who knows what is behind each door, now has to open one of the two remaining doors, and the door he opens must have a worthless prize behind it. If both remaining doors have worthless prizes behind them, he chooses one at random. After the host opens a door with a worthless prize, he will ask you to decide whether you want to stay with your original choice or to switch to the other closed door. In order to win the valuable prize, is it better for you to stay or to switch?

Table 2.2 presents the solution that shows the three possible arrangements of one valuable prize and two worthless prizes behind three doors and the result of staying or switching after initially picking a door in each case.

A player who stays with the initial choice wins in only one out of three equally likely possibilities, while a player who switches wins in two out of three. Valuable prize has a $\frac{1}{3}$ chance of being behind the player's pick and a $\frac{2}{3}$ chance of being behind one of the other two doors. The host opens a door, the odds for these two probabilities do not change but the odds move to 0 for the open door and $\frac{2}{3}$ to the other closed door. The fact that the host subsequently reveals a worthless prize in one of the unchosen doors changes nothing about the initial probability, after all the host always opens a door with a worthless prize. The host is saying in effect, you can keep your one door or you can have the other two doors. By opening his door, the host is saying to you that there are two doors you did not choose, and the probability that the prize is behind one of them is $\frac{2}{3}$.

Example 2.9

Using the probability theory, prove that in the Monty Hall problem, by switching, you can increase your chances of winning from $\frac{1}{3}$ to $\frac{2}{3}$.

Solution

We define the following events, where they all are equally likely:

$$C1 : \text{The car is behind door } 1 \rightarrow P(C1) = \frac{1}{3}$$

$$C2 : \text{The car is behind door } 2 \rightarrow P(C2) = \frac{1}{3}$$

$$C3 : \text{The car is behind door } 3 \rightarrow P(C3) = \frac{1}{3}$$

Since the first choice of the player is independent of the position of the car, we assume the player initially choosing door 1, as described by the event *D1*. The conditional probabilities are as follows:

$$P(C1 \mid D1) = \frac{1}{3} \quad \text{and} \quad P(C2 \mid D1) = \frac{1}{3} \quad \text{and} \quad P(C3 \mid D1) = \frac{1}{3}$$

With the host opening door 2, described by *H2*, and with the host opening door 3, described by *H3*, we have the following probabilities:

$$P(H2 \mid C1, D1) = 0.5 \quad \text{and} \quad P(H3 \mid C1, D1) = 0.5$$

$$P(H2 \mid C2, D1) = 0 \quad \text{and} \quad P(H3 \mid C2, D1) = 1$$

$$P(H2 \mid C3, D1) = 1 \quad \text{and} \quad P(H3 \mid C3, D1) = 0$$

Then, if the player initially selects door 1, and the host opens door 3, the conditional probability of winning by switching is

$$
\begin{aligned}
P(C2 \mid H3, D1) &= \frac{P(H3, C2, D1)}{P(H3, D1)} \\[2mm]
&= \frac{P(H3 \mid C2, D1)P(C2, D1)}{P(H3, C1, D1) + P(H3, C2, D1) + P(H3, C3, D1)} \\[2mm]
&= \frac{P(H3 \mid C2, D1)P(C2 \mid D1)P(D1)}{(P(H3 \mid C1, D1)P(C1, D1) + P(H3 \mid C2, D1) \\ \times P(C2, D1) + P(H3 \mid C3, D1)P(C3, D1))} \\[2mm]
&= \frac{P(H3 \mid C2, D1)P(C2 \mid D1)P(D1)}{(P(H3 \mid C1, D1)P(C1 \mid D1)P(D1) + P(H3 \mid C2, D1) \\ \times P(C2 \mid D1)P(D1) + P(H3 \mid C3, D1)P(C3 \mid D1)P(D1))} \\[2mm]
&= \frac{P(H3 \mid C2, D1)}{P(H3 \mid C1, D1) + P(H3 \mid C2, D1) + P(H3 \mid C3, D1)} \\[2mm]
&= \frac{1}{0.5 + 1 + 0} = \frac{2}{3}
\end{aligned}
$$

∎

2.7   Digital Transmission Error

In a digital transmission system, the input is a stream of random binary digits (bits), and the output is also a stream of bits. An error occurs when a received bit is not the same as the transmitted bit. The error occurs due to an array of communications channel impairments, such as distortion, noise, interference, and attenuation. As the ultimate measure of performance in a digital transmission system is the bit error rate, the calculation and minimization of the bit error rate is of great importance in digital communications systems and networks.

Example 2.10
In a binary symmetric communication channel, the input bits transmitted over the channel are either 0 or 1 with probabilities p and $1 - p$, respectively. As shown in Figure 2.3, the channel is assumed to be symmetric, which means the probability of receiving 1 when 0 is transmitted is the same as the probability of receiving 0 when 1 is transmitted. The conditional probability of error ε is assumed to be $0 \le \varepsilon \le 1$. Determine the bit error rate as well as the a posteriori probabilities.

Solution
Let X and Y denote the channel input and output, respectively. Note that the effect of the transmission is to alter the probability of each possible input from its a priori probability to its a posteriori probability. The a priori probabilities of bits are as follows:

$$\begin{cases} P(X = 0) = p \\ P(X = 1) = 1 - p \end{cases}$$

With the transition probabilities of

$$P(Y = 1 \mid X = 0) = P(Y = 0 \mid X = 1) = \varepsilon$$

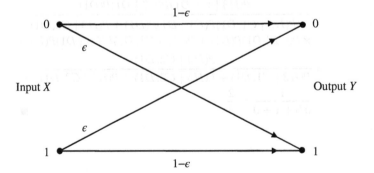

Figure 2.3 Transition probability diagram of binary symmetric channel.

and applying the law of total probability, the bit error rate can be calculated as follows:

$$P(e) = P(Y = 1, X = 0) + P(Y = 0, X = 1)$$
$$= P(Y = 1 \mid X = 0)P(X = 0)$$
$$+ P(Y = 0 \mid X = 1)P(X = 1) = \varepsilon p + \varepsilon(1 - p) = \varepsilon$$

Using Bayes' rule, the a posteriori probabilities are then as follows:

$$P(X = 0 \mid Y = 0) = \frac{P(Y = 0 \mid X = 0)P(X = 0)}{P(Y = 0)}$$
$$= \frac{P(Y = 0 \mid X = 0)P(X = 0)}{P(Y = 0, X = 0) + P(Y = 0, X = 1)}$$
$$= \frac{P(Y = 0 \mid X = 0)P(X = 0)}{(P(Y = 0 \mid X = 0)P(X = 0)}{+ P(Y = 0 \mid X = 1)P(X = 1))}$$
$$= \frac{(1 - \varepsilon)p}{(1 - \varepsilon)p + \varepsilon(1 - p)} = \frac{p - \varepsilon p}{p + \varepsilon - 2\varepsilon p}$$

and

$$P(X = 1 \mid Y = 1) = \frac{P(Y = 1 \mid X = 1)P(X = 1)}{P(Y = 1)}$$
$$= \frac{P(Y = 1 \mid X = 1)P(X = 1)}{P(Y = 1, X = 1) + P(Y = 1, X = 0)}$$
$$= \frac{P(Y = 1 \mid X = 1)P(X = 1)}{(P(Y = 1 \mid X = 1)P(X = 1)}{+ P(Y = 1 \mid X = 0)P(X = 0))}$$
$$= \frac{(1 - \varepsilon)(1 - p)}{(1 - \varepsilon)(1 - p) + \varepsilon p} = \frac{1 - p - \varepsilon + \varepsilon p}{1 - p - \varepsilon + 2\varepsilon p}$$

The following interesting observations can now be made:

- For $\varepsilon = 0$, i.e. when the channel is ideal, both a posteriori probabilities are one.
- For $\varepsilon = \frac{1}{2}$, the a posteriori probabilities are the same as the a priori probabilities.
- For $\varepsilon = 1$, i.e. when the channel is most destructive, both a posteriori probabilities are zero.
- For $p = \frac{1}{2}$, both a posteriori probabilities are $1 - \varepsilon$. ∎

2.8 How to Make the Best Choice Problem

This is a famous problem, also known as selecting-a-partner (marriage) problem or hiring (secretary) problem. The basic form of *how to make the best choice problem* is as follows: You are faced with a choice among n candidates, where n is a positive integer. How do you choose among them, or in probabilistic terms, what decision rule should you use to maximize the probability of choosing the best candidate? Note that the basic assumptions are as follows:

- There are n rankable candidates to be assessed, and the value of n is known in advance.
- The order of arrival of candidates is random.
- A score is assigned to the candidate who is under evaluation.
- No two candidates have the same score, i.e. ties do not exist.
- The best choice is the candidate with the highest score.
- The candidates are assessed one at a time sequentially, i.e. one by one.
- A decision is made about each candidate immediately right after the individual evaluation has been performed and right before the next candidate is assessed.
- A decision on a candidate is made solely based on the relative ranks of the candidates assessed previously.
- A decision to accept or reject a candidate is final and absolute, i.e. the decision is irrevocable.
- Once a candidate is accepted as the best choice, there is no possibility to assess the remaining candidates, i.e. the quality of yet unseen candidates and the absolute ranks of all candidates will ever remain unknown.
- If a decision has not been made by the time the nth candidate (i.e. the last candidate) is being assessed, then that last, i.e. the nth, candidate is chosen.

Some of these assumptions may not be reasonable in real-life applications. For instance, the fact that the value of n is known in advance or there are no two ties may not be reasonable in some cases. The difficulty of this problem is that the decision must be made immediately right after the individual evaluation. The problem has an elegant solution when the following strategy is employed:

(i) Assess the first $k < n$ candidates, but reject all k candidates, regardless of the quality of the k candidates.
(ii) Select the very first candidate who is better than all of the previous candidates, or select the last candidate if no candidate has been selected before.

Let X_n denote the number (arrival order) of the best candidate, and S denote the event of success (the best candidate is selected). The order of arrival of individuals is random and uniformly distributed on $\{1, 2, \ldots, n\}$, hence we have

$P(X_n = i) = \frac{1}{n}, i = 1, 2, \ldots, n$. The strategy is to assess and reject the first k candidates, we thus have $P(S \mid X_n = i) = 0, i = 1, \ldots, k$. Moreover, if the arrival order of the best candidate is $i > k$, then the strategy will succeed if and only if one of the first k candidates (the ones that are automatically rejected) is the best among the first i candidates (i.e. the best candidate in $[1, i]$ is the same best candidate in $[1, k]$). Using the law of total probability, we have the following:

$$P(S) = \sum_{i=1}^{n} P(\text{candidate } i \text{ is selected and candidate } i \text{ is the best})$$

$$= \sum_{i=1}^{n} P(\text{candidate } i \text{ is selected} \mid \text{candidate } i \text{ is the best})$$

$$\times P(\text{candidate } i \text{ is the best})$$

$$= \left[\sum_{i=k+1}^{n} P(\text{the best of the first } i - 1 \text{ candidates is in the first}\right.$$

$$\left. k \text{ candidates} \mid \text{candidate } i \text{ is the best}) \right] \times \frac{1}{n} \qquad (2.9a)$$

$$P(S) = \sum_{i=1}^{n} P(S, X_n = i) = \sum_{i=1}^{n} P(S \mid X_n = i) P(X_n = i)$$

$$= \frac{1}{n} \sum_{i=k+1}^{n} P(S \mid X_n = i)$$

$$= \frac{1}{n} \sum_{i=k+1}^{n} P(X_{i-1} \le k) = \frac{1}{n} \sum_{i=k+1}^{n} \frac{k}{i-1} = \frac{k}{n} \sum_{i=k+1}^{n} \frac{1}{i-1} = \frac{k}{n} \sum_{j=k}^{n-1} \frac{1}{j}$$

$$\qquad (2.9b)$$

Letting n tend to infinity, the above sum can be approximated as follows:

$$\lim_{n \to \infty} P(S) \approx \frac{k}{n} \ln\left(\frac{n}{k}\right) = -\frac{k}{n} \ln\left(\frac{k}{n}\right) \qquad (2.10)$$

Assuming n is very large, it can be shown that the maximum value of $P(S)$ is $\frac{1}{e}$, where $e = 2.7182\ldots$ is the base of the natural logarithm, and it occurs when $\frac{k}{n} = \frac{1}{e} \cong 36.78\%$. Therefore, the best strategy is when k is the largest integer less than $\frac{n}{e}$. In fact, for any value of n, the probability of selecting the best candidate when using the optimal technique is at least $\frac{1}{e} \cong 36.78\%$. Table 2.3 shows the optimal probability $P(S)$ of finding the best candidate as a function of n, where the last column also reflects the probability of randomly selecting the best candidate, but with no particular strategy. As n increases, the impact of the optimal selection over the random selection becomes more prominent.

Table 2.3 Optimal probability $P(S)$ of finding the best candidate as a function of n.

n	k	k/n	P(S)	versus	1/n
1	0	0	1	=	1
2	0	0	0.5	=	0.5
3	1	0.333	0.5	>	0.333
4	1	0.250	0.458	>	0.250
5	1	0.200	0.433	>	0.200
6	2	0.333	0.428	>	0.167
7	2	0.286	0.414	>	0.143
8	2	0.250	0.410	>	0.125
9	3	0.333	0.406	>	0.111
10	3	0.300	0.399	>	0.100
20	7	0.350	0.384	>	0.050
50	18	0.360	0.374	>	0.020
100	36	0.360	0.371	>	0.010
500	183	0.366	0.369	>	0.002
1000	367	0.367	0.368	>	0.001

Example 2.11

Suppose there are three candidates with three different scores of A (highest score), B, and C (lowest score), what is the strategy to make the best choice?

Solution

Since $n = 3$, we have $k = 1$, as k is the largest integer less than $\frac{n}{e}$. The first step of the strategy is to asses the first of three candidates, but the first candidate is not selected as the best candidate. The second step is as follows: if the score of the second candidate is higher than that of the first one, select the second candidate; otherwise, select the third candidate as it is the last candidate. For $n = 3$, we have the following six distinct possible scenarios, where the order of candidates are different:

Order of candidates	First candidate	Best choice
A, B, C	A	C
A, C, B	A	B
B, A, C	B	A
B, C, A	B	A
C, B, A	C	B
C, A, B	C	A

With this strategy, there is a 50% chance we get to select A, had we not chosen this strategy, the probability of selecting A would have been $\frac{1}{3}$. ∎

2.9 The Viterbi Algorithm

The **Viterbi algorithm** is a dynamic programming algorithm to optimization problems. The algorithm has found many applications in digital communications and signal processing as well as telecommunications and transportation networks.

The Viterbi algorithm finds the optimum path, it may be the most likely path, the shortest path, the least costly path, or any other criterion of interest. It is based on the principle that any portion (sub-path) of the overall optimum path must be the optimum path between the end points of that portion (sub-path). This is obvious, since if it were not true, the portion (sub-path) could be replaced to yield a more likely sub-path or a shorter sub-path or a less costly sub-path or a sub-path that better meets the criterion of interest, a contradiction. The salient feature of the Viterbi algorithm is that it reduces the exponential complexity of a brute-force approach, which considers all possible paths, to an approach whose complexity grows linearly. The operation of the Viterbi algorithm can be best illustrated by means of a trellis diagram.

Example 2.12

Using the Viterbi algorithm, determine the most likely path between the nodes a and h in the trellis diagram shown in Figure 2.4a, where a number next to a dotted line represents the transition (conditional) probability from a node to an adjacent node on its right.

Solution

The most likely path between a and h must go through either b or c. If it goes through b, then the probability of getting up to b is $\frac{1}{2}$ and up to c is $\frac{1}{2}$, where the metrics at b and c are then $\frac{1}{2}$ and $\frac{1}{2}$, respectively, as shown in Figure 2.4b.

If the most likely path between a and h must go through d, then it must go through either b or c to get to d. The probability of getting from a to d via b is $\frac{3}{8}$ $\left(=\frac{1}{2}\times\frac{3}{4}\right)$ and that via c is $\frac{1}{3}$ $\left(=\frac{1}{2}\times\frac{2}{3}\right)$. Since we have $\frac{3}{8}>\frac{1}{3}$, it is obvious that the path from a to d via c is not part of the overall optimum path between a and h, and will therefore not be included in subsequent calculations, and the surviving path is from a to b to d. If the most likely path between a and h must go through e, then it must go through either b or c to get to e. The probability of getting from a to e via b is $\frac{1}{8}$ $\left(=\frac{1}{2}\times\frac{1}{4}\right)$ and that via c is $\frac{1}{6}$ $\left(=\frac{1}{2}\times\frac{1}{3}\right)$. Since we have $\frac{1}{6}>\frac{1}{8}$, it is obvious that the path from a to e via b is not part of the overall optimum path between a and h, and will therefore not be included in subsequent calculations, and the surviving path is from a to c to e. Note that the metrics at d and e are $\frac{3}{8}$ and $\frac{1}{6}$, respectively, as shown in Figure 2.4c.

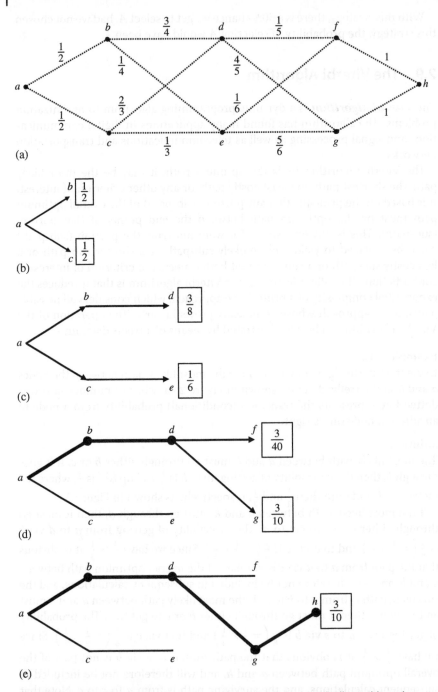

Figure 2.4 Steps in the Viterbi algorithm. (a) The trellis diagram with all possible paths.
(b) The initial possible paths. (c) The initial surviving paths. (d) The intermediate surviving
paths. (e) The optimum path.

If the most likely path between a and h must go through f, then it must go through either d or e to get to f. The probability of getting from a to f via d is $\frac{3}{40}$ $\left(= \frac{3}{8} \times \frac{1}{5}\right)$ and that via e is $\frac{1}{36}$ $\left(= \frac{1}{6} \times \frac{1}{6}\right)$. Since we have $\frac{3}{40} > \frac{1}{36}$, it is obvious that the path from a to f via e is not part of the overall optimum path between a and h, and will therefore not be included in subsequent calculations, and the surviving path is from a to b to d to f. If the most likely path between a and h must go through g, then it must go through either d or e to get to g. The probability of getting from a to g via d is $\frac{3}{10}$ $\left(= \frac{3}{8} \times \frac{4}{5}\right)$ and that via e is $\frac{5}{36}$ $\left(= \frac{1}{6} \times \frac{5}{6}\right)$. Since we have $\frac{3}{10} > \frac{5}{36}$, it is obvious that the path from a to g via e is not part of the overall optimum path between a and h, and will therefore not be included in subsequent calculations, and the surviving path is from a to b to d to g. It is already evident that the most likely path between a and h, whether it goes through f or g, will not include the path going through e. In other words, we already know that the path from a to b to d is part of the optimum path between a and h, where the metrics at f and g are $\frac{3}{40}$ and $\frac{3}{10}$, respectively, as shown in Figure 2.4d.

The probability of getting from a to h via f is $\frac{3}{40}$ $\left(= \frac{3}{40} \times 1\right)$ and that via g is $\frac{3}{10}$ $\left(= \frac{3}{10} \times 1\right)$. Since we have $\frac{3}{10} > \frac{3}{40}$, it is obvious that the path from a to h via f is not the overall optimum path between a and h and the overall optimum path is thus from a to b to d to g to h, as shown in Figure 2.4e.

In this example, between a and h, we have three sets of two states, ((b, c), (d, e), (f, g)), i.e. there are $2^3 = 8$ possible paths. With N sets of M states, the complexity of a brute force approach is M^N, whereas the complexity associated with the Viterbi algorithm is MN. The Viterbi algorithm thus yields linear complexity, whereas the brute force algorithm, which compares all possible path combinations to find the most likely path, gives rise to exponential complexity. ∎

2.10 All Eggs in One Basket

You can put all your eggs in one basket if you are certain (i.e. with the probability of one) that the basket will not break or will not be dropped. In the real world, however, absolute certainty almost never exists, and if the basket is dropped, then all is lost. So spreading your eggs among multiple baskets is a better way to make sure that you are still left with some eggs. The famous expression "do not put all your eggs in one basket" implies that you do not make everything dependent on only one thing or do not concentrate all your prospects or resources or investments in one thing, as you could lose everything you have. This in turn highlights the importance of diversification.

Suppose the probability that a basket is dropped is q, and when there are more than one basket, dropping baskets are statistically independent events. If there is only one basket, then the probability of losing all eggs is q and the probability of keeping all eggs is $1-q$. If there are two baskets, the probability of losing all eggs is $q \times q = q^2$, the probability of keeping all eggs is $(1-q) \times (1-q) = (1-q)^2$, and the probability that the eggs in one basket are all lost but those in the other basket are all kept is $q \times (1-q) + (1-q) \times q = 2q(1-q)$. Assuming n represents the number of baskets, the following table summarizes various possibilities with their corresponding probabilities:

Number of baskets	→	1	2	$\cdots$	n
Probability that all eggs are kept	→	$1-q$	$(1-q)^2$	$\cdots$	$(1-q)^n$
Probability that some eggs are kept	→	0	$2q(1-q)$	$\cdots$	$(1-q^n)-(1-q)^n$
Probability that all eggs are lost	→	q	q^2	$\cdots$	q^n

As reflected in the above table, an increase in the number of baskets, from 1 to $n > 1$, will reduce the probability of keeping all eggs, from $(1-q)$ to $(1-q)^n$, as well as the probability of losing all eggs, from q to q^n, but will increase the probability of keeping some eggs from zero to $(1-q^n)-(1-q)^n$.

Example 2.13

Suppose there are $n \geq 1$ individuals in a group each shooting once at the same target, where each individual can hit the target with the probability of p, and can of course miss it with the probability of $1-p$. Determine the probability that at least one of them hits the target. Comment on the results.

Solution

Since the shootings are independent of one another, the probability that none of the n individuals in the group hit the target is $(1-p)^n$. Therefore, the probability that at least one of the individuals hits the target is as follows:

$$P = 1 - (1-p)^n \rightarrow n = \frac{\log(1-P)}{\log(1-p)}$$

For given values of P and p, we can find n. Since n must be an integer, n is the smallest integer greater than $\frac{\log(1-P)}{\log(1-p)}$. In fact, for a rather small probability p and a rather large probability P, surprisingly n does not need to be extremely large. For instance, with $p = 50\%$ and $P = 98\%$, we have $n = 6$, and with $p = 30\%$ and $n = 20$, we have $P = 99.9\%$. Figure 2.5 shows for various values of n, the probability of the target being hit, i.e. the likelihood of the group success P, as a function of p the probability that a single individual can hit the target.

It is imperative to highlight the fact that no matter how low p (the likelihood of an individual success) may be, with a large enough number of individuals n, P (the likelihood of the group success) can be as high as it needs to be. After all, there is strength in numbers. ∎

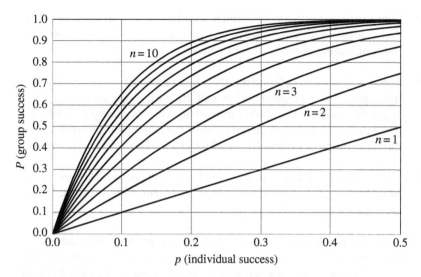

Figure 2.5 Probability of group success versus probability of individual success.

2.11 Summary

In this chapter, some major probability applications in science and engineering, such as systems reliability, medical diagnostic testing, and Bayesian spam filtering, as well as some interesting probability applications in life, such as odds and risk, gambler's ruin problem, and how to make the best choice problem, were briefly discussed. Specific examples were provided to highlight the applications introduced in this chapter.

Problems

2.1 On a multiple-choice test, a student either knows the answer with probability p or guesses with probability of $1 - p$. As m is the number of choices in a question, a student who guesses at the answer will be correct with probability $\frac{1}{m}$. Determine the conditional probability that a student knew the answer to a question, given that the student answered it correctly.

2.2 At a certain stage of a criminal investigation, with probability p, the officer in charge is convinced of the guilt of a certain suspect. Suppose a new piece of evidence that shows the criminal has a certain characteristic (such as left-handedness, blue eyes, above-average height, short hair) is uncovered and we know that r percent of the general population possesses this characteristic. Assuming the suspect has this characteristic, determine how certain of the guilt of the suspect should the officer now be.

2.3 Suppose there are five female students and three male students in a room. The students will have to leave one by one. Determine the most likely order of leaving in terms of their gender.

2.4 Suppose that we have found out that the word *money* occurs 500 of 4000 messages known as spam and in x of 2000 messages known not to be spam. Suppose our threshold for rejecting a message as spam is 90% and it is equally likely that an incoming message is spam or not spam. Determine the range of values of x for which the messages will be rejected.

2.5 Suppose there are four doors in the Monty Hall problem. Determine the probabilities of staying or switching. Comment on the results, if there are n doors.

2.6 A system consists of n subsystems in series. Each of the n subsystems consists of parallel components, where the first subsystem consists of n_1 parallel components, the second subsystem consists of n_2 parallel components, and finally, the nth subsystem consists of n_n parallel components. Assume that the probability of failure of a component is p and the components fail independently. Noting that $n, n_1, n_2, \ldots, n_n$ are all positive integers and may be different from one another, determine the probability that the system functions.

2.7 A system consists of m subsystems in parallel. Each of the m subsystems consists of components in series, where the first subsystem consists of m_1 components in series, the second subsystem consists of m_2 components in series, and finally the mth subsystem consists of m_m components in series. Assume that the probability of failure of a component is p and the components fail independently. Noting that $m, m_1, m_2, \ldots, m_m$ are all positive integers and may be different from one another, determine the probability that the system functions.

2.8 Algorithms that make random choices at one or more steps are called probabilistic algorithms. A particular class of probabilistic algorithms is Monte Carlo algorithms. Monte Carlo algorithms always produce answers to decision problems, but a small probability remains that these answers may be incorrect. A Monte Carlo algorithm uses a sequence of tests and the probability that the algorithm answers the decision problem correctly increases as more tests are carried out. Suppose there are $n \gg 0$ items in a batch, and the probability that an item is defective is p when random testing is done. To decide all items are good, n tests are required to guarantee that none of the items are defective. However, a Monte Carlo algorithm can determine whether all items are good as long as some probability of error is acceptable. A Monte Carlo algorithm proceeds by successively selecting items at random and testing them one by one, where the maximum number of items being tested is a pre-determined $k \ll n$. When a defective item is encountered, the algorithm stops to indicate that out of the n items in a batch, there is at least one defective. If a tested item is good, the algorithm goes on to the next item. If after testing k items, no defective item is found, the algorithm concludes that all n items are good, but with a modest probability of error that is independent of n. Assuming the events of testing different items are independent, determine that the probability not even one item is defective after testing k items. Analyze the impact of n, k, and p on the probability of finding not even a defective item. As a particular case, suppose there are 1 000 000 cell-phones in a factory, where the probability that a cell-phone is in a perfect condition is 0.99. Based on the Monte Carlo algorithm, determine the minimum number of cell-phones that needs to be tested so the probability of finding not even a defective cell-phone among those tested is less than one in a million.

2.9 Suppose there are four candidates with four different scores of A, B, C, and D. What is the strategy to make the best choice?

2.10 In a country with a population of hundreds of millions, where capital punishment is legal, 10 000 individuals are charged and tried for murder every year, and those convicted of murder are put to death. Past records indicate that out of those tried, 95% are truly guilty and 5% are truly innocent. Out of those who are truly guilty, 95% are convicted and 5% are wrongly set free, and out of those who are truly innocent, 5% are wrongly convicted, and 95% are set free. Determine the probability that a person who is tried is truly innocent, but wrongly convicted of murder. Comment on the results.

3

Counting Methods and Applications

Counting problems are ubiquitous. *Enumeration*, the counting of objects with certain properties one after another, can solve a variety of problems. In a random experiment with finite sample space and equiprobable outcomes, we need to know how many outcomes there are in the sample space, as the probability of an event is the ratio of the number of outcomes in the event to the total number of outcomes in the sample space. However, problems arise where direct counting becomes a practical impossibility. The mathematical theory of counting is formally known as *combinatorial analysis*. The counting methods can determine the number of outcomes in the sample space of a random experiment, without actually listing the outcomes. Overcounting is a common enumeration error, for sometimes counting is not straightforward. As a sanity check, it is thus important to use small numbers for which it is easy to find the answer to the counting problem, and then compare the count with the formulas using small numbers.

3.1 Basic Rules of Counting

Suppose a task (experiment) can be broken down into a sequence of k independent subtasks, where k is a positive integer. Any one subtask is thus done regardless of how the other $k-1$ subtasks are done. Assuming $n_1, n_2, ..., n_k$ are all positive integers, the first subtask can be accomplished in n_1 ways, the second subtask in n_2 ways, ..., and the kth subtask in n_k ways. The *fundamental principle of counting*, also known as the *product rule of counting* or the *multiplication rule of counting*, states that there are a total of $n_1 \times n_2 \times \cdots \times n_k$ distinct ways to accomplish the task.

Probability, Random Variables, Statistics, and Random Processes: Fundamentals & Applications,
First Edition. Ali Grami.
© 2020 John Wiley & Sons, Inc. Published 2020 by John Wiley & Sons, Inc.
Companion Website: www.wiley.com/go/grami/PRVSRP

Example 3.1
How many positive three-digit integers are there that are not multiples of five?

Solution:
In a three-digit integer, the first (the most significant) digit cannot be zero. There are therefore nine possible ways for the first digit, from 1 to 9 inclusive. The second digit can be any one of the 10 possible digits, from 0 to 9 inclusive. The third (the least significant) digit cannot be 0 or 5, otherwise the integer is divisible by 5. Therefore, there are eight possible ways for the third digit, 1 to 4 inclusive and 6 to 9 inclusive. Using the product rule, there are a total of $9 \times 10 \times 8 = 720$ positive three-digit integers that are not multiples of 5. ∎

Suppose a task can be done in k mutually exclusive sets of ways, where k is a positive integer. The ways in any one set thus excludes the ways in the other $k-1$ sets. Assuming n_1, n_2, ..., n_k are all positive integers, the task can be accomplished in one of n_1 ways in set 1, in one of n_2 ways in set 2, ..., and in one of n_k ways in set k, where the set of n_1 ways, the set of n_2 ways, ..., and the set of n_k ways are all pairwise disjoint finite sets. The **sum rule of counting**, also known as the **addition rule of counting**, states that there are a total of $n_1 + n_2 + \ldots + n_k$ distinct ways to accomplish the task.

Example 3.2
How many two-digit prime numbers and two-digit even numbers are there?

Solution:
The set of prime numbers between 10 and 99 inclusive, which consists of 21 numbers, and the set of even numbers between 10 and 99 inclusive, which contains 45 numbers, are mutually exclusive. In other words, we cannot find a two-digit number that is a prime number as well as an even number. Using the sum rule, there are, therefore, $21 + 45 = 66$ prime numbers and even numbers between 10 and 99 inclusive. ∎

Suppose a task can be accomplished in k sets of ways, where k is a positive integer. Assuming n_1, n_2, ..., n_k are all positive integers, the task can be accomplished in one of n_1 ways in set 1, in one of n_2 ways in set 2, ..., and in one of n_k ways in set k, where the set of n_1 ways, the set of n_2 ways, ..., and the set of n_k ways are not pairwise disjoint finite sets. In other words, some of the ways to do the task are thus common and must not be counted more than once. The **subtraction rule of counting**, also widely known as the **principle of inclusion–exclusion**, states that the number of distinct ways to accomplish the task is $n_1 + n_2 + \ldots + n_k$ minus the number of common ways that have been overcounted, so no common way is counted more than once.

Example 3.3
How many positive two-digit integers are there that are multiples of 5 or 7 or both?

Solution:
The set of two-digit integers that are multiples of 5 and the set of integers that are multiples of 7 are not mutually exclusive, as 35 and 70 are the integers that are divisible by both 5 and 7. The number of two-digit integers that are multiples of 5 is 18, and the number of two-digit integers that are multiples of 7 is 13. Using the subtraction rule, there are therefore $18 + 13 - 2 = 29$ two-digit integers that are multiples of 5 or 7 or both. ∎

Example 3.4
In a survey of 120 people, it was found that 65 have shares of A stock, 45 have shares of G stock, and 42 have shares of Z stock. It was also found that 20 have shares in both A and G stocks, 25 have shares in both A and Z stocks, 15 have shares in both G and Z stocks, and 8 have shares in all three A, G, and Z stocks. Determine the number of people who have shares in at least one of the three stocks, the number of people who have shares exactly in one stock, and the number of people who have no shares at all.

Solution:
Figure 3.1 shows the Venn diagram representing the number of people with 1, 2, or 3 stocks. Using the principle of inclusion–exclusion, the number of people with shares in at least one of the three stocks is 100 $(= 65 + 45 + 42 - 20 - 25 - 15 + 8)$. Since the number of people with shares in all three stocks A, G, and Z is 8, the number of people with shares only in stocks A and Z is 17 $(= 25 - 8)$, the number of people with shares only in stocks G and Z is 7 $(= 15 - 8)$, and the number of people with shares only in stocks A and G is 12 $(= 20 - 8)$. Thus, the number of people with shares exactly in one stock is 56 $(= 100 - 17 - 7 - 12 - 8)$. Moreover, the number of people with no shares at all is 20 $(= 120 - 100)$. ∎

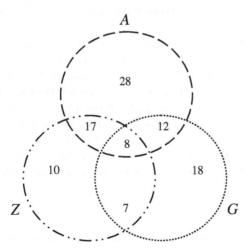

Figure 3.1 Venn diagram.

It is important to note that many counting problems cannot be solved using only one of the above rules, but a combination of them may be required.

Example 3.5
A computer access password consists of four to seven digits chosen from 0 to 9 inclusive, where no digit in a password can be repeated. How many different passwords are possible?

Solution:
Using the product rule and noting that a digit in a password cannot be repeated, the number of passwords consisting of four different digits is $10 \times 9 \times 8 \times 7 = 5040$, the number of passwords consisting of five different digits is $10 \times 9 \times 8 \times 7 \times 6 = 30\,240$, the number of passwords consisting of six different digits is $10 \times 9 \times 8 \times 7 \times 6 \times 5 = 151\,200$, and the number of passwords consisting of seven different digits is $10 \times 9 \times 8 \times 7 \times 6 \times 5 \times 4 = 604\,800$. By the sum rule, the total number of all passwords is thus $5040 + 30\,240 + 151\,200 + 604\,800 = 791\,280$. ∎

There are also some counting problems that cannot be directly solved using any of the above rules. For instance, certain problems, which require tree diagrams with asymmetric structures to solve, cannot easily use these rules, since there are some conditions in these problems that must be met.

The outcomes of a finite sequential experiment can be represented by a **tree diagram**. A tree structure is a methodical way to keep systematic track of all possibilities in scenarios in which events occur in order, but in a finite number of ways. A tree consists of a root, a number of branches leaving the root, and some additional branches leaving the nodes (the endpoints of other branches). The number of branches leaving a node may vary. In order to employ trees in counting, a branch is used to represent each possible choice, and the leaves, which are the nodes not having other branches starting at them, are used to represent the possible outcomes. The number of branches between the root and a leaf in a tree may vary. The number of branches that originate from a node represents the number of events that can occur, given that the event represented by that node occurs. The probability of an outcome of an event is the product of the probabilities of branches going from the root of the tree to a leaf. Probability tree diagrams are thus a way of solving problems involving conditional probabilities through diagrams where each conditional route is represented by a branch.

Example 3.6
How many different sequences of five bits do not have two consecutive zeros?

Solution:
There are $2^5 = 32$ sequences of five bits, however, only some of them can meet the requirement of having no two zeros in a row. Figure 3.2 shows the tree diagram for all sequences of five bits without two connective zeros, where an upper branch indicates a 0 and a lower branch indicates a 1, and no further branching occurs when two zeros occur in a row. Note that after a 0, there can always be only a 1, but after a 1, either a 0 or a 1 can occur, until there are five bits in the sequence. As shown on the tree, there are 13 such bit sequences. ■

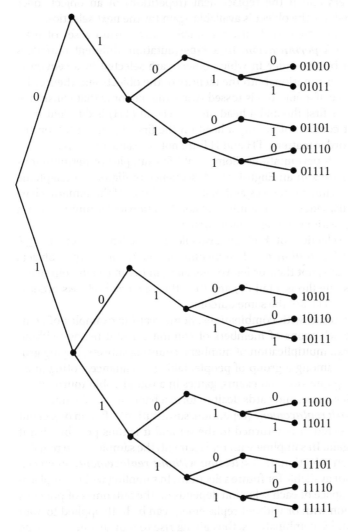

Figure 3.2 Tree diagram for bit strings of length 5 without two consecutive 0s.

3.2 Permutations and Combinations

Combinatorics, which is the study of systematic counting methods, determines the cardinalities of various sets that arise in probability. Permutations and combinations arise when a subset is selected from a set. In a counting problem, we basically determine the number of ways to select a sequence of k objects from a set of n distinguishable objects. In the process of selection, it is important to know in advance if the order of selecting the objects in a sequence matters and if the replacement (repetition) of an object, once selected, is allowed, i.e. the object is available again for the next selection.

An ordered arrangement of k distinguishable objects from a set of $n \geq k$ objects is called a *k-permutation*. In a k-permutation, different outcomes are distinguished by the order in which objects are selected in a sequence. Therefore, in a k-permutation, both the identity of the objects and their order matter. For instance, if a fair coin is tossed twice, the outcome that consists of getting a tail on the first flip and a head on the second (TH) is different from the outcome that consists of getting a head on the first flip and a tail on the second (HT). As order matters, TH and HT are not the same outcome.

Order counts with permutations. Some real-life examples of permutations include order of letters in an English word, sequence of digits in a telephone number, Olympic winners in races and matches, order of alphanumeric characters in passwords, sequence of numbers in combination locks, multiplication of matrices, and positions in hierarchical systems.

An unordered selection of k distinguishable objects from a set of $n \geq k$ objects is called a *k-combination*. In a k-combination, the identity of objects in a sequence matters, not their order. For instance, in tossing a fair coin twice, all that matters is one flip is a tail (T) and the other is a head (H). As order is immaterial, TH and HT are the same outcome.

Order does not count with combinations. Some real-life examples of combinations include selections of members of committees and teams (without counting positions), multiplication of numbers, counting subsets, buying groceries, handshakes among a group of people, taking attendance, voting in an election, answering questions on exams, games in a round-robin tournament, dice rolled in a dice game, and cards dealt to form a hand in a card game.

In a *selection with replacement* (repetition, substitution), after an object out of n objects is selected, it is returned to the set and it is thus possible that it will be selected again. In sampling with replacement, the sample space remains the same after each selection. In a *selection without replacement*, an object, once selected, is not available for future selections. In sampling without replacement, the sample space of each selection depends on the outcomes of previous selections. Selection with or without replacement can be both applied to both k-permutations and k-combinations, thus giving rise to four distinct counting methods.

Before introducing the permutations and combinations formulas, we need to define the symbol $\binom{m}{r}$. The symbol $\binom{m}{r}$, read as *m* choose *r*, where *m* and *r* are integers with $0 \le r \le m$, is called a **binomial coefficient**. Note that we have $\binom{m}{r} \triangleq \frac{m!}{r!(m-r)!}$, where *m*!, read as **m factorial**, is defined as $m! \triangleq m \times (m-1)! = m \times (m-1) \times (m-2) \times \cdots \times 2 \times 1$, and $0! \triangleq 1$. As a useful tool in the calculation of *m* factorial when *m* is very large, the factorial of *m* can be asymptotically approximated by the **Stirling's formula** $m! \approx \sqrt{2\pi m}\left(\frac{m}{e}\right)^m$, as $m \to \infty$, where $\pi = 3.141\,59\ldots$ and $e = 2.718\,28\ldots$

3.2.1 Permutations without Replacement

If *n* and *k* are integers, such that $0 \le k \le n$, then the number of ways to make ordered arrangements of *k* objects from a set of *n* distinct objects, but without repetition, is as follows:

$$n(n-1)(n-2)\cdots(n-k+1) = \frac{n!}{(n-k)!} \quad k \ne 0 \tag{3.1}$$

Note that when $k = 0$, there is one unique way to order zero objects, and that is a list with no objects in it, i.e. the empty list, and when $k \ne 0$, the total number of ways to make these *k* choices is the product of the number of ways to make the individual choices.

Example 3.7

How many different ways are there to select a gold medalist, a silver medalist, and a bronze medalist from 195 athletes with different nationalities who have entered a world sports competition?

Solution:

Since it matters which athlete wins which medal and no athlete can win more than one medal, it is a selection with ordering and without replacement. Noting that $n = 195$ and $k = 3$, the number of ways is thus $\frac{195!}{192!} = 195 \times 194 \times 193 = 7\,301\,190$. ∎

3.2.2 Combinations without Replacement

If *n* and *k* are integers, such that $0 \le k \le n$, then the number of ways to make unordered selections of *k* objects from a set of *n* distinct objects, but without repetition, is as follows:

$$\binom{n}{k} = \frac{n!}{k!(n-k)!} \tag{3.2}$$

Example 3.8
There are 15 players on a sports team. The starting lineup consists of only five players. How many possible starting lineups are there, assuming what positions they play are of no concern?

Solution:
Since the order of the selection of the players is immaterial and no player can be selected more than once, it is a selection without ordering and without replacement. Noting that $n = 15$ and $k = 5$, the number of ways is thus $\frac{15!}{10!5!} = 3003.$ ∎

3.2.3 Permutations with Replacement

If n and k are integers, such that $0 \le k$ and $1 \le n$, then the number of ways to make ordered arrangements of k objects from a set of n objects, when repetition of objects allowed, is as follows:

$$n \times n \times \cdots \times n = n^k \tag{3.3}$$

Example 3.9
How many two-letter words in the English language, from the letters A to Z inclusive, can be made, noting that the words do not need to have meaning?

Solution:
This is a selection with ordering and with replacement, as the order of letters in a word matters and a letter can be used in a word more than once. Noting that $n = 26$ and $k = 2$, the number of words is thus $26^2 = 676.$ ∎

3.2.4 Combinations with Replacement

If n and k are integers, such that $0 \le k$ and $1 \le n$, then the number of ways to make unordered selections of k objects from a set of n objects, when repetition of objects allowed, is as follows:

$$\binom{n+k-1}{k} = \frac{(n+k-1)!}{k!(n-1)!} \tag{3.4}$$

Example 3.10
An ice cream parlor has 20 different flavors of ice creams. In how many ways can a selection of five ice creams be chosen?

Solution:
This is a selection without ordering, but with replacement, as the order of the selection does not matter and a flavor can be selected more than once. With $n = 20$ and $k = 5$, the number of ways is thus $\frac{24!}{5!19!} = 42\,504.$ ∎

Table 3.1 Permutations and combinations, with and without replacement.

	Permutations (ordered arrangements)	Combinations (unordered selections)
Without replacement	$\dfrac{n!}{(n-k)!}$	$\dfrac{n!}{k!(n-k)!}$
With replacement	n^k	$\dfrac{(n+k-1)!}{k!(n-1)!}$

Table 3.1 summarizes the formulas for the numbers of ordered arrangements, i.e. permutations, and unordered selections, i.e. combinations, of k objects, with and without repetition (replacement) allowed, from a set of n distinct objects. It is important to note that for a given n, the case of k-permutations with replacement include all the other three cases (k-permutations without replacement, k-combinations with replacement, and k-combinations without replacement). The terms *permutations* (i.e. arrangements with ordering) and *combinations* (i.e. selections without ordering) generally, but not always, imply that there are no replacement (repetition) involved. The following general observations regarding permutations and combinations can be made:

- For $k \neq 1$, the number of permutations is greater than the number of combinations.
- For $k \neq 1$, the number of outcomes with replacement is greater than that without replacement.
- For $k = 1$, the number of permutations and the number of combinations are both n.
- For $k = n$ and without replacement, the number of permutations is $n!$ and the number of combinations is 1.

Example 3.11
There are six balls in a bag, numbered 1, 2, 3, 4, 5, and 6. We randomly pick two balls from it, one by one. Find the number of permutations and combinations, with and without replacement, i.e. consider all four possible cases.

Solution:
Noting that $n = 6$ and $k = 2$, Figure 3.3 lists all possible outcomes. The outcomes can be divided into three mutually exclusive sets: the set A that includes 15 pairs, in each of which the first number is larger than the second number, the set B that includes 15 pairs, in each of which the first number is smaller than

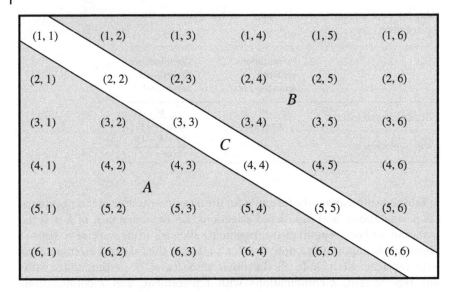

Figure 3.3 List of all possible pairs for $n = 6$ and $k = 2$.

the second number, and the set C that includes 6 pairs, in each of which the first and the second numbers are the same. Using Table 3.1, we have the following four cases:

(i) Combinations without replacement: There are $\frac{6!}{2!4!} = 15$ pairs, which form either the set A or the set B.

(ii) Combinations with replacement: There are $\frac{7!}{2!5!} = 21 \ (= 15 + 6)$ pairs, which form either the sets A and C or the sets B and C.

(iii) Permutations without replacement: There are $\frac{6!}{4!} = 30 \ (= 15 + 15)$ pairs, which form the sets A and B.

(iv) Permutations with replacement: There are $6^2 = 36 \ (= 15 + 15 + 6)$ pairs, which form the sets A, B, and C. ∎

Example 3.12

How many ways are there to select four pieces of fruit from a bowl containing oranges, apples, and bananas if the order in which the pieces are selected does not matter, only the type of fruit matters? Note that there are more than four pieces of each type of fruit in the bowl.

Solution:

This is the case of combinations with replacement. Noting we have $n = 3$ and $k = 4$, there are therefore $\frac{(3+4-1)!}{4!(3-1)!} = 15$ combinations. ∎

Example 3.13

Assuming there are 365 days in a year (i.e. it is not a leap year), and there are 100 students in a class, determine the number of ways that (a) the students can have birthdays, and (b) the students can have distinct birthdays.

Solution:

(a) This is the case of permutations with replacement. Noting that $n = 365$ and $k = 100$, the number of ways is thus $365^{100} \cong 1.7 \times 10^{256}$.

(b) This is the case of permutations without replacement. Noting that $n = 365$ and $k = 100$, the number of ways is thus $\frac{365!}{(365-100)!} = \frac{365!}{265!} \cong 5.2 \times 10^{249}$. ∎

Example 3.14

How many ways are there for three students in a major science competition to perform, if ties are possible, i.e. two or three students may tie?

Solution:

The solution consists of three mutually exclusive cases, depending on the number of ties:

(i) There are no ties, there are therefore $\frac{3!}{(3-3)!} = 6$ different ways for the three students to perform. For instance, one way is that student-1 gets an A, student-2 gets a B, and student-3 gets a C (i.e. ABC), another way is that student-1 gets an A, student-2 gets a C, and student-3 gets a B (i.e. ACB), so on and so forth, the six ways are thus as follows: ABC, ACB, BAC, BCA, CAB, and CBA.

(ii) Two students are tied, but the third one has a different performance. Therefore, there are $\frac{3!}{1!2!} = 3$ different ways to have two students to be tied and there are $\frac{2!}{(2-2)!} = 2$ different ways for the two groups (the pair and the single student) to perform. There are therefore $3 \times 2 = 6$ ways for the students to perform. For instance, one way is that student-1 and student-2 both get an A, but student-3 gets a B (i.e. AAB), another way is that student-1 and student-2 both get a B, but student-3 gets an A (i.e. BBA), so on and so forth, the six ways are thus as follows: AAB, BBA, ABB, BAA, ABA, and BAB.

(iii) There is only one way for all three students to tie: AAA.

The total number of ways is thus 13 ($= 6 + 6 + 1$). ∎

3.3 Multinomial Counting

Suppose we have n objects, out of which there are n_1 indistinguishable objects of type 1, n_2 indistinguishable objects of type 2, ..., and n_k indistinguishable objects of type k, where n as well as n_1, n_2, ..., n_k are all positive integers, and we have $n = n_1 + n_2 + ... + n_k$. The number of distinguishable ways to arrange the n objects, i.e. the number of different permutations of n objects, is as follows:

$$\frac{(n_1 + n_2 + \cdots + n_k)!}{n_1! n_2! \cdots n_k!} = \frac{n!}{n_1! n_2! \cdots n_k!} \tag{3.5}$$

This is known as the **multinomial coefficient**. Note that when $k = 2$, the result is known as the binomial coefficient.

Example 3.15
How many different words can be made by reordering the letters of the word MUMMY, noting that the words do not need to have meaning?

Solution:
In this five-letter word, there are 1 U, 1 Y, and 3 Ms. Noting that $n = 5$, $k = 3$, $n_1 = 1$, $n_2 = 1$, and $n_3 = 3$, the number of distinguishable ways is $\frac{5!}{1!1!3!} = 20$. The list is as follows: MUMYM, MYMUM, MMUMY, MMYMU, MMUYM, MMYUM, UMMYM, YMMUM, MUMMY, MYMMU, MUYMM, MYUMM, UMYMM, YMUMM, MMMUY, MMMYU, UMMMY, YMMMU, UYMMM, and YUMMM. ∎

Example 3.16
How many positive 12-digit integers are there that each contains one 1, two 2s, three 3s, four 4s, one 5, and one 9?

Solution:
Noting that $n = 12$, $k = 6$, $n_1 = 1$, $n_2 = 2$, $n_3 = 3$, $n_4 = 4$, $n_5 = 1$, and $n_6 = 1$, the number of 12-digit integers is thus $\frac{12!}{1!2!3!4!1!1!} = 1\,663\,200$. ∎

Suppose we have n distinguishable objects that is to be partitioned into k distinguishable groups, in such a way that group 1 will have n_1 objects, group 2 will have n_2 objects, ..., and group k will have n_k objects, where k and n as well as n_1, n_2, ..., n_k are all positive integers and we have $n = n_1 + n_2 + ... + n_k$. In other words, the n distinct objects are distributed among k distinct non-overlapping groups, where each group contains distinguishable objects. Note that the order of objects within a group is of no importance, but the order of the k groups matters. The number of ways to accomplish this partitioning can be calculated using the multinomial coefficient as well.

Example 3.17
In a card game, how many ways are there to distribute hands of five cards to each of six players from a standard deck of 52 cards?

Solution:
With $n = 52$, $n_1 = 5$, $n_2 = 5$, $n_3 = 5$, $n_4 = 5$, $n_5 = 5$, $n_6 = 5$, and $n_7 = 52 - 5 \times 6 = 22$, the number of ways is thus $\frac{52!}{5!5!5!5!5!5!22!} \cong 2.4 \times 10^{34}$.   ∎

Example 3.18
There are four workers, say A, B, C, and D, who can be assigned to three different chores, say X, Y, and Z. Note that one worker is needed to do chore X, one worker is needed to do chore Y, but two workers are needed to do chore Z. In how many ways can the workers be formed?

Solution:
With $n = 4$, $n_1 = 1$, $n_2 = 1$, and $n_3 = 2$, the number of ways is thus $\frac{4!}{1!1!2!} = 12$. Noting that both the members of the set (the workers) and the subsets (the chores) are distinguishable, the list of all possibilities is as follows:

	1	2	3	4	5	6	7	8	9	10	11	12
$X \rightarrow$	A	A	A	B	B	B	C	C	C	D	D	D
$Y \rightarrow$	B	C	D	A	C	D	A	B	D	A	B	C
$Z \rightarrow$	C,D	B,D	B,C	C,D	A,D	A,C	B,D	A,D	A,B	B,C	A,C	A,B

∎

3.4 Special Arrangements and Selections

Sometimes we need to find the number of ways to select objects from various groups. Suppose $n_1 > k_1$, $n_2 > k_2$, ..., $n_m > k_m$ are all positive integers, and we want to make an unordered selection without replacement of k_1 objects from a group of n_1 objects, k_2 objects from a group of n_2 objects,..., k_m objects from a group of n_m objects. Note that $n = n_1 + n_2 + ... + n_m$ is the total number of objects and $k = k_1 + k_2 + ... + k_m$ is the total number of objects selected in an unordered fashion and without replacement. The number of ways to make such a particular selection is the product of m binomial terms. The **hypergeometric probability** is thus defined as follows:

$$p = \frac{\prod_{i=1}^{m} \binom{n_i}{k_i}}{\binom{n}{k}} = \frac{\frac{n_1!}{k_1!(n_1 - k_1)!} \times \cdots \times \frac{n_m!}{k_m!(n_m - k_m)!}}{\frac{n!}{k!(n - k)!}} \tag{3.6}$$

Example 3.19
Suppose we have 8 red balls, 12 green balls, and 18 blue balls in a bag. Determine the number of ways to make an unordered selection without replacement

of 6 red balls, 4 green balls, and 2 blue balls, and also the corresponding hyper-geometric probability.

Solution:
As we have $n_1 = 8$, $n_2 = 12$, $n_3 = 18$, $k_1 = 6$, $k_2 = 4$, and $k_3 = 2$, the number of ways is thus $\frac{8!12!18!}{6!2!4!8!2!16!} = 2\,120\,580$, which is obviously much less than $\frac{38!}{12!26!} = 2\,707\,475\,148$. The hypergeometric probability is thus $\frac{2\,120\,580}{2\,707\,475\,148} \cong 0.08\%$. ∎

Example 3.20
In a class, there are seven male students and eight female students. Determine the number of ways for each of the following cases: (a) select a six-member committee from the students, (b) select a six-member committee with three male students and three female students, (c) select a six-member committee with a given student as a chairperson, and (d) elect a president, a vice president, and a treasurer.

Solution:
(a) This is a combination with no replacement. Therefore, there are $\binom{15}{6} = 5005$ ways.
(b) This is a combination with no replacement. Therefore, there are $\binom{7}{3}\binom{8}{3} = 1960$ ways.
(c) This is a combination with no replacement. We have to select 5 students out of 14 students to form the committee. Therefore, there are $\binom{14}{5} = 2002$ ways.
(d) This concerns permutations, not combinations, for order does matter. Moreover, repetition is not allowed. Therefore, there are $15 \times 14 \times 13 = 2730$ ways. ∎

If n objects are arranged in a circle, there are $(n-1)!$ permutations of objects around the circle. This is known as the ***circular permutation***. Note that by considering one object in a fixed position and arranging the other $n-1$ objects, we have $(n-1)! = \frac{n!}{n}$ ways, this is also known as the ***division rule of counting***. One application is to seat n people around a round table, where two arrangements are considered different when each person does not have the same left neighbor and the same right neighbor.

Example 3.21
How many different ways are there to seat six people around a circular table?

Solution:
With $n = 6$, there are thus $5! = 120$ permuations. ∎

Sometimes we need to find the number of **_integer solutions of equations_**. To this effect, it is important to note that there are $\binom{n+r-1}{r-1}$ distinct nonnegative integer-valued solutions satisfying the equation $x_1 + x_2 + \ldots + x_r = n$, where r and $n \geq r$ are both positive integers.

Example 3.22
How many distinct nonnegative integer-valued solutions of $x_1 + x_2 = 4$ are possible?

Solution:
With $r = 2$ and $n = 4$, there are $\binom{5}{1} = 5$ distinct nonnegative integer-valued solutions, namely $(x_1 = 0, x_2 = 4)$, $(x_1 = 1, x_2 = 3)$, $(x_1 = 2, x_2 = 2)$, $(x_1 = 3, x_2 = 1)$, and $(x_1 = 4, x_2 = 0)$. ∎

3.5 Applications

In this section, some interesting applications of counting methods are highlighted.

3.5.1 Game of Poker

Poker is a popular game using a deck of cards. A deck of cards contains 52 cards, for each of the four suits (spades, clubs, hearts, and diamonds), there are 13 different ranks of cards – twos, threes, fours, fives, sixes, sevens, eights, nines, tens, jacks J, queens Q, kings K, and aces A. The subset of cards held at one time by a player during a game is commonly called a hand. These hands are compared using a hand ranking system that is standard across almost all variants of poker. In poker, the player with the highest-ranking hand wins that particular deal. The relative ranking of the various hand categories is based on the probability of being randomly dealt such a hand from a well-shuffled deck. A poker hand consists of five cards. In games where more than five cards are available to each player, the best five-card combination of those cards is generally played. Individual cards are ranked A (highest), $K, Q, J, 10, 9, 8, 7, 6, 5, 4, 3$, and 2 (lowest). Aces can also appear low (as if having a value of 1). In most variants, suits play no part in determining the ranking of a hand. Hands are ranked first by category, then by individual card ranks. Even the lowest hand that qualifies in a certain category defeats all hands in all lower categories. Between two hands in the same category, card ranks are used to break ties.

Since the order of cards dealt does not matter, we have $\frac{52!}{5!47!} = 2\,598\,960$ possible combinations in five-card poker. Using hypergeometric probabilities, Table 3.2 lists the probabilities of hand categories. Probability analysis can

Table 3.2 Hands in poker.

Category	Description	Probability
Royal flush	It contains ace, king, queen, jack, and a 10, all of the same suit, the highest possible hand	$\dfrac{\binom{4}{1}}{\binom{52}{5}} = \dfrac{4}{2\,598\,960} \cong 0.000\,15\%$
Straight flush	It contains five cards of sequential rank, all of the same suit, but not royal flush	$\dfrac{\binom{10}{1}\binom{4}{1} - \binom{4}{1}}{\binom{52}{5}} = \dfrac{36}{2\,598\,960} \cong 0.001\,4\%$
Four of a kind	It contains four cards of one rank and one other card	$\dfrac{\binom{13}{1}\binom{4}{4}\binom{12}{1}\binom{4}{1}}{\binom{52}{5}} = \dfrac{624}{2\,598\,960} \cong 0.024\%$
Full house	It contains three matching cards of one rank and two matching cards of another rank	$\dfrac{\binom{13}{1}\binom{4}{3}\binom{12}{1}\binom{4}{2}}{\binom{52}{5}} = \dfrac{3\,744}{2\,598\,960} \cong 0.144\%$
Flush	It contains five cards of the same suit, but not in sequence	$\dfrac{\binom{13}{5}\binom{4}{1} - \binom{10}{1}\binom{4}{1}}{\binom{52}{5}} = \dfrac{5\,108}{2\,598\,960} \cong 0.197\%$
Straight	It contains five cards of sequential rank, but in more than one suit	$\dfrac{\binom{10}{1}\binom{4}{1}^5 - \binom{10}{1}\binom{4}{1}}{\binom{52}{5}} = \dfrac{10\,200}{2\,598\,960} \cong 0.392\%$
Three of a kind	It contains three cards of the same rank, plus two unmatched (not of this rank nor the same as each other) cards	$\dfrac{\binom{13}{1}\binom{4}{3}\binom{12}{2}\binom{4}{1}^2}{\binom{52}{5}} = \dfrac{54\,912}{2\,598\,960} \cong 2.11\%$
Two pairs	It contains two cards of the same rank, plus two cards of another rank (that match each other, but not the first pair) and one unmatched (not of either rank) card	$\dfrac{\binom{13}{2}\binom{4}{2}^2\binom{11}{1}\binom{4}{1}}{\binom{52}{5}} = \dfrac{123\,552}{2\,598\,960} \cong 4.75\%$

(continued)

Table 3.2 (Continued)

Category	Description	Probability	
One pair	It contains two cards of the same rank, plus three other unmatched (not of this rank nor the same as each other) cards	$\dfrac{\binom{13}{1}\binom{4}{2}\binom{12}{3}\binom{4}{1}^3}{\binom{52}{5}}$	$= \dfrac{1\,098\,240}{2\,598\,960} \cong 42.26\%$
High card	It contains no two cards of the same rank, the five cards are not in sequence, and the five cards are not all the same suit	$\dfrac{\left(\binom{13}{5}-10\right)\left(\binom{4}{1}^5-4\right)}{\binom{52}{5}}$	$= \dfrac{1\,302\,540}{2\,598\,960} \cong 50.12\%$

provide the valuable knowledge about the probabilities of an array of hands, a necessary but not a sufficient requirement to win in poker. In poker, there are also other important factors involved, such as the betting sequence, other players' hands, and above of all the playing decisions made by the other players.

3.5.2 Birthday Paradox

Birthday Paradox is a famous puzzle asking the minimum number of people who need to be in a room so that it is more likely than not that at least two of them have the same birthday. As briefly discussed later, the solution to the birthday problem leads to the solution of secure communications using message authentication.

This problem is now addressed in a more general way, i.e. determine the probability that, in a set of k randomly chosen people in a room, there is at least one pair of people who have the same birthday. Although a year may have 366 days and more people are born on some days of the year than others, we assume that a year is not a leap year, i.e. there are 365 days in a year, and each day of the year is equally probable for a birthday. We further assume that the birthdays of the people in the room are independent, and in particular, there are no twins, triplets, and so on.

The occurrence of one pair of people to have the same birthday seems unlikely unless k is quite large, and in fact, by the pigeonhole principle, the probability reaches 100% when the number of people reaches 366, i.e. $k = 366$. The requirement of at least one pair simplifies the problem conceptually. However, the direct approach warrants a very significant level of probability computation. For the direct approach, we have to consider the probability that two people have the same birthday on 1 January, the probability that three people have the same birthday on 1 January, and the probability that k people have the same birthday on 1 January. We then have to consider probabilities that two or three or k people share a common birthday on 2 January, on 3 January, up to and including on 31 December. This warrants the indirect approach, in which no two people share a common birthday. In other words, what needs to be done is to count the number of ways that k people can have distinct birthdays.

The first selected birthday could be any day, with the probability of $\frac{365}{365}$. The probability that a randomly selected person whose birthday is different from the first birthday is $\frac{364}{365}$. The probability that a randomly selected person whose birthday is different from both birthdays, i.e. the birthdays of the first two persons, is $\frac{363}{365}$. In general, the ith person, with $2 \leq i \leq 365$, has a birthday different from the birthdays of $i-1$ people already given that these $i-1$ people have different birthdays is $\frac{365-(i-1)}{365} = \frac{366-i}{365}$. We can thus conclude that the probability that k people in the room have different birthdays is the multiplication of k independent probabilities, or equivalently, the probability is the number of ways of making a permutation of k days taken from 365 without replacement (i.e. $\frac{365!}{(365-k)!}$) divided by the number of ways making an ordered with replacement selection of k days from 365 (i.e. 365^k), i.e. we have

$$p_k = \left(\frac{366-1}{365}\right)\left(\frac{366-2}{365}\right)\cdots\left(\frac{366-k}{365}\right) = \frac{\frac{365!}{(365-k)!}}{365^k}$$

$$= \frac{365!}{365^k(365-k)!} \tag{3.7}$$

Note that the probability that among k people in a room at least two people having the same birthday is $1 - p_k$. Table 3.3 lists, for various values of k, the probabilities that at least two people share a common birthday. It is interesting to note that the minimum number of people needed so that the probability that at least two people have the same birthday is greater than 50% is only 23. With only 50 people, the probability is greater than 97%, and with only 70 people, the probability is greater than 99.9%, surprising results indeed.

The probabilities are quite high simply because every pair of people is potential matches, and as the number of people increases, the number of pairs increases much faster. It is thus a key point to highlight the fact that in the birthday problem, neither of the two people is chosen in advance. For instance, in a room of 30 people, the probability 70.63% does not imply that at least one other person shares the birthday of a specific person in the room. The

Table 3.3 Probabilities that at least two out of k people share a common birthday.

k	$1 - p_k$	k	$1 - p_k$	k	$1 - p_k$
1	0%	9	9.46%	40	89.12%
2	0.27%	10	11.69%	50	97.04%
3	0.82%	15	25.29%	60	99.41%
4	1.64%	20	41.14%	70	99.916%
5	2.71%	23	50.73%	80	99.991 4%
6	4.05%	25	56.87%	90	99.999 3%
7	5.62%	30	70.63%	100	99.999 97%
8	7.43%	35	81.43%	366	100%

probabilities are for some collection of two or more people, and we cannot specify any of the people ahead of time.

Assuming we have n possibilities (instead of 365 days) and also $n \gg k$, it can be shown that p_k can be closely approximated as follows:

$$
\begin{aligned}
p_k &= \frac{n!}{n^k (n-k)!} \\
&= \frac{n(n-1)\cdots(n-k+1)}{n^k} = \frac{n}{n} \times \frac{(n-1)}{n} \times \cdots \times \frac{(n-k+1)}{n} \\
&\cong \exp\left(-\frac{k(k-1)}{2n}\right) \qquad \rightarrow \qquad k \cong \sqrt{-2\ln p_k}\sqrt{n}
\end{aligned}
\tag{3.8}
$$

We can thus determine the smallest value of n given a value of k such that the probability of no collision is greater than a particular threshold. For instance, when $p_k = 0.5$, we have $k \cong 1.177\,41\sqrt{n}$. As reflected above, for $n = 365$, we have $k = 23$.

In any message authentication or digital signature mechanism, there must be some sort of function that produces an authenticator, a value to be used to authenticate a message. Authenticators may be grouped into message encryption (cipher-texting of the entire message), message authentication code (computing a fixed-length value from the message and a secret key), and hash function (mapping a message of any length into a fixed-length hash function). In the context of secure communications using message authentication code, we have $n = 2^m$, where m, the number of bits in the authenticator, is typically 128, 196, and 256. For instance, with $m = 128$, we have $n = 2^{128} \cong 3.403 \times 10^{34}$, and with $p_k = 0.5$, to find any two messages that their authenticators match, we must then have $k \cong 1.177\sqrt{2^{128}} \cong 2.171 \times 10^{19}$ messages. With

$1 - p_k = 10^{-12}$ and $m = 128$, we then have $k = 2.608 \times 10^{13}$ messages. Note that an increase in m and/or a decrease in p_k can bring about an increase in k.

3.5.3 Quality Control

There are K items, out of which $k \leq K$ are defective, i.e. $K - k$ items work properly. $M \leq K$ items are chosen at random and tested, i.e. $K - M$ remain untested. It is important to determine the probability that m of the M tested items are found defective, where we have $m \leq k$ and $m \leq M$.

It is an unordered sampling without replacement. There are $\binom{k}{m}$ ways to choose the m defective items from the total of k defective items and $\binom{K-k}{M-m}$ ways to choose the $M - m$ nondefective items from the total of $K - k$ nondetective items. Hence, there are $\binom{k}{m}\binom{K-k}{M-m}$ possible ways to make such a selection. However, the number of ways to select M items out of K items at random is $\binom{K}{M}$. Using the hypergeometric probability, we have

$$p(K, k, M, m) = \frac{\binom{k}{m}\binom{K - k}{M - m}}{\binom{K}{M}} \tag{3.9}$$

Table 3.4 provides some insight into some special cases.

3.5.4 Best-of-Seven Championship Series

In a sports championship, the Eastern and Western teams play against one another in a best-of-seven series that is the first team that wins four games wins the series and becomes the champion. We use the tree diagram to find the number of ways that the championship can occur. Assuming p is the probability that the Eastern team wins a game and $q = 1 - p$ is the probability that the

Table 3.4 Probability that m of M tested items are defective.

Parameters	Probability
$m = 0, k = 1$	$\dfrac{K - M}{K}$
$m = k$	$\dfrac{M(M - 1)\cdots(M - k + 1)}{K(K - 1)\cdots(K - k + 1)}$
$m = M$	$\dfrac{k(k - 1)\cdots(k - M + 1)}{K(K - 1)\cdots(K - M + 1)}$

Western team wins a game, and the games are independent of one another, we determine the probabilities that the series can end in four games, five games, six games, or seven games.

The tree for all cases is quite large, we thus show only half, namely, the half corresponding to the Eastern team having won the first game, as shown in Figure 3.4. Due to symmetry, the final answer will be twice the number calculated with this tree. Note that an upper branch indicates a win by the Eastern team and a lower branch indicates a win by the Western team. No further

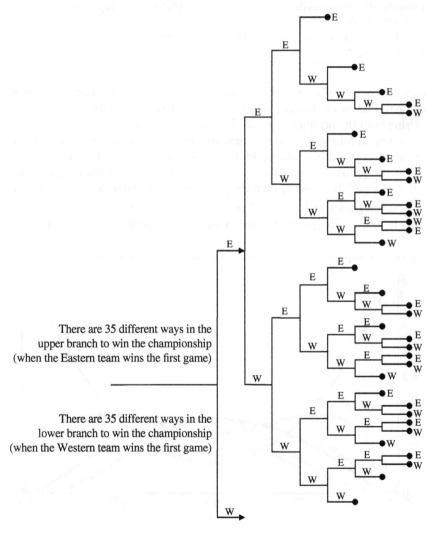

Figure 3.4 Tree diagram for championship series.

Table 3.5 Probabilities of winning in the best-of-seven championship series.

	Eastern team wins	Western team wins	One of the two teams wins
Probability that the series can be won in four games	p^4	q^4	$p^4 + q^4$
Probability that the series can be won in five games	$4p^4q$	$4q^4p$	$4p^4q + 4q^4p$
Probability that the series can be won in six games	$10p^4q^2$	$10q^4p^2$	$10p^4q^2 + 10p^2q^4$
Probability that the series can be won in seven games	$20p^4q^3$	$20q^4p^3$	$20p^3q^3$

branching occurs when a team wins four games altogether. The number of ways on the tree shown in Figure 3.4 is 35, and the final answer is thus 70.

Using both the product rule and the sum rule, we can find the probabilities of winning in four games, five games, six games, and seven games in term of probabilities p and q, as shown in Table 3.5. If $p \rightarrow 0$ (i.e. $q \rightarrow 1$) or $q \rightarrow 0$ (i.e. $p \rightarrow 1$), the probability that the series is won in four games then approaches 1. If $p \rightarrow \frac{1}{2}$ and thus $q \rightarrow \frac{1}{2}$, then the probability that the series can be won in six or seven games each approaches $\frac{5}{16}$, which is higher than those to win the series in four or five games. Figure 3.5 shows P (the probability of winning the

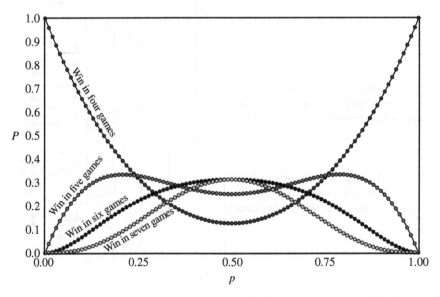

Figure 3.5 Probability of winning in championship series in terms of the probability p.

championship) in terms of p (the probability that the Eastern team wins a game) under various scenarios.

3.5.5 Lottery

Lottery is a form of gambling that involves the drawing of numbers for prizes. It is a type of gambling where skills cannot play a role and who wins what is solely decided by luck. The amount of money an average person loses in a lottery is not significant, and except for a very few people, a very large number of people lose. In a typical lottery game, a player chooses n distinct numbers from 1 to N inclusive, where clearly n and N are both positive integers. At the lottery drawing, identical balls numbered from 1 to N are mixed, and $n \leq N$ balls are randomly picked from a device, one by one.

Assuming that once a ball is drawn, it is not put back in the device, we need to determine the probability that $0 \leq k \leq n$ of the n balls picked match the player's choices. This is a combination without replacement, in which k numbers are drawn from n numbers and $n - k$ numbers are drawn from $N - n$ numbers. Using hypergeometric probability, we can have the following probability:

$$p = \frac{\binom{n}{k}\binom{N-n}{n-k}}{\binom{N}{n}} \tag{3.10}$$

Assuming we have $N = 49$ and $n = 6$, Table 3.6 highlights the probabilities of winning for various values of k.

Table 3.6 Probabilities of winning six-forty-nine lottery.

k	$p\ (N = 49, n = 6)$
0	$\dfrac{6\,096\,454}{13\,983\,816} \cong 43.596\,5\%$
1	$\dfrac{5\,775\,588}{13\,983\,816} \cong 41.301\,9\%$
2	$\dfrac{1\,851\,150}{13\,983\,816} \cong 13.237\,8\%$
3	$\dfrac{246\,820}{13\,983\,816} \cong 1.765\,0\%$
4	$\dfrac{13\,545}{13\,983\,816} \cong 0.096\,9\%$
5	$\dfrac{258}{13\,983\,816} \cong 0.001\,8\%$
6	$\dfrac{1}{13\,983\,816} \cong 0.000\,007\%$

3.6 Summary

In this chapter, we briefly discussed basic rules of counting and selections of k objects from n objects, with ordering (permutations) and without ordering (combinations) as well as with and without replacement. Also, some special selections, arrangements, and interesting applications, such as birthday paradox, game of poker, quality control, and lottery, were highlighted.

Problems

3.1 How many even four-digit numbers are there, if only the six digits 0, 1, 2, 3, 4, and 5 can be used, but no digit more than once?

3.2 Noting that a string of numbers or letters whose reversal is identical to the string is called a palindrome, i.e. it reads the same backward or forward, how many bit strings of length n, where n is an even integer, are palindromes?

3.3 A drawer has six black socks, four brown socks, and two gray socks. Determine the number of ways two socks can be picked from the drawer for each of the following cases: (a) the two socks can be any color, and (b) the two socks must be the same color.

3.4 Teams A and B play against one another in a championship series. The first team that wins three games or wins two games in a row wins the tournament. Determine the number of ways the tournament can occur.

3.5 Given the integers from 1 to 9 inclusive, how many additions of four different digits can be formed?

3.6 There are nine chairs in a row, numbered from 1 to 9 inclusive, and there are also five men and four women. It is required to seat the women in even seats and men in odd seats. How many arrangements are possible?

3.7 How many different salads can be made from lettuce, tomatoes, cucumbers, green peppers, olives, and onions, assuming lettuce must always be used?

3.8 From 21 consonants and 5 vowels in the English language, how many words can be formed consisting of two different consonants and two different vowels? Note that the words do not need to have meaning.

3.9 How many ways are there for four students in a major science competition to perform, if ties are possible?

3.10 In a bag, there are three black balls, four white balls, and five red balls. Two balls are picked from the bag. How many ways are there if the two balls are not of the same color?

3.11 How many four-digit integers are there using the digits 0, 1, 2, 3, 4, 5, where no digit is repeated and the integer is a multiple of 3.

3.12 Eight married couples (four males and four females) are to sit around a round table. (a) How many arrangements are possible? (b) In how many ways of these will the four males sit next to one another?

3.13 In a group of 15 athletes, there are 2 soccer players, 3 volleyball players, and 10 basketball players. How many ways can one choose a committee of 6 athletes if (a) there are no constraints, (b) the 2 soccer players must be included, (c) the 2 soccer players must be excluded, (d) there must be at least 3 basketball players, and (e) there must be at most 2 basketball players?

3.14 In a small party, there are 10 males and 10 females who want to dance. (a) How many different couples can be formed, where a couple is defined as a male and a female? (b) Suppose one of the males has 3 sisters among the 10 females, and he would not accept any of them as a partner. How many different couples can be formed?

3.15 How many ways are there to seat four people around a circular table where two seatings are considered the same when everyone has the same two neighbors without regard to whether they are right or left neighbors?

3.16 How many ways are there to seat 4 people of a group of 10 around a circular table where 2 seatings are considered the same when everyone has the same immediate left and immediate right neighbors?

3.17 A 12-person jury is to be selected from a group of 24 potential jurors of which 16 are men and 8 are women. (a) How many 12-person juries are there with 6 men and 6 women? (b) What is the probability of selecting exactly 6 men and 6 women to form a 12-person jury?

3.18 In how many ways can 10 people be assigned to 1 triple room, 2 double rooms, and 3 single rooms?

3.19 Bridge is a popular card game in which 52 cards are dealt to four players, each having 13 cards. The order in which the cards are dealt is not important; just the final 13 cards each player ends up with are of importance. How many different ways are there to deal hands of 13 cards to each of 4 players?

3.20 In a 10-bit string, how many strings are there that can have 7 ones and 3 zeros?

3.21 A three-member committee has to be formed from six women and four men. The committee must include at least two women. In how many ways can this be done?

3.22 How many permutations, without repetition, of the letters $UVWXYZ$ contain the letters XYZ as a block?

3.23 Determine the number of ways that six-letter words can be formed using the letters of the word CANADA.

3.24 A farmer buys four cows, three pigs, two hens, and one horse from another farmer who has nine cows, seven pigs, five hens, and three horses. Determine the total number of choices.

3.25 Find the number of ways that eight people can arrange themselves (a) in a row of chairs, and (b) around a circular table.

3.26 How many different words can be made by reordering the letters of the word TELL? Note that the words do not need to have meaning.

3.27 Someone has six different books from author A, four different books from author B, two different books from author C, and wants to arrange these books on a shelf. How many different arrangements are possible if (a) the books by each author must all stand together, (b) only the books from author B must stand together?

3.28 In how many ways can four people be seated at a round table if 2 people must not sit next to one another?

3.29 Note that n and r are both positive integers, where $1 \leq r \leq n$. Assume no repetition (i.e. no replacement) is allowed. If the number of r-permutations of a set with n distinct elements is six times the number of r-combinations of a set with n distinct elements, then determine the possible values of r and n.

3.30 Derive all the probabilities for all categories of hands in the game of poker.

3.31 Students in a class, which consists of 18 females and 12 males, write an exam. Noting no two students have the same mark, the students are ranked according to their marks. (a) How many different rankings are possible? (b) Assuming all males are ranked among themselves and all females are ranked among themselves, how many different rankings are possible?

3.32 A mother has 20 candies to distribute among her 4 kids. (a) How many different distributions are possible? (b) What if not all the candies need to be distributed?

3.33 A student must answer 8 out of 12 questions in an exam. (a) How many choices does the student have? (b) How many choices are there if the student must answer at least four of the first six questions?

3.34 If 3 balls are randomly drawn from a bowl containing six white balls and five black balls, what is the probability that one of the balls is white and the other two are black?

3.35 There are n balls in a box; they are all of the same color but one. If r of these balls are taken one at a time, what is the probability that the ball with the different color is chosen?

3.36 How many ways are there to go in xyz space from the point (i, j, p) to the point (l, m, q) by taking steps one unit in the positive x, positive y, and positive z direction, where $i, j, p, l, m,$ and q are all nonnegative integers, and $l > i, m > j$, and $q > p$?

Part II
Random Variables

4

One Random Variable: Fundamentals

The focus of this chapter and the remainder of this book is on probability models that assign real numbers to the random outcomes in the sample space. A numerical representation of the outcome of a random experiment is referred to as a random variable. Before proceeding to describe all fundamental aspects of a random variable, it may be important to note that the term *random variable* is a misleading nomenclature, as it is not random nor is it a variable. In fact, it is a well-defined deterministic function on the points of a sample space. Nevertheless, the term random variable is ubiquitously used, as it has no substitute. A thorough and detailed discussion of all major concepts associated with a single random variable and methods for calculating probabilities of various events involving a random variable are provided in this chapter.

4.1 Types of Random Variables

The outcome of a random experiment may be numerical or descriptive, and the primary interest may not be in the outcome itself, but rather in the numerical attribute of the outcome. In our notation, the name of a random variable is always an uppercase letter, such as X, and the corresponding lowercase letter, such as x, denotes a possible value of the random variable. A *random variable* X is a deterministic function that assigns a real number to each outcome in the sample space S, i.e. a mapping from the sample space to the set of real numbers R, $X : S \rightarrow R$, as shown in Figure 4.1. The sample space S is the *domain of the random variable* and the set of all values taken on by the random variable, denoted by S_X, is the *range of the random variable*. The range S_X is a subset of all real numbers $(-\infty, \infty)$. Note that two random variables are said to be equal if and only if the probability of the set of points in the sample space on which they differ has zero probability.

There are fundamentally two distinct types of random variables: discrete random variables and continuous random variables. If the range of the random

Probability, Random Variables, Statistics, and Random Processes: Fundamentals & Applications,
First Edition. Ali Grami.

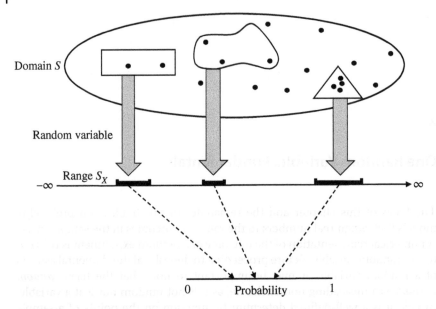

Figure 4.1 Illustration of the relationship among sample space, random variable, and probability.

variable assumes values from a countable set, it is then a ***discrete random variable***. The defining characteristic of a discrete random variable is that the set of possible values in the range can all be listed, where it may be a finite list or a countably infinite list. If the range can take infinitely many real values, i.e. takes on values that vary continuously within one or more intervals of real numbers, it is then a ***continuous random variable***. The determining feature of a continuous random variable is that the set of possible values in the range cannot all be enumerated, as it is an uncountably infinite list. During a given time interval, the number of telephone calls received is an example of a discrete random variable, whereas the duration of a call is an example of a continuous random variable. Randomly selecting a nonnegative integer out of a countably infinite list, i.e. {0, 1, 2, ...}, is an example of a discrete random variable, whereas randomly selecting a nonnegative real number out of an uncountably infinite list, i.e. [0, ∞), is an example of a continuous random variable.

It is important to note that there are random variables that are neither discrete nor continuous, but are a mixture of both. A ***mixed random variable*** has a range with both continuous and discrete parts. In other words, the range of a mixed random variable has at least one value with nonzero probability as well as values that include at least one interval of real numbers. For instance, the waiting time for a person who wants to buy a train ticket is a mixed random variable, assuming the probability that there is no one ahead of her in line (i.e. there is

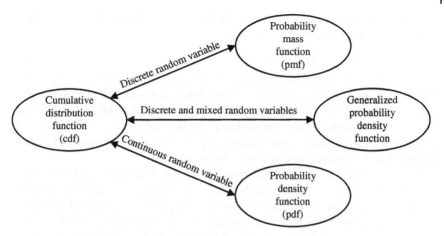

Figure 4.2 Illustration of the relationship among cdf, pmf, and pdf.

no waiting time) is $p > 0$, thus resembling a discrete random variable, and the probability that there are some people ahead of her in line (i.e. there is some waiting time) is $(1 - p) > 0$, thus behaving like a continuous random variable.

Any random variable has a cumulative distribution function (cdf), through which the probabilistic properties of the random variable can be fully summarized. However, a discrete random variable has a probability mass function (pmf) and a continuous random variable has a probability density function (pdf). Figure 4.2 shows the relationship between the cdf on the one hand and the pmf or pdf on the other hand, as they will be all discussed in this chapter. The knowledge of either the pmf of a discrete random variable or the pdf of a continuous random variable allows the cdf of the random variable to be completely determined, and vice versa. In most cases, the pmf or pdf is a more effective and intuitive way of providing the information contained in the cdf. The phrase *distribution function* is usually used exclusively for the cdf, whereas the word *distribution* is often used in a broader sense to refer to the pmf or pdf.

4.2 The Cumulative Distribution Function

Every random variable has a cdf. The cdf of a random variable contains all the information required to calculate probability for any event involving the random variable. The notation for cdf is $F_X(x)$, where we use the uppercase letter F, with a subscript corresponding to the name of the random variable X, as a function of a possible value of the random variable, denoted by the corresponding lowercase letter x. The cdf $F_X(x)$ is thus a function of x, not of the random variable X. The *cumulative distribution function* of a random variable X expresses the complete probability model of a random experiment as the following mathematical function:

$$F_X(x) = P(X \le x) \quad -\infty < x < \infty \tag{4.1}$$

The event $\{X \le x\}$ and its probability may vary, as x is varied, i.e. $F_X(x)$ is a function of the variable x. For any real number x, the cdf is the probability that the random variable X is no larger than x, that is the cdf is the probability that the random variable X takes on a value in the interval $(-\infty, x]$. The definition of the cdf thus contains a loose inequality (i.e. less than or equal to), which means that the function is continuous from the right.

If the cdf $F_X(x)$ is a piecewise flat function of $x \in S_X$ with discontinuous jumps, then X is a discrete random variable. In other words, for a discrete random variable X, $F_X(x)$ has zero slope everywhere except at values of x with nonzero probabilities, and at these points, the cdf has a discontinuity in the form of a jump of magnitude $P(X = x) \ne 0$. The cdf for a discrete random variable is thus in the form of a staircase of finite or countably infinite number of steps. Using the unit step function $u(x)$, defined as $u(x) = 0$ for $x < 0$ and $u(x) = 1$ for $x \ge 0$, the cdf for a discrete random variable can be written in terms of the unit step functions, where the number of the unit step functions corresponds to the number of nonzero probabilities.

If the cdf $F_X(x)$ is a continuous nondecreasing function of $x \in S_X$, then X is a continuous random variable. When X is a continuous random variable, we have $P(X = x) = 0$ for $x \in S_X$. This is of course an example of an event with probability zero that is not necessarily an impossible event. For continuous random variables, the probabilities of intervals rather than points are of importance.

If the cdf $F_X(x)$ increases continuously over at least one interval of values of x, and also possesses at least one discontinuity, then X is a mixed random variable. In other words, the cdf of a mixed random variable possesses features of both the cdf of a continuous random variable and that of a discrete random variable.

Figure 4.3 shows cdf examples for discrete, continuous, and mixed random variables. The cdf of a random variable always exists, and has the following important properties:

(i) $0 \le F_X(x) \le 1$

(ii) $a < b \rightarrow F_X(a) \le F_X(b)$

(iii) $\lim\limits_{x \to -\infty} F_X(x) = 0$

(iv) $\lim\limits_{x \to +\infty} F_X(x) = 1$

(v) $F_X(b) = F_X(b^+)$

(vi) $P(a < X \le b) = F_X(b) - F_X(a)$

(vii) $P(X = b) = F_X(b^+) - F_X(b^-)$

(viii) $P(X > x) = 1 - F_X(x) \tag{4.2}$

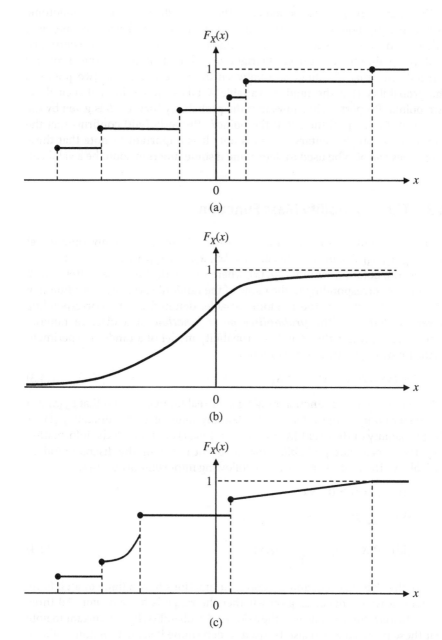

Figure 4.3 cdf examples: (a) discrete random variable; (b) continuous random variable; (c) mixed random variable.

The first four properties indicate that the cdf $F_X(x)$ is a monotonic nondecreasing bounded function that starts at zero and ends at one, as x increases from $-\infty$ to ∞. Property (v) implies that at points of discontinuity, a cdf is equal to the limit from the right, i.e. cdf is right-continuous. Property (vi) indicates that the difference between the cdf evaluated at two points is the probability that the random variable X takes on a value between these two points. Property (vii) shows that the probability that $X = b$ is given by the height of the jump of the cdf at the point b. Property (viii) confirms that the sum of two complementary events is one. It is important to note that these properties can also be used to determine if some function could be a valid cdf.

4.3 The Probability Mass Function

A discrete random variable assumes values from a countably infinite set $S_X = \{x_1, x_2, x_3, \ldots\}$ or a finite set $S_X = \{x_1, x_2, \ldots, x_n\}$, where n is a positive integer. The notation for pmf is $p_X(x)$, where we use the lowercase letter p, with a subscript corresponding to the name of the random variable X, as a function of a possible value of the random variable, denoted by the corresponding lowercase letter x. The **probability mass function** of a discrete random variable X expresses the complete probability model of a random experiment as the following mathematical function:

$$p_X(x) = P(X = x) \quad x \in S_X \tag{4.3}$$

Note that $p_X(x)$ is a function ranging over real numbers x and that $p_X(x)$ can be nonzero only at the values $x \in S_X$. For any value of x, the function $p_X(x)$ is the probability of the event $\{X = x\}$. The pmf $p_X(x)$ contains all the information required to calculate probability for any event involving the discrete random variable X. The pmf must satisfy the following important properties:

(i) $p_X(x) \geq 0 \quad \forall x$

(ii) $\displaystyle\sum_{x \in S_X} p_X(x) = 1$

(iii) $\displaystyle P(X \in B) = \sum_{x \in B \subseteq S_X} p_X(x)$ (4.4)

Note that B is an event and the above summations have a finite or an infinite number of terms, depending on whether the range is finite or not. All three properties are consequences of the axioms of probability. It is important to note that these properties can also be used to determine if some function could be a valid pmf. The graph of pmf $p_X(x)$ of a discrete random variable has vertical lines of height $p_X(x)$ at the values x in S_X. The relative values of pmf at different points give an indication of the relative likelihoods of occurrence.

The cdf and pmf of a discrete random variable X are related as follows:

$$F_X(x) = \sum_{u \leq x} p_X(u) \quad x \in S_X \tag{4.5}$$

In other words, the value of $F_X(x)$ is evaluated by simply adding together the probabilities $p_X(u)$ for all values of u that are no larger than x.

Example 4.1

Suppose we have a fair coin. Let X be the number of heads in three coin tosses. Find and sketch the pmf $p_X(x)$ and cdf $F_X(x)$ of the discrete random variable X.

Solution:

In a coin toss, there are two possibilities, a head and a tail. In tossing a coin three times, there are a total of 8 ($=2^3$) different possible outcomes, as reflected below:

$$\{TTT, TTH, THT, THH, HTT, HTH, HHT, HHH\}$$

Figure 4.4 pmf of the discrete random variable X and its cdf.

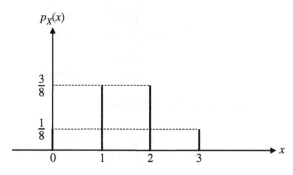

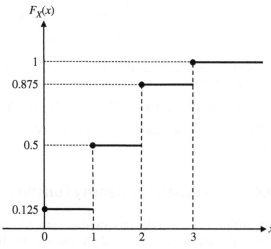

As the coin is fair, the likelihood of a tail is the same as the likelihood of a head. It is a reasonable assumption that the coin tosses are independent. After tossing a fair coin three times, the probability of each of the eight outcomes is $\frac{1}{8}\left(=\frac{1}{2}\times\frac{1}{2}\times\frac{1}{2}\right)$. In short, we have eight equally likely outcomes, and the pmf of X is therefore as follows:

$$P(X = 0) = P(TTT) = \frac{1}{8}$$

$$P(X = 1) = P(TTH) + P(THT) + P(HTT) = \frac{3}{8}$$

$$P(X = 2) = P(THH) + P(HHT) + P(HTH) = \frac{3}{8}$$

$$P(X = 3) = P(HHH) = \frac{1}{8}$$

or equivalently

$$p_X(x) = P(X = x) = \begin{cases} 0.125 & x = 0 \\ 0.375 & x = 1 \\ 0.375 & x = 2 \\ 0.125 & x = 3 \\ 0 & \text{otherwise} \end{cases}$$

The cdf $F_X(x)$ is thus as follows:

$$F_X(x) = P(X \leq x) = \begin{cases} 0 & x < 0 \\ 0.125 & 0 \leq x < 1 \\ 0.5 & 1 \leq x < 2 \\ 0.875 & 2 \leq x < 3 \\ 1 & 3 \leq x \end{cases}$$

Using the unit step function $u(x)$, the cdf can also be written in the following compact form:

$$F_X(x) = 0.125\, u(x) + 0.375\, u(x-1) + 0.375\, u(x-2) + 0.125\, u(x-3)$$

Figure 4.4 shows the pmf $p_X(x)$ and cdf $F_X(x)$ of the discrete random variable X. ∎

4.4 The Probability Density Function

An analog for a continuous random variable to the concept of the pmf developed for a discrete random variable is the concept of the pdf. A continuous random variable assumes values from an uncountably infinite set S_X. The notation for the pdf is $f_X(x)$, where we use the lowercase letter f, with a subscript corresponding to the name of the random variable X, as a function of a possible value of the random variable, denoted by the corresponding lowercase letter x. The **probability density function** of a continuous random variable X, if it exists, expresses the complete probability model of a random experiment as the following mathematical function:

$$f_X(x) = \frac{dF_X(x)}{dx} \qquad (4.6)$$

This derivative, when it exists that is when $F_X(x)$ is differentiable at x, is non-negative, as the cdf is a non-decreasing function. There may, however, be places where the derivative is not defined. For example, a continuous cdf with corners (points of abrupt change in slope) has a derivative with step-type discontinuities. The definition of pdf allows placing a Dirac delta function of weight $P(X = x)$ at the point x where the cdf is discontinuous. The important properties associated with the pdf are as follows:

(i) $f_X(x) \geq 0 \quad \forall x$

(ii) $\int_{-\infty}^{\infty} f_X(x)\,dx = 1$

(iii) $P(a \leq X \leq b) = \int_{a}^{b} f_X(x)\,dx \qquad (4.7)$

All these three properties are consequences of the axioms of probability and can also be used to determine if some function could be a valid pdf. Note that the properties (i) and (ii) are referred to as the nonnegativity and normalization properties, respectively. According to the property (iii), if the width of the interval shrinks to zero, i.e. $a = b$, then the area under the pdf also goes to zero. Hence, the probability of a sample point in the case of a continuous random variable is zero, i.e. a distinction is not made between the probabilities $P(X < x)$ and $P(X \leq x)$ when X is a continuous random variable. This highlights the fact that for a continuous random variable, the probability of a possible event can be zero. For instance, there are an infinite number of real numbers between, say, 1 and 2, the probability of randomly selecting a certain real number between 1 and 2 is thus zero, yet it can be selected. Figure 4.5 shows for a continuous random variable the connections between possible and impossible events on the one hand and zero and nonzero probabilities on the other hand.

The cdf and pdf of a continuous random variable X are related as follows:

$$F_X(x) = \int_{-\infty}^{x} f_X(\tau)\,d\tau \qquad (4.8)$$

The pdf is generally more useful than the cdf, for the graph of the pdf provides a good indication of likely values of observations. Note that pdf must be defined

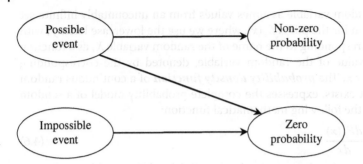

Figure 4.5 Probabilities of possible and impossible events for a continuous random variable.

for all real values of x, if X does not take on values in some interval, the pdf is then set equal to zero in that interval. The pdf has larger values in regions of high probability and smaller values in regions of low probability.

Example 4.2

Let X be a continuous random variable whose pdf is as follows:

$$f_X(x) = \begin{cases} kx & 0 \le x \le 1 \\ 0 & \text{otherwise} \end{cases}$$

where k is a nonrandom constant. Determine the value of k, the cdf $F_X(x)$, and $P(0.25 \le X \le 2)$.

Solution:

We solve the following equation for k:

$$\int_{-\infty}^{\infty} f_X(x)\, dx = \int_0^1 kx\, dx = 1 \to k = 2$$

We thus have

$$F_X(x) = \int_{-\infty}^{x} f_X(\tau)\, d\tau = \begin{cases} 0 & x < 0 \\ x^2 & 0 \le x < 1 \\ 1 & 1 \le x \end{cases}$$

We can determine $P(0.25 \le X \le 2)$ in two different ways. Using the cdf, we have

$$P(0.25 \le X \le 2) = F_X(2) - F_X(0.25) = 1 - \left(\frac{1}{4}\right)^2 = \frac{15}{16}$$

and using the pdf, we have

$$P(0.25 \le X \le 2) = \int_{0.25}^{2} 2x\, dx = \int_{0.25}^{1} 2x\, dx = \frac{15}{16}$$ ∎

If we use the Dirac delta function, also known as the unit impulse function $\delta(x)$, defined as $\delta(x) = 1$ for $x = 0$ and $\delta(x) = 0$ for $x \neq 0$, we could then define the *generalized pdf* for discrete random variables as follows:

$$f_X(x) = \sum_{x_k \in S_X} p_X(x_k)\delta(x - x_k) \tag{4.9}$$

where $S_X = \{x_1, x_2, \dots\}$ or $S_X = \{x_1, x_2, \dots, x_n\}$, assuming n is a positive integer, is the range of the discrete random variable and $p_X(x)$ represents its pmf. Therefore, the probability of $X = x_k$ is given by the coefficient of the corresponding delta function $\delta(x - x_k)$. The advantage of using a generalized pdf is that all types of random variables can then have a pdf. If the generalized pdf of a random variable has only delta functions, it is then a discrete random variable, if it does not include any delta functions, it is then a continuous random variable, and if it has both delta functions and nondelta functions, it is then a mixed random variable.

Example 4.3
Let X be a random variable whose cdf is as follows:

$$F_X(x) = \begin{cases} 0 & x < 0 \\ 0.75 - 0.5e^{-x} & 0 \leq x < 1 \\ 1 - 0.5e^{-x} & 1 \leq x \end{cases}$$

Determine the generalized pdf of X, and specify what type of random variable it is.

Solution:
Noting that the cdf has discontinuities at $x = 0$ and $x = 1$, as we have $F_X(0^-) \neq F_X(0^+)$ and $F_X(1^-) \neq F_X(1^+)$, the pdf, which is the derivative of the cdf, is then as follows:

$$f_X(x) = 0.25\,\delta(x) + 0.25\,\delta(x - 1) + 0.5\,e^{-x}u(x)$$

where $u(x)$ is the unit step function. Since the generalized pdf of the random variable X is a mix of delta and nondelta functions, it is a mixed random variable. ∎

4.5 Expected Values

Expectation is an important operation on a random variable, as it can bring about meaningful insight into the behavior of a random variable.

4.5.1 Mean of a Random Variable

The expected value or the mean of a random variable X represents a real number $(-\infty, \infty)$, and in a very large number of observations of X, can correspond to the average of X or the sample mean or the arithmetic mean. The expected value of a random variable is a significant measure, which in a limited, but effective way, can help characterize the random variable. Nevertheless, not all random variables have expected values, as there are few random variables whose means do not exist (not defined), such as the well-known Cauchy random variable.

The **expected value of a random variable** X, denoted by $E[X]$ or μ_X, is defined as follows:

$$E[X] = \mu_X = \begin{cases} \int_{-\infty}^{\infty} x f_X(x)\, dx = \int_{x \in S_X} x f_X(x)\, dx & \text{Continuous } X \\ \sum_{k=-\infty}^{\infty} x_k\, P(X = x_k) = \sum_{x \in S_X} x p_X(x) & \text{Discrete } X \end{cases} \quad (4.10)$$

In summary, the expected value is obtained by multiplying each possible value by its respective probability and then summing these products over all the values that have nonzero probability. Assuming a continuous random variable has an expected value, it is possible to observe an outcome that is equal to its expected value. However, for a discrete random variable whose expected value is defined, it may not be possible to observe an outcome that is equal to its expected value. For instance, the expected value of the random outcomes of a fair die is 3.5, but none of its outcomes can ever be 3.5.

Let X be a random variable and let $Y = g(X)$ denote a real-valued deterministic function of X. Therefore, Y is also a random variable. The **expected value of a function of a random variable** is defined as follows:

$$E[g(X)] = \begin{cases} \int_{-\infty}^{\infty} g(x) f_X(x)\, dx & \text{Continuous } X \\ \sum_{k=-\infty}^{\infty} g(x_k)\, P(X = x_k) & \text{Discrete } X \end{cases} \quad (4.11)$$

It is interesting to note that the expected value of a derived random variable, such as $g(X)$, can be calculated without having its distribution. This is known as the **law of the unconscious statistician.** In general, the expected value of a function of a random variable is not equal to the function of the expected value of the random variable, i.e. $E[g(X)] \neq g(E[X])$, as the expectation operator does not commute.

Note that the statistical expectation is a linear operation, that is the expected value of a linear combination of n random functions, where n is a positive integer, is equal to the same linear combination of n expected values of the random functions, simply put, we have the following:

$$E\left[\sum_{i=1}^{n} a_i g_i(X) \right] = \sum_{i=1}^{n} a_i E[g_i(X)] \quad (4.12)$$

where $\{a_1, a_2, \ldots, a_n\}$ are some nonrandom constants. In other words, the mean value of a weighted sum of functions of a random variable equals the weighted sum of the mean values of individual functions of the random variable.

Example 4.4

Let X be a continuous random variable whose pdf is as follows:

$$f_X(x) = \begin{cases} |x| & -1 \le x \le 1 \\ 0 & \text{otherwise} \end{cases}$$

and Y is a derived random variable, where $Y = X^3 - X^2 - X + 1$. Determine the expected values of the continuous random variables X and Y.

Solution:

We have

$$E[X] = \mu_X = \int_{-1}^{1} x|x|\, dx = \int_{-1}^{0} -x^2\, dx + \int_{0}^{1} x^2\, dx$$

$$= -\frac{1}{3}x^3 \Big|_{-1}^{0} + \frac{1}{3}x^3 \Big|_{0}^{1} = 0$$

and

$$E[Y] = \int_{-\infty}^{\infty} (x^3 - x^2 - x + 1) f_X(x)\, dx = \int_{-1}^{1} (x^3 - x^2 - x + 1)\,|x|\, dx$$

$$= \int_{-1}^{0} -(x^4 - x^3 - x^2 + x)\, dx + \int_{0}^{1} (x^4 - x^3 - x^2 + x)\, dx$$

$$= \frac{23}{60} + \frac{7}{60} = \frac{1}{2}$$

∎

Example 4.5

Let X be the random outcome in rolling a fair die. Suppose Y is a derived random variable, where $Y = \cos\left(\frac{X\pi}{6}\right)$. Determine the expected values of the discrete random variables X and Y.

Solution:

We have

$$E[X] = \mu_X = \sum_{k=1}^{6} k \times \frac{1}{6} = \frac{1}{6}(1 + 2 + 3 + 4 + 5 + 6) = 3.5$$

and

$$E[Y] = \sum_{k=1}^{6} \cos\left(\frac{k\pi}{6}\right) \times \left(\frac{1}{6}\right) = \frac{1}{6}\left(\frac{\sqrt{3}}{2} + \frac{1}{2} + 0 - \frac{1}{2} - \frac{\sqrt{3}}{2} - 1\right)$$

$$= -\frac{1}{6}$$

∎

4.5.2 Variance of a Random Variable

As the expected value of a random variable fails to show the spread of random values in its distribution, a measure to highlight its dispersion is essential. If we have $g(X) \triangleq (X - E[X])^2$, $E[g(X)]$, i.e. the expected value of $g(X)$, is then called the ***variance of the random variable*** X. The variance of X, denoted by σ_X^2, is thus the mean square of the difference between a random variable X and its mean $E[X]$. In summary, the variance is obtained by multiplying the squared distance of each possible value from the expected value by its respective probability and then summing these products over all the values that have nonzero probability. Note that not all random variables have variances, as there are few random variables whose variances do not exist (not defined), such as the well-known Cauchy random variable. The variance of a random variable, if it is defined, is always nonnegative.

The variance of a random variable provides essentially a measure of the effective width of the pdf or pmf of the random variable. The variance of a random variable in some sense is a measure of the variable's randomness, as it indicates the variability of the outcomes. For instance, a large variance indicates the random variable is quite spread out and it is thus more unpredictable, whereas a small variance shows the random variable is concentrated around its mean and it is thus less random. In fact, when the variance is zero, the variable is no longer random, meaning there is no uncertainty at all.

The square root of the variance of X, denoted by σ_X, is called the ***standard deviation of the random variable*** X, and is a positive quantity with the same unit as X. For instance, the random variable X and its standard deviation σ_X may be both in volts, meters, dollars or kilograms. The importance of the standard deviation of a random variable lies in the fact that it brings context to the mean value. There is an often-used principle, which is true for many, but certainly not all, random variables, that states the area under the pdf or pmf of a random variable within two standard deviations, i.e. $2\sigma_X$, of the mean μ_X is approximately 95%. In many applications, most of the observations of a random variable are within one standard deviation of the expected value. For instance, a student, who has a test mark of 8 points above the test mean, is likely to be in the middle of the class, if the standard deviation of test marks is 16 points, however, she is likely to be near the top of the class, if the standard deviation is 4 points.

As statistical expectation is a linear operation, the variance of the random variable X can be equivalently expressed as follows:

$$\sigma_X^2 \triangleq E[(X - E[X])^2] = E[X^2] - (E[X])^2 = E[X^2] - \mu_X^2 \tag{4.13}$$

In order to determine the variance, we first need to find the mean, as the variance of a random variable is the difference between its mean square and square of its mean. Note that variance, in contrast to mean, is a nonlinear operator. In other words, the variance of a weighted sum of random variables, in general, is not equal to the weighted sum of the variances of the random variables.

Example 4.6
Suppose we have a fair coin. Let X be the number of tails in four coin tosses. Determine the variance of the discrete random variable X.

Solution:
The pmf of the discrete random variable X is as follows:

$$p_X(x) = P(X = x) = \begin{cases} \dfrac{1}{16} & x = 0 \\[2mm] \dfrac{4}{16} & x = 1 \\[2mm] \dfrac{6}{16} & x = 2 \\[2mm] \dfrac{4}{16} & x = 3 \\[2mm] \dfrac{1}{16} & x = 4 \end{cases}$$

The mean is thus as follows:

$$E[X] = \mu_X = (0)\left(\frac{1}{16}\right) + (1)\left(\frac{4}{16}\right) + (2)\left(\frac{6}{16}\right) + (3)\left(\frac{4}{16}\right)$$
$$+ (4)\left(\frac{1}{16}\right) = 2$$

which intuitively makes sense, as out of four tosses of a fair coin, on average two are tails. The mean square is obtained as follows:

$$E[X^2] = (0)^2\left(\frac{1}{16}\right) + (1)^2\left(\frac{4}{16}\right) + (2)^2\left(\frac{6}{16}\right) + (3)^2\left(\frac{4}{16}\right)$$
$$+ (4)^2\left(\frac{1}{16}\right) = 5$$

The variance is therefore as follows:

$$\sigma_X^2 = 5 - (2)^2 = 1 \qquad \blacksquare$$

Example 4.7
We have $X = A\cos(2\pi f_c t + \Theta)$, where the amplitude $A \neq 0$ and the frequency f_c are both nonrandom constants, and the initial phase Θ is a continuous random variable whose pdf is as follows:

$$f_\Theta(\theta) = \begin{cases} \dfrac{1}{2\pi} & 0 \leq \theta \leq 2\pi \\[2mm] 0 & \text{otherwise} \end{cases}$$

Find the variance of the continuous random variable X.

Solution:
The mean is obtained as follows:

$$E[X] = \mu_X = \int_0^{2\pi} A\cos(2\pi f_c t + \theta)\left(\frac{1}{2\pi}\right) d\theta$$

$$= \left(\frac{A}{2\pi}\right)\int_0^{2\pi} \cos(2\pi f_c t + \theta)\, d\theta = 0$$

Using a trigonometric power relation, the mean square is as follows:

$$E[X^2] = \int_0^{2\pi} A^2(\cos(2\pi f_c t + \theta))^2 \left(\frac{1}{2\pi}\right) d\theta$$

$$= \left(\frac{A^2}{2\pi}\right)\int_0^{2\pi} \frac{(1 + \cos(4\pi f_c t + 2\theta))}{2}\, d\theta$$

$$= \left(\frac{A^2}{4\pi}\right)\left(\int_0^{2\pi} d\theta + \int_0^{2\pi} \cos(4\pi f_c t + 2\theta)\, d\theta\right) = \frac{A^2}{2}$$

We thus obtain $\sigma_X^2 = \frac{A^2}{2}$. Note that the variance of a sinusoidal waveform is independent of its frequency, and is a function of only its amplitude. ∎

Example 4.8
Suppose $Y = aX + b$, where X is a random variable with mean μ_X and variance σ_X^2, and a and b are both nonrandom constants. Determine the variance of the random variable Y in terms of a, b, μ_X, and σ_X^2. Comment on the results.

Solution:
We first determine the mean of the random variable Y:

$$\mu_Y = E[Y] = E[aX + b] = a\,E[X] + b = a\,\mu_X + b$$

This in turn means the mean value of Y is linearly related to the mean value of X in the same way that the random variables Y and X are linearly related. We first find the mean square value of Y:

$$E[Y^2] = E[(aX + b)]^2 = E[a^2X^2 + b^2 + 2abX] = a^2E[X^2] + b^2 + 2abE[X]$$

We then determine the variance of Y:

$$\sigma_Y^2 = E[Y^2] - (E[Y])^2 = (a^2E[X^2] + b^2 + 2abE[X]) - (a\,E[X] + b)^2$$
$$= a^2(E[X^2] - (E[X])^2) = a^2\sigma_X^2$$

The mean and variance of Y can be thus independently specified by using the appropriate values of a and b. Note that for certain values of a and b, we can make the following interesting observations about the mean and variance of the random variable Y:

- If $b = 0$, then Y is directly proportional to X and we have $\mu_Y = a\,\mu_X$ and $\sigma_Y^2 = a^2\sigma_X^2$. The mean and variance of Y are both functions of a, and their values are thus not independent of one another.
- If $b = -a\,\mu_X + \mu_X$, then b is a particular linear function of a. We thus have $\mu_Y = \mu_X$ and $\sigma_Y^2 = a^2\sigma_X^2$. The mean of Y remains the same as the mean of X, but its variance is then a function of a.
- If $a = 1$, then Y is a shifted version of X. We thus have $\mu_Y = \mu_X + b$ and $\sigma_Y^2 = \sigma_X^2$. The variance of Y remains the same as the variance of X, but its mean is then the shifted version of the mean of X by b.
- If $a = 0$, then Y is not a random variable, and we have $\mu_Y = b$ and $\sigma_Y^2 = 0$. The variance of Y becomes zero, and the value of Y is always b. ∎

The mean and variance in probability have comparable concepts in other fields. Examples include (i) in physics, if a distribution of a random variable corresponds to a mass distribution, then the mean is the center of mass and the variance is the moment of inertia about the mean, (ii) in electrical engineering, if a distribution of a random variable corresponds to a signal amplitude, then the mean is the DC value and the variance is the AC power, and (iii) in statistics, to estimate a random variable in the absence of observations, the minimum mean square error estimate of a random variable is its mean, and the mean square error associated with the estimate is its variance.

4.5.3 Moments of a Random Variable

The mean and variance are important parameters to provide valuable insights about the distribution of a random variable. However, the expected values of all the powers of a random variable, known as the moments of the random variable, can completely specify the distribution of the random variable. For the random variable X, assuming n is a positive integer, the **nth moment**, denoted by $E[X^n]$, and the **nth central moment**, denoted by $E[(X - E[X])^n]$, are respectively, defined as follows:

$$E[X^n] = \begin{cases} \int_{-\infty}^{\infty} x^n f_X(x)\,dx & \text{Continuous } X \\ \sum_{k=-\infty}^{\infty} (x_k)^n P(X = x_k) & \text{Discrete } X \end{cases}$$

$$E[(X - E[X])^n] = \begin{cases} \int_{-\infty}^{\infty} (x - E[X])^n f_X(x)\,dx & \text{Continuous } X \\ \sum_{k=-\infty}^{\infty} (x_k - E[X])^n P(X = x_k) & \text{Discrete } X \end{cases} \tag{4.14}$$

If the mean value of a random variable is zero, its nth moment and nth central moment are then equal. Note that the mean and variance represent the first moment and the second central moment of a random variable, respectively. As it will be discussed in the context of descriptive statistics, the third central moment is used to determine the skewness, a measure of the asymmetry of a distribution about its mean, and the fourth central moment is used to describe the tailedness of a distribution, a measure of outliers in a distribution.

Example 4.9

Suppose X is a continuous random variable whose pdf is as follows:

$$f_X(x) = \begin{cases} 2x & 0 \le x \le 1 \\ 0 & \text{otherwise} \end{cases}$$

Determine its nth moment and nth central moment as well its mean and variance.

Solution:

We find the nth moment as follows:

$$E[X^n] = \int_0^1 x^n (2x)\, dx = 2\int_0^1 x^{n+1}\, dx = \frac{2}{n+2}$$

Note that when $n = 1$, we have $E[X] = \frac{2}{3}$. We now obtain the nth central moment as follows:

$$\begin{aligned}
E[(X - E[X])^n] &= \int_0^1 \left(x - \frac{2}{3}\right)^n (2x)dx \\
&= \frac{2}{n+2}\left(\left(\frac{1}{3}\right)^{n+2} - \left(-\frac{2}{3}\right)^{n+2}\right) \\
&+ \frac{4}{3(n+1)}\left(\left(\frac{1}{3}\right)^{n+1} - \left(-\frac{2}{3}\right)^{n+1}\right)
\end{aligned}$$

Note that the right-hand side of the above equation was obtained by first using a change of variable and then carrying out integration. Note that when we have $n = 2$, we have $\sigma_X^2 = \frac{1}{18}$. ∎

The expected values, such as the mean, variance and moments of a random variable, summarize valuable information about the random variable. These parameters in some applications of practical significance are a lot easier to measure and estimate than the corresponding cdf (equivalently pmf or pdf).

4.5.4 Mode and Median of a Random Variable

There are two other simple measures that can provide further insights into the possible values of a random variable, they are, namely, the mode and the median. The mode and the median both make the most sense when a large sample space must be described.

The *mode of a random variable* is that value that occurs most often or has the greatest probability of occurring or when its pdf or pmf is a maximum. Sometimes, a random variable has more than one mode, thus naming it a *multi-modal random variable*.

The *median of a random variable* is that particular value for which the probability of having a value greater than that and the probability of having a value less than that are equal, that is each probability is equal to 0.5. If a random variable has a pmf or pdf that is symmetric, then the point of symmetry represents the median of the random variable. A continuous random variable always has a median, but the median of a discrete random variable may not exist.

Example 4.10

Determine the mode of the continuous random variable X, whose pdf is as follows: $f_X(x) = \left(\frac{x}{\alpha^2}\right) \exp\left(-\frac{x^2}{2\alpha^2}\right) u(x)$, where $\alpha > 0$ and $u(x)$ is the unit step function.

Solution:

In order to find the mode of the random variable X, we need to find the derivative of its pdf and set it equal to zero to find the maximum value, we thus have

$$\frac{d}{dx} f_X(x) = \frac{d}{dx}\left(\frac{x}{\alpha^2}\right) \exp\left(-\frac{x^2}{2\alpha^2}\right)$$

$$= \frac{1}{\alpha^2} \exp\left(-\frac{x^2}{2\alpha^2}\right) - \frac{2x^2}{2\alpha^4} \exp\left(-\frac{x^2}{2\alpha^2}\right) = 0 \rightarrow x = \alpha \qquad \blacksquare$$

Example 4.11

Determine the median of the continuous random variable X, whose pdf is as follows: $f_X(x) = \lambda e^{-\lambda x} u(x)$, where $\lambda > 0$ and $u(x)$ is the unit step function.

Solution:

In order to find the median of the random variable, we need to have the following two probabilities equal to one another:

$$\int_0^x \lambda e^{-\lambda t} dt = \int_x^\infty \lambda e^{-\lambda t} dt \rightarrow -(e^{-\lambda t})|_0^x = -(e^{-\lambda t})|_x^\infty \rightarrow$$

$$2e^{-x\lambda} = 1 \rightarrow x = \frac{\ln 2}{\lambda} \qquad \blacksquare$$

In summary, the expected values have the following interesting properties:

- Not all random variables have expected values.
- The expected value of a random variable can be any finite real number $(-\infty, \infty)$.
- More than one random variable may have the same expected value.
- The expected value of a nonrandom constant is the constant itself, i.e. $E[c] = c$.
- Expected values scale by multiplicative constants, i.e. $E[cg(X)] = cE[g(X)]$.
- Expected values are additive, i.e. $E[g_1(X) + g_2(X)] = E[g_1(X)] + E[g_2(X)]$.
- Expected values are not multiplicative, i.e. $E[g_1(X)g_2(X)] \neq E[g_1(X)]E[g_2(X)]$.
- The expected value does not generally indicate the most probable value.
- The expected value of a nonnegative random variable is nonnegative.
- Expected values commute with derivatives and integrals of a nonrandom parameter.

4.6 Conditional Distributions

A conditional distribution can incorporate partial knowledge about the outcome of an experiment in the evaluation of probabilities of events. If there is some information about a random variable, then its conditional distribution needs to incorporate that.

Suppose the event B, defined as $\{X \leq b\}$, is given, and we have $P(B) = P(X \leq b) > 0$. Using Bayes' theorem, the **conditional cdf** of X given the event B is then defined as follows:

$$F_X(x \mid B) = P(X \leq x \mid B) = \frac{P(X \leq x, X \leq b)}{P(B)} \tag{4.15}$$

There are now two mutually exclusive cases, depending on whether x or b is larger. If $b \leq x$, then the event $\{X \leq b\}$ is contained in the event $\{X \leq x\}$, and we thus have $P(X \leq x, X \leq b) = P(X \leq b) = P(B)$. However, if $x \leq b$, then the event $\{X \leq x\}$ is contained in the event $\{X \leq b\}$, and we thus have $P(X \leq x, X \leq b) = P(X \leq x)$. As a result, the above conditional cdf can be simplified as follows:

$$F_X(x \mid B) = \begin{cases} \dfrac{P(B)}{P(B)} = 1 & x \geq b \\[2ex] \dfrac{P(X \leq x)}{P(B)} & x \leq b \end{cases} \tag{4.16}$$

Based on the conditional cdf, we can thus obtain the following **conditional pdf** and **conditional pmf** of X given the event B, respectively:

$$f_X(x \mid B) = \begin{cases} 0 & x \notin B \\ \dfrac{f_X(x)}{P(B)} & x \in B \end{cases} \qquad \text{Continuous } X$$

$$p_X(x \mid B) = \begin{cases} 0 & x \notin B \\ \dfrac{p_X(x)}{P(B)} & x \in B \end{cases} \qquad \text{Discrete } X \qquad (4.17)$$

Note that dividing the pdf $f_X(x)$ or the pmf $p_X(x)$ by $P(B) < 1$ ensures that for $x \leq b$, the integral of the conditional pdf $f_X(x \mid B)$ or the summation of the conditional pmf $p_X(x \mid B)$, over the entire range of interest, is one, an essential requirement for any pdf or pmf to meet.

Example 4.12

Let X be a random variable whose pdf is as follows: $f_X(x) = \left(\dfrac{1}{\sqrt{2\pi}}\right) \exp\left(-\dfrac{x^2}{2}\right)$. Determine the conditional pdf $f_X(x \mid |X| \leq 1)$.

Solution:

We first find the probability of the conditioning event:

$$P(|X| \leq 1) = P(-1 \leq X \leq 1) = P(-1 \leq X) - P(1 \leq X)$$

$$= \int_{-1}^{\infty} \left(\frac{1}{\sqrt{2\pi}}\right) \exp\left(-\frac{x^2}{2}\right) dx$$

$$- \int_{1}^{\infty} \left(\frac{1}{\sqrt{2\pi}}\right) \exp\left(-\frac{x^2}{2}\right) dx$$

The above definite integrals cannot be evaluated analytically, and must thus be found numerically. Using $Q(x) \triangleq \int_x^{\infty} \left(\frac{1}{\sqrt{2\pi}}\right) \exp\left(-\frac{x^2}{2}\right) dx$, we can numerically find the probability of interest:

$$P(|X| \leq 1) = Q(-1) - Q(1) = (1 - Q(1)) - Q(1) = 1 - 2Q(1) \cong 0.683$$

The conditional pdf is thus as follows:

$$f_X(x \mid |X| \leq 1) = \begin{cases} 0 & x < -1 \text{ or } 1 < x \\ \dfrac{f_X(x)}{0.683} \cong 1.464\, f_X(x) & -1 \leq x \leq 1 \end{cases}$$

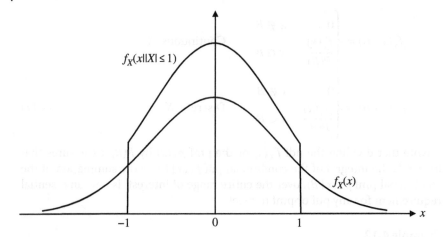

$f_X(x||X| \le 1)$

$f_X(x)$

x

−1   0   1

Figure 4.6 pdf of the continuous random variable X and its conditional pdf.

Figure 4.6 shows the pdf of X and its conditional pdf. Note that for the range of the random variable X that the condition is not met, the conditional pdf is zero, and for the range that the condition is met, the conditional pdf is the scaled up version of the pdf of X so as to ensure the area under the conditional pdf is unity. ■

Example 4.13
We have an unfair (biased) die with the possible outcomes 1, 2, 3, 4, 5, and 6. In rolling this die, we know that the probability of a 6 is six times the probability of a 1, the probability of a 5 is five times the probability of a 1, the probability of a 4 is four times the probability of a 1, the probability of a 3 is three times the probability of a 1, and the probability of a 2 is twice the probability of a 1. Determine the conditional pmf $p_X(x \mid B)$, where B is the set of outcomes that are prime numbers, i.e. $B = \{2, 3, 5\}$.

Solution:
Assuming $y = P(X = 1)$, we then have

$$P(X = 1) + P(X = 2) + P(X = 3) + P(X = 4) + P(X = 5)$$
$$+ P(X = 6) = 1 \rightarrow y + 2y + 3y + 4y + 5y + 6y = 1 \rightarrow$$
$$21y = 1 \rightarrow y = \frac{1}{21}$$

We then obtain the probability of the conditioning event:

$$P(B) = P(X = 2) + P(X = 3) + P(X = 5) = \frac{2}{21} + \frac{3}{21} + \frac{5}{21} = \frac{10}{21}$$

The conditional pmf is thus as follows:

$$p_X(x \mid B) = \begin{cases} 0 & x = 1 \\ \dfrac{\left(\dfrac{2}{21}\right)}{\left(\dfrac{10}{21}\right)} = \dfrac{2}{10} & x = 2 \\ \dfrac{\left(\dfrac{3}{21}\right)}{\left(\dfrac{10}{21}\right)} = \dfrac{3}{10} & x = 3 \\ 0 & x = 4 \\ \dfrac{\left(\dfrac{5}{21}\right)}{\left(\dfrac{10}{21}\right)} = \dfrac{5}{21} & x = 5 \\ 0 & x = 6 \end{cases}$$

Figure 4.7 shows the pmf of X and its conditional pmf. As expected, the sum of the conditional probabilities over all possible values of B is 1. ∎

The **conditional expected value** of X given the event B is as follows:

$$E[X \mid B] = \begin{cases} \int_{-\infty}^{\infty} x f_X(x \mid B) dx & \text{Continuous } X \\ \sum_{k=-\infty}^{\infty} x_k P(X = x_k \mid B) & \text{Discrete } X \end{cases} \tag{4.18}$$

Example 4.14
Suppose the pdf of the random variable X representing the duration of a telephone call in minutes is as follows: $f_X(x) = \frac{1}{4} \exp\left(-\frac{x}{4}\right) u(x)$, where $u(x)$ is the unit step function. Determine the conditional expected value for calls that last at least five minutes.

Solution:
The probability of the conditioning event $X \geq 5$ is as follows:

$$P(B) = P(X \geq 5) = \int_5^{\infty} \frac{1}{4} \exp\left(-\frac{x}{4}\right) dx = \exp\left(-\frac{5}{4}\right)$$

The conditional pdf of X given $X \geq 5$ is thus as follows:

$$f_X(x \mid B) = \frac{\dfrac{1}{4} \exp\left(-\dfrac{x}{4}\right)}{\exp\left(-\dfrac{5}{4}\right)} = \frac{1}{4} \exp\left(-\frac{x-5}{4}\right) \quad x \geq 5$$

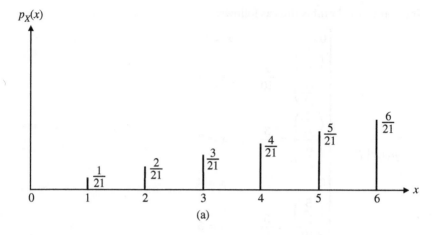

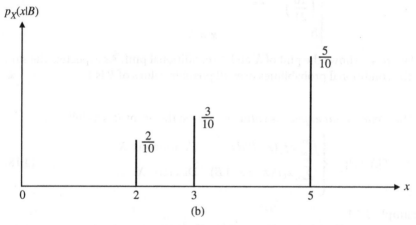

Figure 4.7 Discrete random variable: (a) marginal pmf; (b) conditional pmf.

Using the method of integration by parts, the conditional expected value can then be obtained as follows:

$$E[X \mid B] = \int_5^\infty \frac{x}{4} \exp\left(-\frac{x-5}{4}\right) dx = 9$$

∎

4.7 Functions of a Random Variable

Let X be a random variable and $g(X)$ be a real-valued function of X. If $Y = g(X)$, then Y is also a random variable. The cdf of Y depends on both the deterministic function $g(X)$ and the cdf of the random variable X. The probability of an event

C involving Y is equal to the probability of the equivalent event B involving X such that $g(X)$ is in C, i.e. we have $P(Y \text{ in } C) = P(g(X) \text{ in } C) = P(X \text{ in } B)$.

4.7.1 pdf of a Function of a Continuous Random Variable

In order to determine the ***pdf of a function of a continuous random variable***, i.e. the pdf of Y, where Y is related to the continuous random variable X by the equation $Y = g(X)$ or equivalently $X = g^{-1}(Y) = h(Y)$, we first need to solve it for x in terms of y. Suppose there are $n \geq 1$ solutions $\{x_1, x_2, ..., x_n\}$, i.e. $y = g(x_1) = g(x_2) = ... = g(x_n)$, and for each solution x_i, $i = 1, 2, ..., n$, the derivative $g'(x_i)$ exists and is also nonzero. It can then be shown that the pdf of the random variable $Y = g(X)$ is as follows:

$$f_Y(y) = \sum_{i=1}^{n} \frac{f_X(x_i)}{|g'(x_i)|} \tag{4.19}$$

We now focus on $n = 1$, in other words, suppose the function $g(X)$ is a continuous monotonic increasing or decreasing function, that is $Y = g(X)$ is a one-to-one function and has the inverse function $X = g^{-1}(Y) = h(Y)$. The pdf of Y, which is a one-to-one differentiable function of the random variable X, is then simply as follows:

$$f_Y(y) = \frac{f_X(x)}{|g'(x)|} = \frac{f_X(h(y))}{\left|\frac{dy}{dh(y)}\right|} = \left|\frac{dh(y)}{dy}\right| f_X(h(y)) \tag{4.20}$$

where $y = g(x)$, $g'(x) = \frac{dy}{dx}$, and $x = h(y)$. To find the distribution of the random variable Y, it is important to first find the possible values of Y, i.e. the sample space S_Y, and then determine the pdf of Y.

Example 4.15

Let X be a continuous random variable whose pdf $f_X(x)$ is as follows: $f_X(x) = \left(\frac{1}{\sqrt{2\pi}}\right) \exp\left(-\frac{x^2}{2}\right)$, where $-\infty < x < \infty$. Assuming $Y = X^2$, determine the pdf of the random variable Y, i.e. $f_Y(y)$.

Solution:

We first solve the equation $y = x^2$, and thus obtain $x = \pm\sqrt{y}$, assuming $y > 0$. We then find the derivative $y' = 2x = \pm 2\sqrt{y}$, and thus have the following result:

$$f_Y(y) = \begin{cases} \dfrac{f_X(\sqrt{y})}{|2\sqrt{y}|} + \dfrac{f_X(-\sqrt{y})}{|-2\sqrt{y}|} = \left(\dfrac{1}{\sqrt{2\pi y}}\right)\exp\left(-\dfrac{y}{2}\right) & y > 0 \\ 0 & y \leq 0 \end{cases}$$

∎

Example 4.16
The continuous random variables X and Y are linearly related, i.e. $Y = aX + b$, where $a \neq 0$ and b are both nonrandom constants. Determine the cdf and pdf of Y in terms of the cdf and pdf of X.

Solution:
We have

$$y = g(x) = ax + b \rightarrow x = h(y) = \frac{y-b}{a} \rightarrow \frac{dh(y)}{dy} = \frac{1}{a}$$

Since $Y = aX + b$ is a one-to-one function, the pdf of y is as follows:

$$f_Y(y) = \left| \frac{dh(y)}{dy} \right| f_X(h(y)) = \frac{1}{|a|} f_X \left(\frac{y-b}{a} \right)$$

There is another method to get the pdf of interest. Noting that the sign of the constant a plays a role in the following inequality, the cdf of Y can be determined as follows:

$$F_Y(y) = P(Y \leq y) = P(aX + b \leq y)$$

$$= \begin{cases} P\left(X \leq \dfrac{y-b}{a} \right) = F_X \left(X \leq \dfrac{y-b}{a} \right) & a > 0 \\[3mm] P\left(X \geq \dfrac{y-b}{a} \right) = 1 - F_X \left(X \leq \dfrac{y-b}{a} \right) & a < 0 \end{cases}$$

We differentiate the above equation with respect to y to obtain the pdf of Y:

$$f_Y(y) = \begin{cases} \left(\dfrac{1}{a} \right) f_X \left(\dfrac{y-b}{a} \right) & a > 0 \\[3mm] \left(\dfrac{1}{-a} \right) f_X \left(\dfrac{y-b}{a} \right) & a < 0 \end{cases}$$

We can thus have it in the following compact form, as derived earlier:

$$f_Y(y) = \frac{1}{|a|} f_X \left(\frac{y-b}{a} \right) \qquad \blacksquare$$

Example 4.17
The continuous random variables X and Y are related by $Y = \tan X$. Find the pdf of the random variable Y, i.e. $f_Y(y)$, if $f_X(x)$ is as follows:

$$f_X(x) = \begin{cases} \dfrac{1}{\pi} & -\dfrac{\pi}{2} \leq x \leq \dfrac{\pi}{2} \\[3mm] 0 & \text{otherwise} \end{cases}$$

Solution:
The equation $Y = \tan X$ has infinitely many solutions for any y, as we have $x_i = \tan^{-1}y$, $i = 0, \pm 1, \pm 2, \dots$. We have the following:

$$y = g(x) = \tan x \rightarrow x_i = \tan^{-1}y \text{ and } g'(x_i) = 1 + (\tan x_i)^2 = 1 + y^2$$

We thus have

$$f_Y(y) = \sum_{i=1}^{n} \frac{f_X(x_i)}{|g'(x_i)|} = \frac{1}{1+y^2} \sum_{i=-\infty}^{\infty} f_X(x_i)$$

Since we have $f_X(x_0) = \frac{1}{\pi}$ and all other terms in the summation are zero, Y has the following pdf:

$$f_Y(y) = \frac{1}{\pi(1+y^2)}$$

∎

4.7.2   pmf of a Function of a Discrete Random Variable

In order to determine the *pmf of a function of a discrete random variable*, i.e. the pmf of Y, where Y is related to the discrete random variable X by the equation $Y = g(X)$, we need to have the following:

$$p_Y(y_i) = \sum_{\{j:g(x_j)=y_i\}} p_X(x_j) \tag{4.21}$$

The pmf of the random variable Y is obtained when we sum up the probabilities for all the values of $X = x_j$ that are mapped into $Y = y_i$. Note that if $Y = g(X)$ is a monotonic function, then there is a one-to-one correspondence between X and Y, as such there is no need to sum up.

Example 4.18
Suppose X is a discrete random variable represented by the outcome in rolling a fair die. If the random variable Y is related to X by the function $Y = \sin\left(\frac{X\pi}{6}\right)$, determine the pmf of Y.

Solution:
The sample space for the discrete random variable X is $S_X = \{1, 2, 3, 4, 5, 6\}$, we thus have the following:

$$X = 1 \rightarrow Y = \sin\left(\frac{\pi}{6}\right) = \frac{1}{2}$$

$$X = 2 \rightarrow Y = \sin\left(\frac{2\pi}{6}\right) = \frac{\sqrt{3}}{2}$$

$$X = 3 \rightarrow Y = \sin\left(\frac{3\pi}{6}\right) = 1$$

$$X = 4 \rightarrow Y = \sin\left(\frac{4\pi}{6}\right) = \frac{\sqrt{3}}{2}$$

$$X = 5 \rightarrow Y = \sin\left(\frac{5\pi}{6}\right) = \frac{1}{2}$$

$$X = 6 \rightarrow Y = \sin\left(\frac{6\pi}{6}\right) = 0$$

Consequently, the sample space for the random variable Y is as follows:

$$S_Y = \left\{ 0, \frac{1}{2}, \frac{\sqrt{3}}{2}, 1 \right\}$$

Noting that $p_X(x_j) = \frac{1}{6}, j = 1, 2, \ldots, 6$, and Y is not a one-to-one function, $p_Y(y_i)$ is determined by summing the probabilities of all x_j's that map into y_i using $g(x_j) = y_i$, for $i = 1, 2, 3, 4$:

$$y_1 = 0 \rightarrow p_Y(0) = p_X(6) = \frac{1}{6}$$

$$y_2 = \frac{1}{2} \rightarrow p_Y\left(\frac{1}{2}\right) = p_X(1) + p_X(5) = \frac{1}{6} + \frac{1}{6} = \frac{1}{3}$$

$$y_3 = \frac{\sqrt{3}}{2} \rightarrow p_Y\left(\frac{\sqrt{3}}{2}\right) = p_X(2) + p_X(4) = \frac{1}{6} + \frac{1}{6} = \frac{1}{3}$$

$$y_4 = 1 \rightarrow p_Y(1) = p_X(3) = \frac{1}{6}$$ ■

4.7.3 Computer Generation of Random Variables

As part of studying probability models, computers are extensively used to perform calculations, simulations, and graphing. To this end, random numbers need to be generated according to prescribed distributions. To generate a random variable with a specified distribution, it is reasonable to assume that values of a random variable X with uniform distribution, i.e. equally distributed, on the interval (0, 1) can be easily generated. The problem is thus to find the transformation $T(X)$ that will give rise to a random variable Y of prescribed distribution when X has a uniform distribution over (0, 1). The inverse cdf of Y is thus as follows:

$$F_Y(y) = x \rightarrow y = F_Y^{-1}(x) = T(x) \quad 0 < x < 1 \tag{4.22}$$

Since the cdf of Y is a nondecreasing function, its inverse is also nondecreasing. Therefore, for a given distribution $F_Y(y)$ for the random variable Y, we find the inverse function by solving $F_Y(y) = x$ for y. The result is $y = T(x)$. If the inverse cannot be found analytically, then the required inverse is found numerically and stored for a large array of points (y, x) and the simulation can then use interpolation between computed points to obtain values of y for any value of x.

Example 4.19
Determine the transformation required to generate a random variable whose cdf is as follows: $F_Y(y) = (1 - \exp(-y^2))\, u(y)$, where $u(y)$ is the unit step function.

Solution:
We have

$$F_Y(y) = 1 - \exp(-y^2) = x \quad 0 < x < 1$$

We solve it for y, and thus have

$$y = T(x) = \sqrt{-\ln(1-x)} \quad 0 < x < 1$$

∎

4.8 Transform Methods

The transform methods are very powerful computational aids to solve equations involving derivatives and integrations, to determine the moments of a random variable, to examine convergence of probability distributions, and to help determine the distribution of the sum of independent random variables. There are various transform methods, however, our focus is on the moment generating function (mgf), which is the Laplace transform with s replaced by $-v$, and the characteristic function (cf), which is the Fourier transform with the sign of exponent reversed. These transforms can also be easily defined for multiple random variables, resulting in joint mgfs and joint cfs.

4.8.1 Moment Generating Function of a Random Variable

The ***moment generating function*** of the random variable X is as follows:

$$M_X(v) = E[e^{vX}] = \begin{cases} \int_{-\infty}^{\infty} e^{vx} f_X(x)\, dx & \text{Continuous } X \\ \sum_{i=-\infty}^{\infty} e^{vx_i} p_X(x_i) & \text{Discrete } X \end{cases} \quad (4.23)$$

where v is a real number, i.e. $-\infty < v < \infty$. Note that the $M_X(v)$ may not exist for all random variables and all values of v. In general, $M_X(v)$ will exist only for those values of v for which the above integral or sum converges absolutely. Also, note that we have $M_X(0) = 1$, which is a useful fact for checking the mgf calculation. The mgf uniquely determines the pdf or pmf. If $M_X(v)$ exists for all values of v in the neighborhood of $v = 0$, it can then be shown that the moments of the random variable X are as follows:

$$E[X^k] = \frac{d^k}{dv^k} M_X(v) \bigg|_{v=0} \quad k = 1, 2, \dots \quad (4.24)$$

In other words, the kth moment of the random variable X is the kth derivative of the mgf $M_X(v)$ with respect to v, and then evaluated at $v = 0$.

Example 4.20
Suppose X is a continuous random variable whose pdf is as follows: $f_X(x) = \lambda e^{-\lambda x} u(x)$, where $\lambda > 0$ and $u(x)$ is the unit step function. Using the mgf, determine the first, second, and kth moment of X.

Solution:
The mgf is as follows:

$$M_X(v) = E[e^{vX}] = \int_0^\infty \lambda e^{-\lambda x} e^{vx} dx = \int_0^\infty \lambda e^{-(\lambda - v)x} dx = \frac{\lambda}{\lambda - v}$$

The first moment at $v = 0$, which is the expected value, is as follows:

$$E[X] = \frac{dM_X(v)}{dv}\bigg|_{v=0} = \frac{\lambda}{(\lambda - v)^2}\bigg|_{v=0} = \frac{1}{\lambda}$$

The second moment at $v = 0$, which is the mean square value, is as follows:

$$E[X^2] = \frac{d^2}{dv^2} M_X(v)\bigg|_{v=0} = \frac{2\lambda}{(\lambda - v)^3}\bigg|_{v=0} = \frac{2}{\lambda^2}$$

By using the mathematical induction, we can show that we have the kth moment as follows:

$$E[X^k] = \frac{d^k}{dv^k} M_X(v)\bigg|_{v=0} = \frac{k!\lambda}{(\lambda - v)^{k+1}}\bigg|_{v=0} = \frac{k!}{\lambda^k}$$

∎

4.8.2 Characteristic Function of a Random Variable

The **characteristic function** of the random variable X is as follows:

$$\Psi_X(\omega) = E[e^{j\omega X}] = \begin{cases} \int_{-\infty}^\infty e^{j\omega x} f_X(x)\, dx & \text{Continuous } X \\ \sum_{i=-\infty}^\infty e^{j\omega x_i} p_X(x_i) & \text{Discrete } X \end{cases} \tag{4.25}$$

where ω is a real number, i.e. $-\infty < \omega < \infty$, and $j = \sqrt{-1}$ represents the imaginary unit number. Note that $\Psi_X(\omega)$ is obtained by replacing v in $M_X(v)$ by $j\omega$, if $M_X(v)$ exists. The cf $\Psi_X(\omega)$ always exists, even if $M_X(v)$ does not. The moments can always be found if both the moments and derivatives of $\Psi_X(\omega)$ exist. It can be shown that the moments of the random variable X are as follows:

$$E[X^k] = \frac{1}{j^k} \frac{d^k}{d\omega^k} \Psi_X(\omega)\bigg|_{\omega=0} \quad k = 1, 2, \ldots \tag{4.26}$$

In other words, the kth moment of the random variable X is the kth derivative of the cf $\Psi_X(\omega)$ with respect to ω, and then evaluated at $\omega = 0$ and divided by j^k.

Example 4.21

Suppose X is a discrete random variable whose pmf is as follows: $p_X(x) = p(1-p)^{x-1}, 0 \leq p \leq 1, x = 1, 2, \ldots$. Using the cf, determine the variance of X.

Solution:

The cf is as follows:

$$\Psi_X(\omega) = E[e^{j\omega X}] = \sum_{i=1}^{\infty} e^{j\omega i} p(1-p)^{i-1} = pe^{j\omega} \sum_{i=1}^{\infty} ((1-p)e^{j\omega})^{i-1}$$

$$= pe^{j\omega} \times \frac{1}{1 - (1-p)e^{j\omega}} = \frac{p}{e^{-j\omega} - (1-p)}$$

Note that the last summation represented a geometric series. The expected value is thus as follows:

$$\mu_X = \frac{1}{j} \frac{d}{d\omega} \Psi_X(\omega) \bigg|_{\omega=0} = \frac{1}{j} \left(\frac{je^{-j\omega}p}{(e^{-j\omega} - (1-p))^2} \right) \bigg|_{\omega=0} = \frac{1}{p}$$

and the mean square value is thus as follows:

$$E[X^2] = \frac{1}{j^2} \frac{d^2}{d\omega^2} \Psi_X(\omega) \bigg|_{\omega=0} = \frac{2-p}{p^2}$$

The variance of the random variable X is thus as follows:

$$\sigma_X^2 = \frac{2-p}{p^2} - \left(\frac{1}{p} \right)^2 = \frac{1-p}{p^2}$$ ∎

4.9 Upper Bounds on Probability

A bound, by definition, encompasses all cases, including the worst-case, therefore, when it is applied to a particular case, it may not be very tight. Our focus is on the Markov, Chebyshev, and Chernoff inequalities, which all provide upper bounds on the probability that a value of a random variable is greater than some number. In principle, using more information about the random variable brings about tighter bounds. The Markov bound uses only the expected value, the Chebyshev bound uses both the expected value and the variance, while the much more accurate Chernoff bound requires the knowledge of the distribution of the random variable.

4.9.1 Markov Bound

The Markov inequality provides an upper bound on the probability that a value of a nonnegative random variable is greater than or equal to some positive constant. More specifically, for a random variable X such that $P(X < 0) = 0$ and thus

$E[X] > 0$, the **Markov inequality** is as follows:

$$P(X \geq c) \leq \frac{E[X]}{c}$$

$$P(X \geq kE[X]) \leq \frac{1}{k} \tag{4.27}$$

where $c > 0$, and thus $k = \frac{c}{E[X]} > 0$. The Markov inequality is valid only for non-negative random variables and refers only to the expected value of a random variable. The Markov inequality provides a loose, but useful, bound. A simple example of an application of the Markov inequality is that no more than 25% of the population can have more than four times the average income, assuming incomes are nonnegative, this result is due to having $k = 4$ and $c = 4E[X]$.

4.9.2 Chebyshev Bound

The Chebyshev inequality states that the probability of a large deviation from the expected value is inversely proportional to the square of the deviation. More specifically, for an arbitrary random variable X, the **Chebyshev inequality** is as follows:

$$P(|X - E[X]| \geq c) \leq \frac{\sigma_X^2}{c^2}$$

$$P(|X - E[X]| \geq k\sigma_X) \leq \frac{1}{k^2} \tag{4.28}$$

where $E[X]$ and σ_X^2 are the mean and variance of the random variable X, respectively, and we have $c > 0$ and $k = \frac{c}{\sigma_X} > 0$. The Chebyshev inequality thus indicates that the probability that a random variable deviates from its mean by more than c in either direction is less than or equal to its variance divided by c^2. This confirms the fact that the probability of an outcome departing from the mean becomes smaller as the variance decreases. In other words, as the variance becomes smaller, the Chebyshev inequality says it is unlikely the random variable is far away from its mean. The significance of the Chebyshev inequality is that it is valid for any random variable, nonnegative or not. The Chebyshev bound is obviously useful when $c > \sigma_X$, i.e. $k > 1$.

4.9.3 Chernoff Bound

For an arbitrary random variable X, and a constant c, the **Chernoff inequality** is as follows:

$$P(X \geq c) \leq e^{-sc}E[e^{sX}] \quad s \geq 0 \tag{4.29}$$

where $E[e^{sX}]$ is the expected value of e^{sX}. The upper bound must hold when s is chosen such that $e^{-sc}E[e^{sX}]$ is minimum. The minimum value is called the Chernoff bound. The Chernoff bound is tighter than the Chebyshev bound, as it requires the evaluation of the expected value of an exponential function of the random variable X, and that, in turn, requires the knowledge of the distribution of the random variable X. The Chernoff bound can be applied to any random variable, nonnegative or not. The Chernoff bound is useful when c is large relative to $E[X]$ and $P(X > c)$ is small.

Example 4.22
Suppose the pdf of the random variable X is as follows: $f_X(x) = \lambda e^{-\lambda x} u(x)$, where $\lambda > 0$ and $u(x)$ is the unit step function. Determine $P\left(X \geq \frac{10}{\lambda}\right)$ as well as an upper bound on it using the Markov inequality, the Chebyshev inequality, and the Chernoff bound.

Solution:
It can be easily shown that the expected value and variance of the random variable X are $E[X] = \frac{1}{\lambda}$ and $\sigma_X^2 = \frac{1}{\lambda^2}$, respectively.
The Markov inequality is as follows:

$$P\left(X \geq \frac{10}{\lambda}\right) \leq \frac{\dfrac{1}{\lambda}}{\dfrac{10}{\lambda}} = 0.1$$

In order to find the Chebyshev inequality, we need to find $P\left(X \geq \frac{10}{\lambda}\right)$:

$$P\left(X \geq \frac{10}{\lambda}\right) = P\left(X - \frac{1}{\lambda} \geq \frac{9}{\lambda}\right) = P\left(\left|X - \frac{1}{\lambda}\right| \geq \frac{9}{\lambda}\right)$$

The right-hand side of the above equation was obtained, as X is nonnegative. The Chebyshev inequality is thus as follows:

$$P\left(X \geq \frac{10}{\lambda}\right) = P\left(\left|X - \frac{1}{\lambda}\right| \geq \frac{9}{\lambda}\right) \leq \frac{\dfrac{1}{\lambda^2}}{\dfrac{81}{\lambda^2}} = \frac{1}{81} \cong 0.012\,345\,7$$

In order to find the Chernoff bound, we need to find the following:

$$P\left(X \geq \frac{10}{\lambda}\right) \leq e^{-\left(\frac{10s}{\lambda}\right)} E[e^{sX}] = e^{-\left(\frac{10s}{\lambda}\right)} \int_0^\infty e^{sx}(\lambda e^{-\lambda x}) dx$$

$$= \frac{\lambda e^{-\left(\frac{10s}{\lambda}\right)}}{\lambda - s} \quad s \geq 0$$

The minimum of the right-hand side is obtained by differentiating it with respect to s and the minimum thus occurs when $s = \frac{9\lambda}{10}$. The Chernoff bound is therefore as follows:

$$P\left(X \geq \frac{10}{\lambda}\right) \leq 10e^{-9} \cong 0.001\,234\,1$$

Note the actual probability is as follows:

$$\int_{10}^{\infty} \lambda e^{-\lambda x}\, dx = e^{-10} \cong 0.000\ 045\ 4$$

∎

Example 4.23

Suppose the pmf of the discrete random variable X is as follows: $p_X(x) = \binom{n}{x} p^x (1-p)^{n-x}$, for $x = 0, 1, 2, \ldots, n$, where n is a positive integer and $p = \frac{1}{4}$. Determine an upper bound on $P\left(X \geq \frac{n}{2}\right)$, using the Markov inequality, the Chebyshev inequality, and the Chernoff bound.

Solution:

It can be easily shown that the expected value and variance of the random variable X are $E[X] = \frac{n}{4}$ and $\sigma_X^2 = \frac{3n}{16}$, respectively.

The Markov inequality is as follows:

$$P\left(X \geq \frac{n}{2}\right) \leq \frac{\left(\frac{n}{4}\right)}{\left(\frac{n}{2}\right)} = \frac{1}{2}$$

In order to determine the Chebyshev inequality, we need to find $P\left(X \geq \frac{n}{2}\right)$ as follows:

$$P\left(X \geq \frac{n}{2}\right) = P\left(X - \frac{n}{4} \geq \frac{n}{4}\right) = P\left(\left|X - \frac{n}{4}\right| \geq \frac{n}{4}\right)$$

The right-hand side of the above equation was obtained, as X is nonnegative. The Chebyshev inequality is thus as follows:

$$P\left(\left|X - \frac{n}{4}\right| \geq \frac{n}{4}\right) \leq \frac{\frac{3n}{16}}{\frac{n^2}{16}} = \frac{3}{n}$$

In order to find the Chernoff bound, we need to find the following:

$$P\left(X \geq \frac{n}{2}\right) \leq e^{-\left(\frac{ns}{2}\right)} E[e^{sX}] = e^{-\left(\frac{ns}{2}\right)} \sum_{k=0}^{n} e^{sk} \binom{n}{k} \left(\frac{1}{4}\right)^k \left(\frac{3}{4}\right)^{n-k}$$

$$= e^{-\left(\frac{ns}{2}\right)} \left(\frac{3 + e^s}{4}\right)^n$$

The minimum of the right-hand side is obtained by differentiating it with respect to s and the minimum thus occurs when $e^s = 3$. The Chernoff bound is therefore as follows:

$$P\left(X \geq \frac{n}{2}\right) \leq \left(\frac{3}{4}\right)^{\frac{n}{2}} = (0.5\sqrt{3})^n$$

The Markov inequality offers the loosest bound, which is a constant and does not change as n increases. The Chebyshev bound is obviously a tighter bound where $\frac{3}{n}$ goes to zero as n approaches infinity. The tightest bound is the Chernoff bound as it exponentially goes to zero with n approaching infinity. ∎

4.10 Summary

In this chapter, we introduced the concept of a random variable and developed procedures for characterizing random variables, including the cdf, as well as the pmf for a discrete random variable and the pdf for a continuous random variable. The important concepts of the expected values, such as the mean, variance, and moments, were introduced. After a discussion of conditional distributions and functions of a random variable, we presented transform methods and upper bounds on probability.

Problems

4.1 Prove the Chebyshev inequality.

4.2 Suppose the average age of people in a town is 36 years. If all people over 60 years old should be vaccinated against a certain disease, determine the maximum fraction of people of the town who should be vaccinated.

4.3 Consider the experiment of rolling a fair die. Find the average number of rolls required in order to obtain a 2.

4.4 The cdf of the random variable X is as follows:

$$F_X(x) = \begin{cases} 0 & x < 0 \\ x^3 & 0 \leq x < 1 \\ 1 & 1 \leq x \end{cases}$$

Determine its pdf and the probability that the random variable is between 0.25 and 0.75.

4.5 Suppose X is a random variable, whose pdf is defined as follows:

$$f_X(x) = 2\,mx(u(x) - u(x - 2))$$

where $u(x)$ is the unit step function. Determine the value of m.

4.6 Assuming X is a discrete random variable whose pmf is as follows: $p(X=0)=0.2$, $p(X=1)=0.3$, $p(X=2)=0.4$ and $p(X=3)=0.1$. Determine the mean and variance of X.

4.7 Consider a random variable X whose pdf is $f_X(x) = 2e^{-b|x|}$, where $-\infty < x < \infty$ and $b > 0$. Determine the value of b and find $P(1 < X \le 2)$.

4.8 Suppose the random variable X that is uniformly (with equal probability) distributed between 0 and 1 with probability $\frac{1}{4}$, takes on the value of 1 with probability p, and is uniformly (with equal probability) distributed between 1 and 2 with probability $\frac{1}{2}$. Determine p as well as the pdf and cdf of the random variable X.

4.9 Determine the constant c if the cdf of the continuous random variable X is as follows:

$$F_X(x) = \begin{cases} cx^2 & 0 < x \le 1 \\ 0 & \text{elsewhere} \end{cases}$$

Determine the pdf of X, i.e. $f_X(x)$, and $P(0.1 < X \le 0.4)$.

4.10 The pdf of a continuous random variable is as follows:

$$f_X(x) = \frac{4x(9 - x^2)}{81} \quad 0 \le x \le 3$$

Determine the mean, median, and mode of this random variable.

4.11 The mean of the random variable Y is twice the mean of the random variable X, and the standard deviation of Y is twice that of X. If $Y = aX + b$, then determine the values of a and b.

4.12 Let the random variable X denote the telephone call duration. With the probability p, a call is not made either because no one answers or the line is busy, where $0 \le p \le 1$, and with the probability of $(1-p)$, a call is made. When a call is made, its random duration ranges between 0 and t minutes with a pdf of $\frac{1}{t}$, where $t > 0$ is a nonrandom constant. Assuming the expected value of X is m, determine a relation between p, t, and m.

4.13 The radial miss distance, measured in meters, of the landing point of a parachuting sky diver from the center of the target area has the following cdf:

$$F_X(x) = 1 - \exp\left(-\frac{x^2}{400}\right)$$

Determine the probability that the sky diver will land within a radius of 20 m from the center of the target area, and the probability of such a landing if the landing is within 40 m from the center of the target area.

4.14 In a bag, there are 12 identical balls, numbered 1–12 inclusive. Let X be the discrete random variable denoting the ball number. Determine the conditional pmf of X given B, where B is the event representing balls with prime numbers.

4.15 Suppose X is a random variable, whose pdf is defined as follows:

$$f_X(x) = \left(\frac{x}{8}\right)(u(x) - u(x-4))$$

where $u(x)$ is the unit step function. Determine the conditional pdf $f_X(x \mid 2 < X \le 3)$ and its mean.

4.16 Determine $E[X \mid X > 0.5]$, if the random variable X has the following pdf:

$$f_X(x) = \begin{cases} -|x| + 1 & 0 < |x| \le 1 \\ 0 & \text{elsewhere} \end{cases}$$

4.17 Suppose X is a random variable, whose pdf is defined as follows:

$$f_X(x) = \left(\frac{2x}{9}\right)(u(x) - u(x-3))$$

where $u(x)$ is the unit step function. Determine the conditional pdf $f_X(x \mid 1 < X \le 2)$ and its mean.

4.18 Let Y be the sum of the outcomes of rolling a pair of fair dice. Determine the pmf of Y and its mean. Also, determine the conditional pmf given that $5 \le Y \le 9$ as well as the mean of this conditional pmf.

4.19 Let the discrete random variable X have values $x = -1, 0, 1,$ and 2 with probabilities 0.1, 0.3, 0.4, and 0.2, respectively. Assuming we have $Y = \frac{X^3}{3} - X^2 + 2$, determine the pmf of the random variable Y.

4.20 Let X be a continuous random variable whose pdf is as follows:

$$f_X(x) = \begin{cases} |x| & -1 \leq x \leq 1 \\ 0 & \text{otherwise} \end{cases}$$

The random variable X is transformed by a piecewise-linear rectifier characterized by the input–output relation:

$$Y = \begin{cases} X & X \geq 0 \\ 0 & X < 0 \end{cases}$$

Determine the pdf of the new random variable Y.

4.21 Let X be the outcome of rolling a fair die. For the random variable $Y = X + X^2$, determine the pdf of Y and $P(10 < Y \leq 24)$.

4.22 Determine the transformation required to generate a random variable whose cdf is as follows:

$$F_Y(y) = 1 - \exp(-\lambda y) \quad \lambda > 0$$

4.23 Suppose X is a continuous random variable whose pdf is as follows:

$$f_X(x) = \begin{cases} 4x^3 & 0 < x \leq 1 \\ 0 & \text{otherwise} \end{cases}$$

Assuming $Y = \frac{1}{X}$, determine the pdf of the random variable Y, i.e. $f_Y(y)$.

4.24 The cf of a continuous random variable X is given by

$$\Psi_X(\omega) = \begin{cases} 1 - |\omega| & |\omega| < 1 \\ 0 & |\omega| > 1 \end{cases}$$

Find the pdf of X.

4.25 Let X be a random variable whose pdf $f_X(x)$ is as follows:

$$f_X(x) = \left(\frac{1}{\sqrt{2\pi}} \right) \exp\left(-\frac{x^2}{2} \right) \quad -\infty < x < \infty$$

Assuming we have

$$Y = \begin{cases} X^2 & X \geq 0 \\ 0 & X < 0 \end{cases}$$

Determine the pdf of the random variable Y, i.e. $f_Y(y)$.

4.26 The number of phone calls made by a cellphone user during a day is a random variable. The mean and variance of this random variable are 15 and 9, respectively. Using the Chebyshev inequality, estimate the probability that the number of calls is more than 5 from the mean.

4.27 For a random variable X, with $P(X < 0) = 0$ and a constant $c > 0$, prove the Markov inequality.

4.28 For an arbitrary random variable X and a constant c, prove the Chernoff inequality.

4.29 Suppose the pdf of the continuous random variable X is as follows:
$$f_X(x) = \begin{cases} 0 & x \leq 0 \\ x & 0 < x \leq 1 \\ -x + 2 & 1 < x \leq 2 \end{cases}$$
Determine the cdf.

4.30 Consider rolling an unfair (biased) die, with the following probabilities for its possible outcomes:
$$P(1) = P(2) = P(3) = 0.1, \quad P(4) = P(5) = 0.2, \quad P(6) = 0.3$$
Determine its mean and variance.

4.31 In a course, the number of students who go to the office hours to ask questions during a semester is a random variable, whose mean and standard deviation are 18 and 2.5, respectively. Determine the probability that there will be more than 8, but fewer than 28 students going to the office hours.

4.32 The pdf of the random variable X is as follows:
$$f_X(x) = \left(\frac{x}{8}\right)(u(x) - u(x - 4))$$
where $u(x)$ is the unit step function. Determine the mean, median, and mode of the conditional pdf $f_X(x \mid 1 < X \leq 3)$.

4.33 Consider the continuous random variable Y with the following pdf:
$$f_Y(y) = \begin{cases} 0 & y \leq 0 \\ 0.4 & 0 < y \leq 1 \\ 0.6 & 1 < y \leq 2 \\ 0 & y > 2 \end{cases}$$

Determine the mean and variance of the continuous random variable Y. Determine the conditional pdf $f_Y\left(y \mid \frac{4}{5} < Y \le \frac{5}{4}\right)$.

4.34 Suppose the random variable $X \ge 0$ has a distribution whose mean is 5.5 and variance is 1. Determine an upper bound on $P(X \ge 11)$ using the Markov and Chebyshev inequalities.

4.35 The random variable X takes on the values -1 and $+1$ with equal probability. Determine the cf of X.

4.36 Determine the mgf of the random variable X whose pmf is as follows:

$$p_X(k) = \frac{e^{-\lambda}\lambda^k}{k!}$$

where k is a nonnegative integer and λ is a nonnegative real number.

5

Special Probability Distributions and Applications

There are infinitely many random variables, defined by their cdfs (equivalently pmfs or pdfs), out of which a few have real-life applications and their own specific probability distributions. The focus of this chapter is on probability distributions that are important enough to have been given names, where there is a random experiment behind each of these distributions. After introducing some of the widely used discrete random variables, some of the well-known continuous random variables are described. However, the Gaussian (normal) distribution, which is the most special random variable, gets its own chapter later. Few widely known applications are also highlighted in this chapter.

5.1   Special Discrete Random Variables

Discrete random variables are usually easy to understand intuitively, but sometimes unwieldy to analyze. There are dozens of discrete random variables, among which some well-known and frequently encountered discrete probability distributions are now briefly discussed.

5.1.1   The Bernoulli Distribution

The Bernoulli distribution is a very simple, yet important, discrete random variable. The Bernoulli random variable X takes the value of 1 with probability p (also known as the probability of success) and the value of 0 with probability $1 - p$ (also known as the probability of failure), where $0 < p < 1$. It is therefore a discrete random variable with the range $\{0, 1\}$. The pmf of a **Bernoulli random variable** is defined as follows:

$$p_X(x) = \begin{cases} 1 - p & x = 0 \\ p & x = 1 \\ 0 & x \neq 0, 1 \end{cases} \quad 0 < p < 1 \tag{5.1}$$

Probability, Random Variables, Statistics, and Random Processes: Fundamentals & Applications,
First Edition. Ali Grami.
© 2020 John Wiley & Sons, Inc. Published 2020 by John Wiley & Sons, Inc.
Companion Website: www.wiley.com/go/grami/PRVSRP

A **Bernoulli trial**, which corresponds to sampling from the Bernoulli distribution, is a random experiment with exactly two possible outcomes, in which the probability of each of the two outcomes remains the same every time the experiment is conducted. In **independent trials**, the outcome of one trial has no influence over the outcome of another trial. The Bernoulli trial is equivalent to the tossing of a biased coin or examining if a component is defective in a system or modeling any other process that has only two possible outcomes, such as taking a pass-fail exam. The Bernoulli trial is a basic building block for some well-known discrete distributions, such as the binomial, geometric, and Pascal distributions. The mean μ_X and variance σ_X^2 of the Bernoulli random variable X, which cannot be independently set as they are both functions of p only are, respectively, as follows:

$$\mu_X = p$$
$$\sigma_X^2 = p(1-p) \tag{5.2}$$

Example 5.1
Determine the maximum and minimum values of the variance of the Bernoulli random variable. Comment on the results.

Solution
The maximum value of the variance is obtained as follows:

$$\sigma_X^2 = p - p^2 \rightarrow \begin{cases} \dfrac{d\sigma_X^2}{dp} = 1 - 2p = 0 \\[2ex] \dfrac{d^2\sigma_X^2}{dp^2} = -2 < 0 \end{cases} \rightarrow p = \frac{1}{2}$$

Therefore, the maximum value of the variance occurs when $p = \frac{1}{2}$, which in turn corresponds to the highest level of uncertainty, as there are two equally likely outcomes. The variance is minimum, i.e. its value is approaching 0, when $p = 0$ or $p = 1$, which in turn implies there is then no uncertainty, as there is always only one possible outcome. ∎

5.1.2 The Binomial Distribution

The binomial random variable X is the number of times 1 (i.e. successes) occurs in n independent Bernoulli trials, where n is a positive integer, and each occurrence of 1 is assumed to have probability p, where $0 < p < 1$. Note that the notation $X \sim B(n, p)$ denotes the random variable X has the binomial distribution with the parameters n and p. The sum of n independent Bernoulli trials is a binomial random variable. Applications of binomial distribution may include estimation of probabilities of the number of times hitting the target or the number of erroneous bits when a packet of data (long sequence of bits) is transmitted over a noisy communication channel. The pmf of a **binomial random variable** is defined as follows:

$$p_X(x) = \begin{cases} \binom{n}{x} p^x (1-p)^{n-x} & x = 0, 1, \ldots, n \\ 0 & \text{otherwise} \end{cases} \quad 0 < p < 1 \quad (5.3)$$

where $\binom{n}{x} \triangleq \frac{n!}{x!(n-x)!}$ is known as the **binomial coefficient**. The binomial distribution is frequently used to model the number of successes when the sampling is performed with replacement, as the draws are independent. If the sampling is carried out without replacement from a finite population, the draws are not independent and the hypergeometric distribution, as discussed later in this chapter, must be employed. Note that when $n = 1$, the binomial random variable is just the Bernoulli random variable, and when $p = 0.5$, the binomial distribution is symmetric for any value of the parameter n. The mean μ_X and variance σ_X^2 of the binomial random variable X, which can be independently set by the appropriate selection of p and n, are, respectively, as follows:

$$\mu_X = np$$
$$\sigma_X^2 = np(1-p) \quad (5.4)$$

Example 5.2
Suppose that a fair coin is tossed n times, where n is an even number, and x represents the number of heads. Determine the expected number of heads and the probability that the number of heads is exactly $x = \frac{n}{2}$, assuming $n = 2, 4, 8, 16, 32, 64, 128$. Comment on the results.

Solution
Since the coin is fair, we have $p = 0.5$. The expected number of heads and the probability that the number of heads is half of the number of times the coin is flipped are as follows:

p	n	x	μ_X	σ_X^2	$\binom{n}{x} p^x (1-p)^{n-x}$
0.5	2	1	1	0.5	50%
0.5	4	2	2	1	37.5%
0.5	8	4	4	2	27.34%
0.5	16	8	8	4	19.64%
0.5	32	16	16	8	13.99%
0.5	64	32	32	16	9.93%
0.5	128	64	64	32	7.04%

The expected number of heads is always equal to half of the number of times the fair coin is tossed. However, the probability that exactly half of the tosses is heads decreases, as the number of times the coin is tossed increases. The reason lies in the fact that an increase in the number of times that the coin is tossed brings about a direct increase in the total number of possibilities, hence an increase in variance. The possibility that the number of heads is exactly half of the tosses is then only one of the many possibilities. Each of these many possibilities warrants a certain probability and the sum of these probabilities becomes larger and in turn the probability of having exactly half of the tosses heads becomes smaller, as the number of tosses increases. ∎

Example 5.3

In a digital communication system, bits are transmitted over a channel in which the average bit error rate is assumed to be 0.0001. The transmitter sends each bit five times, rather than once, and the receiver takes a majority vote of the received bits to determine the transmitted bit. Determine the probability that the receiver will make an incorrect decision. Comment on the results.

Solution

Noting that the bit transmission can be viewed as a Bernoulli trial, the probability of interest is as follows:

$$P(X \geq 3) = \sum_{x=3}^{5} \binom{5}{x} (0.0001)^x (0.9999)^{5-x} \cong 10^{-11}$$

By sending the same bit five times, the bit error rate can be reduced from 10^{-4} to about 10^{-11}, a very significant reduction in bit error rate, but of course at the heavy price of repeating a bit five times. ∎

5.1.3 The Geometric Distribution

In a sequence of independent Bernoulli trials with a success probability p, where $0 < p < 1$, the random variable X that denotes the number of trials performed until the first success occurs is said to have the geometric distribution with probability p. For instance, the geometric distribution could be used to describe the number of candidates to be interviewed until a candidate is accepted or the number of cars to be test driven until a car is bought. The pmf of a *geometric random variable* is defined as follows:

$$p_X(x) = (1-p)^{x-1}p \quad x = 1, 2, \ldots \quad 0 < p < 1 \tag{5.5}$$

This highlights the fact that there are $x - 1$ failures before the first success occurs on the xth trial. Note that the pmf $p_X(x)$ decays geometrically with the value of x, and as the value of p increases, the pmf dies down faster. With

$p > 0$, it is not possible to run Bernoulli trials without ever seeing a success. For instance, in a broader sense, say in life, we can rightly state that no matter how low the probability of success is, with a large enough number of attempts, one can finally succeed. The mean μ_X and variance σ_X^2 of the geometric random variable X, which cannot be independently set as they are both functions of p only, are respectively as follows:

$$\mu_X = \frac{1}{p}$$

$$\sigma_X^2 = \frac{1-p}{p^2} \tag{5.6}$$

The mode of the geometric distribution is always 1 for all values of p. This means that even for a very rare event (e.g. a lottery win), the most probable number of trials necessary to obtain the first success is 1. This is sometimes viewed as a paradox.

Example 5.4

Tickets for a popular event are sold exclusively online on a single website. Suppose that the chance of successfully accessing the website by a fan to buy a ticket is 0.2. Determine the probability that a fan gets through on the sixth attempt, and the probability that a fan has to attempt ten or more times to get through.

Solution

The probability that a fan gets through on the sixth attempt is simply the probability that the first five attempts are unsuccessful, we thus have

$$P(X = 6) = (0.8)^5 \times 0.2 = 0.065\,536$$

The probability that a fan has to attempt 10 or more times to get through is thus as follows:

$$P(X \geq 10) = 1 - P(X \leq 9) = 1 - (1 - (0.8)^9) = (0.8)^9 = 0.134\,218$$

This is simply the probability that the first nine attempts are unsuccessful. ∎

The most important property of the geometric distribution is its memoryless property, and the geometric distribution is the only discrete probability distribution that possesses this property. The **memoryless property**, also known as **forgetfulness property**, states that if a success has not occurred in the first n trials, then the probability of having to perform at least k more trials is the same as the probability of initially having to perform at least k trials. The probability of k additional trials until the first success is independent of how many failures have already transpired. Conversely, if a discrete random variable satisfies the memoryless property, then it has a geometric distribution.

Example 5.5
Prove that the geometric random variable satisfies the memoryless property. In
other words, show that we have

$$P(X \geq (k + n) \mid X > n) = P(X \geq k) \quad \forall k, n > 1$$

Solution
The conditional probability of interest can be written as follows:

$$P(X \geq (k + n) \mid X > n) = \frac{P(X \geq (k + n), X > n)}{P(X > n)}$$

$$= \frac{P(X \geq (k + n))}{P(X > n)} = \frac{P(X \geq (k + n))}{P(X \geq (n + 1))}$$

$$= \frac{\sum_{i=k+n}^{\infty} (1 - p)^{i-1} p}{\sum_{i=n+1}^{\infty} (1 - p)^{i-1} p} = \frac{\sum_{i=k+n}^{\infty} (1 - p)^{i-1}}{\sum_{i=n+1}^{\infty} (1 - p)^{i-1}}$$

By a change of variable, the numerator and denominator can be both simplified as follows:

$$P(X \geq (k + n) \mid X > n) = \frac{\sum_{m=0}^{\infty} (1 - p)^{m+k+n-1}}{\sum_{m=0}^{\infty} (1 - p)^{m+n}}$$

$$= \frac{\sum_{m=0}^{\infty} (1 - p)^{k+n-1}(1 - p)^{m}}{\sum_{m=0}^{\infty} (1 - p)^{n}(1 - p)^{m}}$$

$$= \frac{(1 - p)^{k+n-1} \sum_{m=0}^{\infty} (1 - p)^{m}}{(1 - p)^{n} \sum_{m=0}^{\infty} (1 - p)^{m}} = \frac{(1 - p)^{k+n-1}}{(1 - p)^{n}}$$

$$= (1 - p)^{k-1} = P(X \geq k)$$

which is, as expected, independent of n. ∎

5.1.4 The Pascal Distribution

The Pascal distribution is commonly used in quality control for a product and
represents the number of Bernoulli trials that take place until one of the two
outcomes is observed a certain number of times. For instance, in choosing 12
citizens to comprise a jury, the Pascal distribution could be applied to estimate the number of rejections before the jury selection is completed. The number of trials up to and including the kth success in a sequence of independent
Bernoulli trials with a constant success probability p, where $0 < p < 1$, has a
Pascal distribution. The pmf of a *Pascal random variable*, also known as a
negative binomial random variable, is defined as follows:

$$p_X(x) = \binom{x - 1}{k - 1} p^k (1 - p)^{x-k} \quad 0 < p < 1 \quad x = k, k + 1, \ldots \quad k \in \{1, 2, \ldots\}$$

$$(5.7)$$

Note that for $k = 1$, the Pascal distribution becomes the geometric distribution. The mean μ_X and variance σ_X^2 of the Pascal random variable X, which can be independently set by the appropriate selection of p and k, are respectively as follows:

$$\mu_X = \frac{k}{p}$$

$$\sigma_X^2 = \frac{k(1-p)}{p^2} \tag{5.8}$$

Example 5.6
Suppose an IT company is recruiting two IT specialists, and each applicant interviewed has a probability of 0.4 to be hired. Determine the probability that exactly four applicants need to be interviewed, and the probability that up to and including five applicants need to be interviewed.

Solution
With $p = 0.4$ and $k = 2$, the probability that exactly four applicants need to be interviewed, i.e. $x = 4$, is as follows:

$$P(X = 4) = \binom{3}{1}(0.4)^2(0.6)^2 = 0.1728$$

and the probability that up to and including five applicants need to be interviewed, i.e. $2 \leq x \leq 5$, is as follows:

$$P(2 \leq X \leq 5) = \sum_{x=2}^{5}\binom{x-1}{1}(0.4)^2(0.6)^{x-2} = 0.663\,04 \qquad \blacksquare$$

5.1.5 The Hypergeometric Distribution

The hypergeometric distribution has many applications in quality control and card games. Suppose there is a finite population of N items of which K possess a certain attribute, the hypergeometric distribution then describes the probability that a sample of n items, without replacement, is selected of which x possess the attribute. The pmf of a *hypergeometric random variable* is defined as follows:

$$p_X(x) = \frac{\binom{K}{x}\binom{N-K}{n-x}}{\binom{N}{n}} \qquad \max(0, n+K-N) \leq x \leq \min(n, K)$$

$$0 \leq n \leq N \quad 0 \leq x \leq K \tag{5.9}$$

The mean μ_X and variance σ_X^2 of the hypergeometric random variable X, where each is a function of n, K, and N, are respectively as follows:

$$\mu_X = \frac{nK}{N}$$

$$\sigma_X^2 = \frac{nK}{N}\left(\frac{N-K}{N}\right)\left(\frac{N-n}{N-1}\right) \tag{5.10}$$

Example 5.7
Suppose there are 20 goldfish in a small aquarium, 8 of these fish are red, and the remaining 12 are black. A photo of two fish is taken. Determine the probability that both fish in the photo are red.

Solution
With $N = 20$, $K = 8$, $n = 2$, and $x = 2$, the probability of interest is thus as follows:

$$P(X = 2) = \frac{\binom{8}{2} \times \binom{12}{0}}{\binom{20}{2}} = \frac{28}{190} \cong 0.147 \qquad \blacksquare$$

It is also useful to highlight that if the total number of items N is much larger than the sample size n, then sampling without replacement is very similar to sampling with replacement. This is due to the fact that the reduction of the total population by the items already selected does not significantly change the selection probabilities of additional items. In such cases, the binomial distribution provides an excellent approximation for the distribution of the number of items selected.

5.1.6 The Poisson Distribution

The Poisson distribution represents the number of occurrences of events occurring within certain specified boundaries, such as the number of phone calls received by a mobile user during a day or the number of printing errors in a book or the number of defective units in a sample taken from a production line. The Poisson distribution is applied in counting the number of rare but open-ended events. The Poisson random variable arises in situations where the events occur completely at random in time or space. These events occur with a constant average rate and independently of the time or space associated with the last event. In other words, the occurrence of one event does not affect the probability that a second event will occur. The pmf of a *Poisson random variable* is defined as follows:

$$p_X(x) = \frac{e^{-\lambda}\lambda^x}{x!} \quad x = 0, 1, 2, \dots \quad \lambda > 0 \tag{5.11}$$

where λ is the parameter of the distribution reflecting the average rate of occurrence. The mean μ_X and variance σ_X^2 of the Poisson random variable X are respectively as follows:

$$\mu_X = \lambda$$
$$\sigma_X^2 = \lambda \qquad\qquad (5.12)$$

The mean and variance of a Poisson distribution are both equal to the value of the distribution parameter, and cannot be independently specified. By increasing λ, a larger expected value and a more spread out distribution are expected.

Example 5.8
Suppose that the number of software bugs in a computer program has a Poisson distribution with parameter $\lambda = 5$. Determine the probability that a computer program has no bugs, and the probability that the number of bugs are five or more.

Solution
The probability that a computer program has no bugs is as follows:

$$P(X = 0) = \frac{e^{-5}5^0}{0!} \cong 0.006\,74$$

The probability that the number of bugs are five or more is as follows:

$$P(X \geq 5) = 1 - \sum_{i=0}^{4} P(X = i) = 1 - \sum_{i=0}^{4} \frac{e^{-5}5^i}{i!} \cong 0.5595 \qquad\blacksquare$$

The Poisson distribution can be used to approximate the binomial distribution with parameters n and p, when n is very large and p is very small. A parameter value of $\lambda = np$ should be then used for the Poisson distribution, so that it has the same expected value as the binomial distribution. An increase in the value of n or a decrease in the value of p can improve the approximation. In the Poisson distribution, the discrete random variable theoretically ranges in the interval $[0, \infty)$, but in a binomial distribution, it ranges in the interval $[0, n]$.

Example 5.9
Suppose 8% of the students at a school have a cold. Using the Poisson distribution as an approximation for the binomial distribution, determine the probability of having 3 or more students with a cold in a sample of 20 students. Comment on the results.

Solution
We use the Poisson distribution with mean $\lambda = 20 \times 0.08 = 1.6$. The probability of having three or more students with a cold is approximately as follows:

$$P(X \geq 3) = 1 - P(X = 0) - P(X = 1) - P(X = 2)$$

$$= 1 - \frac{e^{-1.6}1.6^0}{0!} - \frac{e^{-1.6}1.6^1}{1!} - \frac{e^{-1.6}1.6^2}{2!}$$

$$\cong 0.217$$

But we have to use the binomial distribution to get the exact value, with $n = 20$ and $p = 0.08$, the probability of interest is then as follows:

$$P(X \geq 3) = 1 - P(X = 0) - P(X = 1) - P(X = 2)$$

$$= 1 - \frac{20!}{0!20!}(0.08)^0(0.92)^{20} - \frac{20!}{1!19!}(0.08)^1(0.92)^{19}$$

$$- \frac{20!}{2!18!}(0.08)^2(0.92)^{18}$$

$$\cong 0.212$$

The approximation, as expected, is thus very close to the true value. ∎

5.1.7 The Discrete Uniform Distribution

The discrete uniform random variable occurs when outcomes are equally likely, such as rolling a fair die. It takes on values in a set of L positive integers with equal probability. This distribution is generally employed when there are no information at all regarding the possible outcomes, as such they are all assumed to have the same probability. The pmf of a ***discrete uniform random variable*** is defined as follows:

$$P(X = k) = p_X(k) = \frac{1}{L} \quad k = m + 1, \dots, m + L$$

$$-\infty < m < \infty \quad L \in \{1, 2, \dots\} \tag{5.13}$$

The mean μ_X and variance σ_X^2 of the discrete uniform distribution X are, respectively, as follows:

$$\mu_X = m + \frac{L+1}{2}$$

$$\sigma_X^2 = \frac{L^2 - 1}{12} \tag{5.14}$$

Example 5.10

Suppose the mean and variance of a discrete uniform random variable X are both 10. Determine the set of consecutive integers that this discrete uniform random variable can take on.

Solution

We have

$$\sigma_X^2 = \frac{L^2 - 1}{12} = 10 \rightarrow L = 11$$

We can therefore have

$$\mu_X = m + \frac{L+1}{2} = 10 \rightarrow m + \frac{11+1}{2} = 10 \rightarrow m = 4$$

Hence, X takes on values in the set of consecutive integers {5, 6, 7, 8, 9, 10, 11, 12, 13, 14, 15}, each with probability $\frac{1}{11}$. ∎

The discrete random variable is used to generate random numbers in computer simulation models. Note that events that have an equiprobable finite number of outcomes form the basis upon which counting formulas are developed.

5.1.8 The Zipf (Zeta) Distribution

The Zipf distribution is commonly used to provide a close model for the size or rank of an object randomly chosen from certain types of populations. The Zipf random variable has wide applications when a very small number of outcomes occur quite frequently but a very large number of outcomes occur quite rarely. The pmf of a *Zipf random variable*, also known as a *Zeta random variable*, is defined as follows:

$$p_X(x) = x^{-\alpha}(\zeta(\alpha))^{-1} \quad \alpha > 1 \quad x = 1, 2, \dots \tag{5.15}$$

where the well-known *zeta function* $\zeta(\alpha)$ is defined as follows:

$$\zeta(\alpha) = \sum_{k=1}^{\infty} \frac{1}{k^\alpha} \quad \alpha > 1 \tag{5.16}$$

The mean μ_X and variance σ_X^2 of the Zipf random variable X are, respectively, as follows:

$$\mu_X = \frac{\zeta(\alpha - 1)}{\zeta(\alpha)} \quad \alpha > 2$$

$$\sigma_X^2 = \frac{\zeta(\alpha)\zeta(\alpha - 2) - (\zeta(\alpha - 1))^2}{(\zeta(\alpha))^2} \quad \alpha > 3 \tag{5.17}$$

Note that if the number of outcomes in the population is finite and $\alpha = 1$, then the Zipf's law, which provides an empirical observation, results. The *Zipf's law* states that in a large, but finite population of different types, the proportion belonging to the nth most common type is approximately proportional to $\frac{1}{n}$. In short, the rank of an item on the list times its frequency (probability) is approximately constant. In other words, the frequency of the one ranked second is about half of the frequency of the one ranked first, the frequency of the one ranked third is close to one-third of the frequency of the one ranked first, so on and so forth. For instance, the words with the highest frequencies in the English language, which are generally "the," "of," and "to," or the websites most visited on the Internet, which are normally "Google," "YouTube," and "Facebook," follow the Zipf law rather closely.

Example 5.11
Suppose an infinite number of items have been ranked according to their frequencies from the highest probability to the lowest probability, and the probabilities of the items follow the Zipf distribution. Assuming $\alpha = 2$, thus having $\zeta(2) = \frac{\pi^2}{6}$, determine the sum of the probabilities of all those ranked from the fourth most frequent item to the least frequent item inclusive.

Solution
We thus need to determine the following:

$$P(X \geq 4) = 1 - P(X \leq 3) = 1 - \sum_{x=1}^{3} x^{-2}(\zeta(2))^{-1} = 1 - \sum_{x=1}^{3} x^{-2}\left(\frac{\pi^2}{6}\right)^{-1}$$

$$= 1 - \frac{6}{\pi^2}1^{-2} - \frac{6}{\pi^2}2^{-2} - \frac{6}{\pi^2}3^{-2} = 1 - \frac{6}{\pi^2}\left(1 + \frac{1}{4} + \frac{1}{9}\right) \cong 0.173$$ ∎

5.2 Special Continuous Random Variables

Continuous random variables are generally easy to handle analytically, but often hard to understand intuitively. Among dozens of well-known and frequently encountered continuous probability distributions, select few are discussed here. However, the most special continuous random variable, known as the Gaussian (normal) distribution, along with few distributions that are based on the Gaussian distribution, are discussed in Chapter 7.

5.2.1 The Continuous Uniform Distribution

The continuous uniform random variable arises where all values in a known finite interval of the real line are equally likely. The pdf of a *continuous uniform random variable* is defined as follows:

$$f_X(x) = \begin{cases} \dfrac{1}{b-a} & a \leq x \leq b \quad -\infty < a < b < \infty \\ 0 & \text{otherwise} \end{cases} \tag{5.18}$$

The probabilities are simple to calculate, and the probability of falling in an interval of fixed length Δ within $[a, b]$ is a constant equal to $\frac{\Delta}{(b-a)}$, while noting that $\Delta \leq b - a$ and the probability is independent of the location of the interval itself. The mean μ_X and variance σ_X^2 of the continuous uniform random variable X are thus as follows:

$$\mu_X = \frac{a+b}{2}$$

$$\sigma_X^2 = \frac{(b-a)^2}{12} \tag{5.19}$$

The result for the variance indicates that the variance depends only on the length of the interval, and the variance increases as the interval gets larger and the distribution becomes more spread out. A uniform distribution in the interval $[0, 1]$ is known as the **standard uniform distribution**. If X is uniformly distributed over the interval $[a, b]$, then through the linear transformation $Y = \frac{X-a}{b-a}$, the random variable Y has the standard uniform distribution.

The continuous uniform random variable is used when all values in some known finite range seem equally probable or when there is no a priori knowledge of the actual pdf. This random variable is often used to model the initial phase of a sinusoidal signal received in digital communication systems, where a uniform distribution of phase between 0 and 2π is frequently assumed. The continuous uniform random variable plays a critical role in generating random variables of prescribed distributions in computer simulation models.

Example 5.12
Suppose the mean and variance of a continuous uniform random variable are both equal to k, a positive real number. Determine the interval $[a, b]$ of this continuous uniform distribution.

Solution:
We have

$$\begin{cases} \mu_X = \dfrac{a+b}{2} = k \\ \sigma_X^2 = \dfrac{(b-a)^2}{12} = k \end{cases} \rightarrow \begin{cases} a+b = 2k \\ b-a = 2\sqrt{3k} \end{cases} \rightarrow \begin{cases} a = k - \sqrt{3k} \\ b = k + \sqrt{3k} \end{cases}$$
∎

5.2.2 The Exponential Distribution

The exponential distribution arises in many waiting time events, such as how long it takes for something to happen. The pdf of an **exponential random variable** is as follows:

$$f_X(x) = \begin{cases} \lambda \exp(-\lambda x) & x \geq 0 \quad \lambda > 0 \\ 0 & x < 0 \end{cases} \tag{5.20}$$

where the parameter λ is the rate at which events occur. The mean μ_X and variance σ_X^2 of the exponential random variable X are thus as follows:

$$\mu_X = \frac{1}{\lambda}$$

$$\sigma_X^2 = \frac{1}{\lambda^2} \tag{5.21}$$

Note that the standard deviation of the exponential distribution is the same as the expectation of the distribution. This thus indicates that the mean and variance of an exponential distribution cannot be independently specified.

The most important property of the exponential distribution is its memoryless property, and the exponential distribution is the only continuous distribution that possesses this property. The **memoryless property**, also known as **forgetfulness property**, states that if the exponential random variable X measures the time until a certain event occurs and the event has not occurred by time t_0, the additional waiting time for the event to occur beyond t_0 has the very same exponential distribution as X.

Example 5.13
Prove the exponential random variable with parameter λ satisfies the memoryless property. In other words, show that we have

$$P(X > (t_0 + h) \mid X > t_0) = P(X > h) \quad h > 0, t_0 > 0$$

Solution
We have

$$
\begin{aligned}
P(X > (t_0 + h) \mid X > t_0) &= \frac{P((X > (t_0 + h)) \cap (X > t_0))}{P(X > t_0)} \\
&= \frac{P(X > (t_0 + h))}{P(X > t_0)} \\
&= \frac{\exp(-\lambda(t_0 + h))}{\exp(-\lambda t_0)} = \exp(-\lambda h) = P(X > h)
\end{aligned}
$$

which is, as expected, independent of time t_0. ∎

The exponential distribution is often used to model failure or waiting times, such as the time between goals scored in a soccer game, the time between phone calls made to a call center, the time between successive failures of a machine, and the distance between successive potholes on a road. The exponential distribution is an appropriate model if events occur independently and the rate at which events occur remains unchanged, i.e. the rate cannot be higher in some intervals and lower in other intervals.

Example 5.14
The probability of how long a telephone call lasts is often modeled as an exponential distribution. Suppose the mean value is three minutes, determine the probability that a conversation lasts between one and five minutes.

Solution
With $\lambda = 3$, we therefore have the following:

$$P(1 \leq X \leq 5) = \int_1^5 \frac{1}{3} e^{-\frac{x}{3}} dx = e^{-\frac{1}{3}} - e^{-\frac{5}{3}} \cong 0.528$$

∎

5.2.3 The Gamma Distribution

The gamma random variable, which is a nonnegative random variable, has many diverse applications, including queuing theory and reliability theory. The pdf of a *gamma random variable* is defined as follows:

$$f_X(x) = \frac{\lambda^\alpha x^{\alpha-1} e^{-\lambda x}}{\Gamma(\alpha)} \quad x \geq 0, \quad \alpha > 0, \quad \lambda > 0 \tag{5.22}$$

where the well-known *gamma function* $\Gamma(\alpha)$ is defined as follows:

$$\Gamma(\alpha) = \int_0^\infty x^{\alpha-1} e^{-x} dx \quad \alpha > 0 \tag{5.23}$$

The pdf of a gamma random variable takes on a wide variety of shapes depending on the values of α, known as the *shape parameter*, and λ, referred to as the *scale parameter*. There are several widely known random variables that are special cases of the gamma random variable, including $\alpha = 1$, when the gamma distribution becomes the exponential distribution. The mean μ_X and variance σ_X^2 of the gamma random variable X are, respectively, as follows:

$$\mu_X = \frac{\alpha}{\lambda}$$
$$\sigma_X^2 = \frac{\alpha}{\lambda^2} \tag{5.24}$$

Example 5.15
Suppose we have a gamma random variable, with $\alpha = 3$ and $\lambda = 1$. Determine the probability that the random variable falls between 1 and 2.

Solution
With $\alpha = 3$, we have $\Gamma(3) = 2$, and for $\lambda = 1$, the gamma pdf is as follows:

$$f_X(x) = \frac{x^2 e^{-x}}{2}$$

We can now calculate the probability of interest:

$$P(1 \leq X \leq 2) = \int_1^2 \frac{x^2 e^{-x}}{2} dx = (-0.5x^2 - x - 1)e^{-x}\big|_1^2$$
$$= (-2 - 2 - 1)e^{-2} - (-0.5 - 1 - 1)e^{-1}$$
$$= -5e^{-2} + 2.5e^{-1} \cong 0.243$$

∎

5.2.4 The Erlang Distribution

The Erlang distribution is used in system reliability models and queuing systems. The pdf of an **Erlang random variable** is defined as follows:

$$f_X(x) = \frac{\lambda^k x^{k-1} e^{-\lambda x}}{(k-1)!} \quad x > 0 \quad \lambda > 0 \quad k = 1, 2, \dots \tag{5.25}$$

The mean μ_X and variance σ_X^2 of the Erlang random variable X are as follows:

$$\mu_X = \frac{k}{\lambda}$$

$$\sigma_X^2 = \frac{k}{\lambda^2} \tag{5.26}$$

Example 5.16
Show that the sum of k independent exponentially distributed random variables with parameter λ has the Erlang distribution.

Solution
The moment generating function of an exponential distribution with parameter λ is as follows:

$$M_X(v) = E[e^{vX}] = \int_0^\infty e^{vx} \lambda e^{-\lambda x} dx = \frac{\lambda}{\lambda - v}$$

The moment generating function of the sum of k independent exponentially distributed random variables is thus as follows:

$$M_S(v) = \left(\frac{\lambda}{\lambda - v} \right)^k$$

With $v = -s$, we get the corresponding Laplace transform, and using a table of Laplace transforms, we see the inverse transform has the following form:

$$f_X(x) = \frac{\lambda^k x^{k-1} e^{-\lambda x}}{(k-1)!} \quad x > 0$$

which is the pdf of an Erlang distribution. ∎

5.2.5 The Weibull Distribution

The Weibull distribution is often used in reliability theory to model failure. The pdf of a **Weibull random variable** is as follows:

$$f_X(x) = abx^{b-1} e^{-ax^b} \quad x > 0 \quad a > 0 \quad b > 0 \tag{5.27}$$

In the context of reliability theory, different values for b have different implications. A value of $b < 1$ indicates that the failure rate decreases over time, this happens if defective items fail early, and they are then replaced. A value of $b = 1$ indicates that the failure rate is constant over time, and the Weibull distribution reduces to an exponential distribution. This might suggest that random external events are causing failure. A value of $b > 1$ indicates that the failure rate increases with time, this happens if there is an aging process. The mean μ_X and variance σ_X^2 of the Weibull random variable X are as follows:

$$\mu_X = a^{-\frac{1}{b}}\Gamma(1 + b^{-1})$$

$$\sigma_X^2 = a^{-\frac{2}{b}}(\Gamma(1 + 2b^{-1}) - (\Gamma(1 + b^{-1}))^2) \tag{5.28}$$

Example 5.17
Suppose that the Weibull random variable X measures the lifetime of a bacterium, in minutes, at a certain temperature, with $b = 2$ and $a = 0.01$. Determine the time that takes for 95% of the bacteria to die.

Solution
We thus need to find the time t in the following equation:

$$0.95 = \int_0^t 0.02xe^{-0.01x^2}dx \rightarrow 0.95 = 1 - e^{-0.01t^2} \rightarrow t \cong 17.31 \qquad \blacksquare$$

5.2.6 The Beta Distribution

The beta distribution is widely used to model random proportions and in general random variables that range over finite intervals. For instance, the probability of success p in a Bernoulli trial experiment is frequently modeled to be a beta random variable. The pdf of a *beta random variable* is as follows:

$$f_X(x) = \beta(a, b)x^{a-1}(1 - x)^{b-1} \quad 0 < x < 1 \quad a > 0 \quad b > 0 \tag{5.29}$$

where the well-known *beta function*, in terms of the gamma function, is defined as follows:

$$\beta(a, b) = \frac{\Gamma(a + b)}{\Gamma(a)\Gamma(b)} \tag{5.30}$$

The mean μ_X and variance σ_X^2 of the beta random variable X are, respectively, as follows:

$$\mu_X = \frac{a}{a + b}$$

$$\sigma_X^2 = \frac{ab}{(a + b)^2(a + b + 1)} \tag{5.31}$$

Example 5.18
Suppose in a beta distribution, we have $a = b = 1$. Determine the resulting pdf. Comment on the results.

Solution

The values of the gamma function at 1 and 2, are as follows:

$$\Gamma(1) = \int_0^\infty e^{-x} dx = 1$$

$$\Gamma(2) = \int_0^\infty x e^{-x} dx = 1$$

After substituting $a = 1$, $b = 1$, and the above gamma functions into the pdf of the beta distribution, we obtain

$$f_X(x) = 1$$

Therefore, with $a = 1$ and $b = 1$, the beta distribution turns into the standard uniform distribution, whose mean and variance are $\frac{1}{2}$ and $\frac{1}{12}$, respectively. ∎

5.2.7 The Laplace Distribution

The Laplace distribution is also sometimes called the **double exponential distribution**. The Laplace distribution is used for modeling in signal processing including speech amplitudes, various biological processes, and finance. The pdf of a **Laplacian random variable** is given by

$$f_X(x) = \frac{b}{2} \exp(-b|x-a|) \quad b > 0 \quad -\infty < a < \infty \quad -\infty < x < \infty$$

(5.32)

The Laplacian pdf is easily integrated to find the probability of an interval. The mean μ_X and variance σ_X^2 of the Laplacian random variable X are, respectively, as follows:

$$\mu_X = a$$

$$\sigma_X^2 = \frac{2}{b^2}$$

(5.33)

Example 5.19

Suppose that the amplitude of a speech signal has a Laplace distribution with $a = 0$ and $b > 0$. Compute the probability when the absolute value of the amplitude is less than the standard deviation.

Solution

As the standard deviation of the speech amplitude is $\frac{\sqrt{2}}{b}$, we have the following:

$$P\left(-\frac{\sqrt{2}}{b} \leq X \leq \frac{\sqrt{2}}{b}\right) = \int_{-\frac{\sqrt{2}}{b}}^{\frac{\sqrt{2}}{b}} \frac{b}{2} \exp(-b|x|) dx$$

$$= \int_{-\frac{\sqrt{2}}{b}}^{0} \frac{b}{2} \exp(bx)dx + \int_{0}^{\frac{\sqrt{2}}{b}} \frac{b}{2} \exp(-bx)dx$$

$$= 1 - \exp\left(-\frac{\sqrt{2}}{b}\right) \qquad \blacksquare$$

5.2.8 The Pareto Distribution

The Pareto distribution can be viewed as a continuous version of the Zipf discrete random variable. The pdf of a **Pareto random variable**, which is widely known for its very long tail, is defined as follows:

$$f_X(x) = \frac{\alpha k^\alpha}{x^{\alpha+1}} \quad \alpha > 0 \quad \text{and} \quad x \geq k > 0 \tag{5.34}$$

The mean μ_X and variance σ_X^2 of the Pareto random variable X are, respectively, as follows:

$$\mu_X = \frac{\alpha k}{(\alpha - 1)} \quad \alpha > 1$$

$$\sigma_X^2 = \frac{\alpha k^2}{(\alpha - 1)^2(\alpha - 2)} \quad \alpha > 2 \tag{5.35}$$

The Pareto distribution is often described as the basis of the 80/20 rule. The **Pareto principle**, also known as the **80/20 rule**, states that, for many events, roughly 80% of the effects come from 20% of the causes. The Pareto principle has a wide set of applications, such as the study of the distribution of wealth, where about 80% of population generally possess about 20% of the wealth, the study of the Internet behavior, where about 80% of those browsing the Internet usually visit about 20% of the websites, in software quality control, where about 80% of the errors are normally due to about 20% of the code, in occupational health and safety, where about 80% of injuries typically arise from about 20% of hazards, and in oil industry, where about 80% of oil reserves are mostly in few large oil fields and 20% of them are in many smaller oil fields.

Example 5.20
Suppose that the household income in a certain country has a Pareto distribution with $\alpha > 1$. Compute the portion of the population with household incomes that is less than the distribution mean.

Solution
As the minimum and mean of the household income in the country are k and $\frac{\alpha k}{(\alpha-1)}$, respectively, we have the following:

$$P\left(k \leq X \leq \frac{\alpha k}{(\alpha - 1)}\right) = \int_{k}^{\frac{\alpha k}{(\alpha-1)}} \frac{\alpha k^\alpha}{x^{\alpha+1}}dx = 1 - \left(\frac{\alpha - 1}{\alpha}\right)^\alpha \qquad \blacksquare$$

5.3 Applications

The applications of the widely known probability distributions are obviously numerous, a select few, however, with a focus on telecommunications, are highlighted here.

5.3.1 Digital Transmission: Regenerative Repeaters

As a quality control, a degradation level in bit transmission can be initially set and then kept nearly fixed at that value at every step (link) of a digital communication path. This reconstruction of the digital signal, i.e. the regeneration of bits, is done by appropriately spaced regenerative repeaters, which do not allow accumulation of noise and interference. Consider a communication path with $k \geq 1$ identical regenerative repeaters in series.

We assume the average bit error rate incurred in each regeneration of a bit is p, where $0 < p < 1$, and the erroneous decisions on bits are made independently. A bit at the receiver, after being regenerated k times, is correct either if the bit in question is detected correctly over each link or it experiences errors over an even number of links. Noting that $P(c)$ is the average probability of a correct decision on a bit and $P(e)$ is the average bit error rate, then by using the binomial distribution, we can have the following:

$$
\begin{aligned}
P(c) = 1 - P(e) = \ &P(\text{no error over all links}) \\
&+ P(\text{error over two links only}) \\
&+ P(\text{error over four links only}) \\
&+ \cdots + P(\text{error over } m \text{ links only})
\end{aligned}
$$

$$
= (1-p)^k + \sum_{j=2,4,6,\ldots}^{m} \binom{k}{j}(p)^j(1-p)^{k-j}
$$

$$
= \sum_{j=0}^{k} \binom{k}{j}(-p)^j
$$

$$
+ \sum_{j=2,4,6,\ldots}^{m} \binom{k}{j}(p)^j(1-p)^{k-j} \tag{5.36a}
$$

where m is the largest even integer less than or equal to k. With the reasonable assumption that $p \ll 1$, only the first two terms of the first summation and the first term of the second summation are of significance, we thus have

$$
P(c) = 1 - P(e) \cong 1 - kp + \frac{k(k-1)}{2}p^2 \cong 1 - kp \rightarrow P(e) \cong kp \tag{5.36b}
$$

Note that with $kp \ll 1$, the quadratic term was disregarded. We can simply conclude that the average bit error in a system with k links is approximately equal to the bit error rate in a single link times the total number of links in the system.

5.3.2 System Reliability: Failure Rate

It is often important to be able to model the reliability of a system as a function of time. The time interval from the moment the system is put into operation until it fails is called the **time to failure**, which by definition is random. Suppose that the random variable T measures the time that a system fails. We assume the random variable T representing the lifetime of a system has the pdf $f_T(t)$ and the cdf $F_T(t)$. Failure times are typically modeled with the exponential, gamma, Weibull, and log-normal distributions.

Assuming $t = 0$ is the moment the operation of the system resumes, the reliability at time $t > 0$ is defined as the probability $R(t)$ that the system is still functioning at time t, we thus have

$$R(t) = P(T > t) = 1 - P(T \le t) = 1 - F_T(t) \to R'(t) = -f_T(t) \qquad (5.37)$$

where $R'(t)$ is the derivative of $R(t)$ with respect to time. The **mean time to failure** (MTTF) is thus given by the expected value of the random variable T:

$$E[T] = \int_0^\infty t f_T(t) dt = \int_0^\infty (1 - F_T(t)) dt = \int_0^\infty R(t) dt \qquad (5.38)$$

Note that the second integration reflects the mean when a random variable is continuous and nonnegative. Using Bayes' rule, the conditional cdf of T given that $T > t$, i.e. the probability that a system functioning at time t fails prior to time $x > t$, is as follows:

$$F_T(x \mid T > t) = \frac{P(T \le x, T > t)}{P(T > t)} = \frac{F_T(x) - F_T(t)}{1 - F_T(t)}$$

$$\to f_T(x \mid T > t) = \frac{f_T(x)}{1 - F_T(t)} \qquad (5.39)$$

The **failure rate** $r(t)$, also known as the **hazard rate**, measures the instantaneous rate of failure at time t conditional on the system still being operational at time t, in a way, it represents the chance of a system that has not failed by time t suddenly failing. It is thus defined as follows:

$$r(t) = f_T(t \mid T > t) = \frac{f_T(t)}{1 - F_T(t)} = -\frac{R'(t)}{R(t)} \to R(t) = \exp\left(-\int_0^t r(\tau) d\tau\right) \qquad (5.40)$$

Note that the product $r(t) dt$ is the probability that a system functioning at time t fails in the interval $(t, t + dt)$.

A common type of hazard rate for a system is the bathtub shape. The hazard rate has initially high values, for some components fail early due to defective parts. If a system survives this initial time period, then it can be expected to last a reasonable period of time, as the system is stable and has a low failure rate. Eventually, the age of the system leads to an increase in the hazard rate at later times.

Example 5.21
Suppose the random variable T representing the lifetime of a system has the exponential distribution with parameter λ. Determine the mean time to failure as well as the failure rate.

Solution
We have

$$E[T] = \int_0^\infty t\lambda \, \exp(-\lambda t)dt = \frac{1}{\lambda}$$

and

$$r(t) = \frac{f_T(t)}{1 - F_T(t)} = \frac{\lambda \, \exp(-\lambda t)}{1 - (1 - \exp(-\lambda t))} = \lambda$$

The exponential distribution thus gives rise to a constant failure rate, which is the inverse of the mean time to failure. ∎

5.3.3 Queuing Theory: Servicing Customers

Queueing theory provides the solution to a standard problem for the allocation of resources to service customers in a host of practical applications. Some typical applications may include telephone systems, traffic control, hospital management, and time-shared computer systems. In every one of these situations, scarce resources are shared among a number of users. Users demand for these resources at random times, and require their use for random periods of time. Inevitably, requests for the resource may be made while the resource is still in use, as such a mechanism to provide an orderly access to the resource is required. In short, queuing theory deals with the study of waiting times and resource sharing.

The notation $A/B/s/K$ is widely used to classify a queuing system, where A specifies the type of arrival process (e.g. M is used for inter-arrival times with exponential distribution), B denotes the service-time distribution (e.g. M is used for service times with exponential distribution), s highlights the number of servers (e.g. 1 or any other positive integer), and $K > 0$ reflects the maximum number of customers in the system. If K is not specified, then it is assumed that the capacity of the system is unlimited.

We assume customers arrive at the system with inter-arrival times that are independent random variables, and with the same distribution. The average arrival rate λ to the system, measured in customers per second, is the reciprocal of the average inter-arrival time. The time required to service a customer, measured in seconds per customer, is called the **service time** μ. The **traffic load** or intensity $\rho = \frac{\lambda}{s\mu} < 1$, measured in Erlang, implies that the server, on the average, must process the customers faster than their average arrival rate; otherwise, the queue length (the number of customers waiting in the queue) tends to infinity. The following table highlights some basic quantities of queuing systems.

	Average time duration for a customer	Average number of customers
Queue	W_q	$N_q = \lambda W_q$
Service	W_S	$N_S = \lambda W_S$
System	$W_q + W_S$	$N = N_q + N_S = \lambda(W_q + W_S)$

Note that $N = N_q + N_S = \lambda(W_q + W_S)$, which is known as **Little's formula**, states for systems that reach steady sate, the average number of customers in a system is equal to the product of the average arrival rate and the average time spent in the system. It is reasonable to assume that the inter-arrival times are exponential with mean $\frac{1}{\lambda}$, which in turn implies the arrival probability distribution is Poisson with rate λ, and the service time is exponentially distributed with mean μ. It can be shown that in the $M/M/1$ queuing system, we have the following basic quantities:

$$N_q = \frac{\rho^2}{1 - \rho}$$

$$N = \frac{\rho}{1 - \rho}$$

$$W_q = \frac{\rho}{\mu(1 - \rho)}$$

$$W = \frac{1}{\mu(1 - \rho)} \tag{5.41}$$

Note that as $\rho \to 1$, the above values grow without bound, that is perfect scheduling cannot be operated at $\lambda = \mu$.

5.3.4 Random Access: Slotted ALOHA

Random access involves dynamic sharing of the transmission medium among many users so as to better match bursty data traffic. Transmission is random among users, and since there is no scheduled time, access to the medium is based on contention. The random access schemes under light traffic can provide low-delay packet transfer. Under heavy traffic, the throughput and packet delays both can suffer.

In slotted ALOHA, the time is divided into equal time slots, where each time slot is equal to the packet transmission time. In slotted ALOHA, the transmitters keep track of transmission time slots and are allowed to send their packets only at the beginning of a time slot. In slotted ALOHA, there are $n > 1$ independent nodes (transmitters), each transmitting a packet with probability p. However, because there is only one communication channel, simultaneous transmissions from $N \leq n$ different nodes may result in collision. Using the binomial distribution, we have the following three cases:

System is idle $\rightarrow$ $\quad\quad\quad$ $N = 0 \rightarrow P(N = 0) = (1 - p)^n$

Successful transmission $\rightarrow$ $\quad$ $N = 1 \rightarrow P(N = 1) = np(1 - p)^{n-1}$

Packet collision $\rightarrow$ $\quad\quad\quad$ $N \geq 2 \rightarrow P(N \geq 2)$

$$= 1 - (1 - p)^n - np(1 - p)^{n-1} \quad\quad (5.42)$$

Let $\lambda = np$ be the **offered packet rate** or the **average number of packets** generated during one packet transmission time. Using the Poisson distribution as an approximation to the binomial distribution, we have $P(N = x) = \frac{e^{-\lambda}\lambda^x}{x!}$, where $x = 0, 1, 2, \ldots, n$. The throughput $S = P(N = 1)$, that is the average number of successfully transmitted packets, is then as follows:

$$S = \lambda e^{-\lambda} \quad\quad (5.43)$$

The maximum value of the throughput S is $\frac{1}{e} \cong 36.8\%$ that occurs at $\lambda = 1$ and corresponds to a total arrival rate of exactly one packet per slot. Similarly, when $\lambda = 1$, we have $P(N = 0) \cong \frac{1}{e}$. This means that about 36.8% of the time, exactly one node transmits and the packet is successfully transmitted, about 36.8% of the time, no node transmits, and about 26.4% of the time, collisions occur, i.e. two or more than two nodes transmit at the same time.

The probability of a successful packet transmission is $m = P(N = 0) = e^{-\lambda}$, and the probability of packet collision is thus $1 - m = 1 - e^{-\lambda}$. A node involved in collision then waits a random amount of time before sending its packet for the second time. If collision occurs again, the node attempts to transmit for the third time, and so on. Therefore, if the number of attempts needed to transmit the packet is K, then, we have $P(K = 1) = m$, $P(K = 2) = m(1 - m)$, $P(K = 3) = m(1 - m)^2$, and so on. M is thus a geometric random variable whose mean is $\frac{1}{m} = e^{\lambda}$. On average, a packet thus requires e^{λ} attempts before it is successfully transmitted.

5.3.5 Analog-to-Digital Conversion: Quantization

Humans produce and perceive audio and visual signals in an analog form, i.e. a signal at any time could have any real value over the interval $(-\infty, \infty)$. To this effect, in digital transmission of such signals, analog-to-digital conversion in the transmitter is an indispensable component. In an analog-to-digital conversion, first a sampler is employed to convert a continuous-time signal to a discrete-time signal. Afterward, a quantizer is used to process the samples one by one so as to convert a continuous-valued signal to a discrete-valued signal. Quantization is thus a process through which a continuous random variable X is mapped into a discrete random variable, the nearest point to X from a set of 2^R representation values, where R is a positive integer.

Example 5.22
A real number between 0 and $n < \infty$ is randomly selected and rounded-off to an integer. Determine the variance of the random variable X, where we have $X = $ (number selected) $-$ (rounded-off integer), if the number is rounded off to the nearest integer, the nearest lower integer, and the nearest higher integer. Comment on the results.

Solution
In the first case, X varies from -0.5 to 0.5, with all values in the interval equally probable, the distribution of X is thus uniform, and we have

$$\sigma_X^2 = \frac{(0.5 - (-0.5))^2}{12} = \frac{1}{12}$$

In the second case, X varies from 0 to 1, with all values in the interval equally probable, the distribution of X is thus uniform, and we have

$$\sigma_X^2 = \frac{(1 - (0))^2}{12} = \frac{1}{12}$$

In the last case, X varies from -1 to 0, with all values in the interval equally probable, the distribution of X is thus uniform, and we have

$$\sigma_X^2 = \frac{(0 - (-1))^2}{12} = \frac{1}{12}$$

The variance in all cases remain the same, as the variance for uniform distribution depends only on the length of the interval, which is 1. However, the means in these cases are different, so are the mean square values. ∎

5.4 Summary

In this chapter, some widely known discrete and continuous distributions, which have been given special names, were introduced, and their features and applications were discussed. We also highlighted a few well-known applications in which some of these distributions are used.

Problems

5.1 Assuming X is a Bernoulli random variable with the parameter p, determine its mean and variance.

5.2 Assuming X is a binomial random variable with the parameters p and n, determine its mean and variance.

5.3 Determine a recursive formula for the pmf of the binomial distribution.

5.4 Assume 10 000 bits are independently transmitted over a channel in which the bit error rate (probability of an erroneous bit) is 0.0001. Using the binomial distribution, calculate the probability when the total number of errors is less than or equal to 2.

5.5 Determine the cdf of the geometric distribution with parameter p.

5.6 Determine the mean and variance of the geometric distribution.

5.7 Determine the probability of the first error occurring at the 10 000th bit transmission in a digital communication system, where the average probability of bit error is one in a billion.

5.8 If in an accelerator a particle has probability 0.01 of hitting a target material, determine the probability that the first particle to hit the target is the 100th and the probability that the target will be hit first by any of the first 100 particles.

5.9 Show that the sum of all probabilities in a Poisson distribution is 1

5.10 Discuss the values of the Poisson probabilities, including its maximum, in terms of the parameters k and λ.

5.11 Determine the mean and variance of the Poisson random variable.

5.12 Assuming X is an exponential random variable with the parameter λ, determine its mean and variance.

5.13 Suppose $Y = \cos(X)$, where X is uniformly distributed over the interval $[0, 2\pi]$. Determine the pdf of the random variable Y.

5.14 Let X be a discrete random variable with K possible outcomes and probabilities $P(X = i)$, $i = 1, 2, \ldots, K$. Assuming the measure of uncertainty is $\log_2 \frac{1}{P(X=i)}$, the entropy of a random variable X, is then defined as the expected value of the uncertainty of its outcomes. Assuming the geometric random variable, determine its entropy.

5.15   In a game of darts, suppose a dart thrown at random always lands on the circular dart board, i.e. it never misses. Assume all points on the board can be hit with equal probability, i.e. where the dart lands follows a uniform distribution. Determine the probability that it lands in the third quadrant.

5.16   Let X be a continuous random variable with the cdf $F_X(x)$. Show that $Y = F_X(x)$ is a uniform random variable over $[0, 1]$.

5.17   Let $Y = \exp X$. Determine the pdf of Y if X is a random variable that is uniformly (with equal probability) distributed between 0 and 1.

5.18   Assuming X is a continuous uniform random variable in the interval $[a, b]$, determine its mean and variance.

5.19   The Weibull distribution is the most widely used probability distribution for modeling failure times. Note that its cdf with parameters a and λ is as follows:

$$F_T(t) = 1 - \exp(-(\lambda t)^a)$$

Determine the failure rate. Assume $a = 2$ and $\lambda = 0.1$, determine the failure rate. Comment on the results.

5.20   Determine the mean and variance of the beta distribution.

5.21   Determine the mean and variance of the Weibull distribution.

5.22   Determine the mean and variance of the gamma distribution.

5.23   Determine the mean and variance of the Laplace distribution.

5.24   Suppose phone call arrivals at a call center are Poisson and occur at an average rate of 50 per hour. The call center has only one operator. If all calls are assumed to last one minute, determine the probability that a waiting line will occur.

5.25   The variance of the discrete uniform random variable Z, which takes on values in a set of n consecutive integers, is 4. Determine the mean of this random variable.

5.26 Assume 1000 bits are independently transmitted over a digital communications channel in which the bit error rate is 0.001. Determine the probability when the total number of errors is greater than or equal to 998.

5.27 Determine the moment generating functions of the Bernoulli, binomial, and Poisson random variables.

5.28 The number of major earthquakes in the world is represented by a Poisson distribution with a rate of $\lambda = 7.4$ earthquakes in a year. Determine the probability that there are exactly four earthquakes in a year. What is the probability that there are no earthquakes given that there are at most two earthquakes in a year?

5.29 Suppose that $f_T(t) = \exp(-2t)$ is the survival function of a certain type of fish, where the time is measured in months. Determine the expected time for the fish being alive, the probability of survival after the first month, the probability that the survival is as long as the expected value, and the time for which the probability of survival is 1%.

6

Multiple Random Variables

There are experiments that produce a collection of $n \geq 2$ random variables. For most of this chapter, the focus is on two random variables ($n = 2$) and their relationships, as the study of a pair of random variables is sufficient to introduce all relevant concepts and important techniques to calculate the probability of interest for any number of random variables. A discussion of random vectors for $n > 2$ is also briefly provided.

6.1 Pairs of Random Variables

In a random experiment, there may be a need to measure two quantities of interest, and the probability of interest may involve both random quantities. Examples include the transmitted and received signals in a communication system, heights and weights of newborn babies, education level and level of income, cost and reliability, cigarette smoking and lung cancer, texting and car accidents, tax cut and social assistance, and alcohol consumption and drug use, to name just a few. There are many applications of interest in a multitude of fields in science and engineering that can be handled by the theory of two random variables. In practice, quite often, we observe one random variable, but we want to know about the other random variable that is inaccessible.

In our notation, the names of the two random variables are always uppercase letters, such as X and Y, and lowercase letters denote possible values of the two random variables, such as x and y. In an experiment, in which there are two random variables X and Y, each outcome (x, y) is a point in a plane and events are points, straightlines, curves, or areas in the plane.

Let S be the sample space of a random experiment. A *pair of random variables* (X, Y) is a function that assigns a pair of real numbers (x, y) to each outcome in the sample space S. In other words, it is a mapping from the sample space S into the real plane R^2, rather than into the real line R as it is the case for a single random variable.

Probability, Random Variables, Statistics, and Random Processes: Fundamentals & Applications,
First Edition. Ali Grami.
© 2020 John Wiley & Sons, Inc. Published 2020 by John Wiley & Sons, Inc.
Companion Website: www.wiley.com/go/grami/PRVSRP

The sample space S is the ***domain of the pair of random variables*** and the set of all values taken on by the two random variables, denoted by $S_{X,Y}$, is the ***range of the pair of random variables***. The range $S_{X,Y}$ is a subset of all possible points in the real plane with nonzero probability, whereas the range S_X is defined as the set of all values of the random variable X with nonzero probability and the range S_Y as the set of all values of the random variable Y with nonzero probability, we thus have

$$\begin{cases} S_X = \{x \mid P_X(x) > 0\} \\ S_Y = \{y \mid P_Y(y) > 0\} \end{cases} \rightarrow S_{X,Y} = \{(x,y) \mid P_{X,Y}(x,y) > 0\} \tag{6.1}$$

Figure 6.1 shows the ranges S_X, S_Y, and $S_{X,Y}$. If the two random variables are both discrete, then (X, Y) is called a ***pair of discrete random variables***. A random experiment consisting of rolling a pair of two dice, with the pair of numbers (X, Y) facing up, is an example of a pair of discrete random variables.

If the two random variables are both continuous, then (X, Y) is called a ***pair of continuous random variables***. A random experiment consisting of throwing a dart and landing on the board, with the pair of numbers (X, Y) representing the coordinates of the dart on the board, is an example of a pair of continuous random variables.

If one of the two random variables is discrete, while the other is continuous, then (X, Y) is called a ***pair of mixed random variables***. A random experiment consisting of receiving a text message, with the pair of numbers (X, Y) representing the message arrival time and the message length measured in the number of characters, is an example of a pair of mixed random variables.

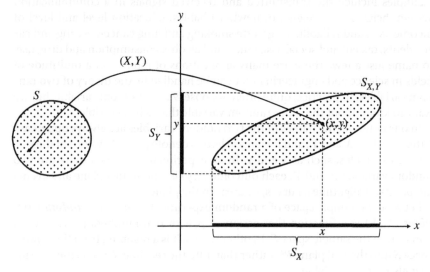

Figure 6.1 A function assigns a pair of real numbers in the real plane to each outcome, a mapping from S to $S_{X,Y}$.

Throughout this chapter, the primary focus is either on a pair of continuous random variables or on a pair of discrete random variables.

6.2 The Joint Cumulative Distribution Function of Two Random Variables

The joint cumulative distribution function (cdf) of a pair of random variables contains all the information required to calculate the probability for any event involving either or both random variables. The notation for joint cdf is $F_{X,Y}(x, y)$, where we use the uppercase letter F – whose subscripts separated by a comma are the names of the random variables X and Y – as a function of possible values of the two random variables, denoted by lowercase letters x and y.

The *joint cdf of two random variables* X and Y, which is a complete probability model, is defined as follows:

$$F_{X,Y}(x, y) = P(X \leq x, Y \leq y) \qquad -\infty < x < \infty \qquad -\infty < y < \infty$$

$$(6.2)$$

The probability of the event $\{X \leq x, Y \leq y\}$ varies as x and y are varied, i.e. $F_{X,Y}(x, y)$ is a function of the variables x and y. The joint cdf is a nondecreasing bounded function of x and y that starts at 0 and ends at 1, as x increases from $-\infty$ to ∞ and y increases from $-\infty$ to ∞. For any real numbers x and y, the cdf is the probability that the random variable X is no larger than x and the random variable Y is no larger than y, that is the cdf is the probability that the random variable X takes on a value in the set $(-\infty, x]$ and the random variable Y takes on a value in the set $(-\infty, y]$. Figure 6.2a shows the area of the real plane corresponding to the joint cdf, where the joint cdf is defined as the probability of the semi-infinite rectangle, i.e. the probability of the area of the plane below and to the left of the point (x, y). The joint cdf has the following important properties:

- $0 \leq F_{X,Y}(x, y) \leq 1$
- $x_1 \leq x_2$ and $y_1 \leq y_2 \rightarrow F_{X,Y}(x_1, y_1) \leq F_{X,Y}(x_2, y_2)$
- $\lim_{x \to a^+} F_{X,Y}(x, y) = F_{X,Y}(a, y)$
- $\lim_{y \to b^+} F_{X,Y}(x, y) = F_{X,Y}(x, b)$
- $F_{X,Y}(x, -\infty) = 0$
- $F_{X,Y}(-\infty, y) = 0$
- $F_{X,Y}(-\infty, -\infty) = 0$
- $F_{X,Y}(\infty, \infty) = 1$

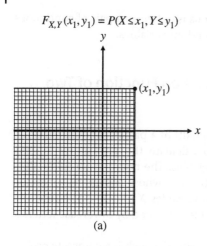

$$F_{X,Y}(x_1, y_1) = P(X \leq x_1, Y \leq y_1)$$

(a)

Figure 6.2 cdfs defined by regions: (a) joint cdf defined by the semi-infinite rectangle; (b) marginal cdf of X defined by the horizontal half-plane; (c) marginal cdf of Y defined by the vertical half-plane.

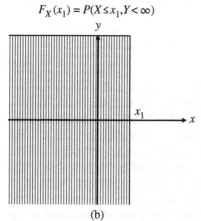

$$F_X(x_1) = P(X \leq x_1, Y < \infty)$$

(b)

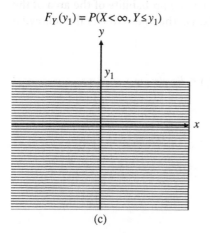

$$F_Y(y_1) = P(X < \infty, Y \leq y_1)$$

(c)

- $P(x_1 < X \le x_2, Y \le y) = F_{X,Y}(x_2, y) - F_{X,Y}(x_1, y)$
- $P(X \le x, y_1 < Y \le y_2) = F_{X,Y}(x, y_2) - F_{X,Y}(x, y_1)$
- $a_1 > b_1$ and $a_2 > b_2 \rightarrow F_{X,Y}(a_1, a_2) \ge F_{X,Y}(a_1, b_2) \ge F_{X,Y}(b_1, b_2)$
- $P(x_1 < X \le x_2, y_1 < Y \le y_2) = F_{X,Y}(x_2, y_2) - F_{X,Y}(x_2, y_1) - F_{X,Y}(x_1, y_2)$
  $+ F_{X,Y}(x_1, y_1)$
- $P(X > x, Y > y) = 1 - F_{X,Y}(x, y)$ (6.3)

It is important to note that these properties can be used either to calculate probability for any event of interest involving either or both random variables or to determine if some function could be a valid joint cdf of two random variables.

6.2.1 Marginal Cumulative Distribution Function

Note that in the context of two random variables, the event $\{X \le x\}$ suggests that Y can have any value so long as the condition on X is met and the event $\{Y \le y\}$ suggests that X can have any value so long as the condition on Y is met. The marginal cdf can be obtained by removing the constraint on one of the two variables. The marginal cdfs are the probabilities of the regions shown in Figure 6.2b,c. The *marginal cdfs* of X and Y are, respectively, defined as follows:

$$F_X(x) = F_{X,Y}(x, \infty) = P(X \le x, Y < \infty) \quad -\infty < x < \infty$$
$$F_Y(y) = F_{X,Y}(\infty, y) = P(X < \infty, Y \le y) \quad -\infty < y < \infty \quad (6.4)$$

From a joint cdf, the corresponding marginal cdfs can be uniquely determined, but in general, the converse is not true, as marginal cdfs can provide only partial information about their joint cdf.

Example 6.1
Suppose the joint cdf of a pair of continuous random variables (X, Y) is as follows:

$$F_{X,Y}(x, y) = \begin{cases} 1 - \exp(-x) - \exp(-y) + \exp(-(x+y)) & x \ge 0 \text{ and } y \ge 0 \\ 0 & \text{elsewhere} \end{cases}$$

Determine $P(X \le 1, Y \le 1)$ and find the marginal cdfs of X and Y.

Solution
The probability of interest is calculated as follows:

$$P(X \le 1, Y \le 1) = F_{X,Y}(1, 1) = 1 - 2\exp(-1) + \exp(-2) \cong 0.3996$$

The marginal cdfs are obtained as follows:

$$F_X(x) = F_{X,Y}(x, \infty) = \begin{cases} 1 - \exp(-x) & x \ge 0 \\ 0 & x < 0 \end{cases}$$

and

$$F_Y(y) = F_{X,Y}(\infty, y) = \begin{cases} 1 - \exp(-y) & y \geq 0 \\ 0 & y < 0 \end{cases}$$ ∎

Although the definition of the joint cdf is simple, we hardly work with it to study probability models, as it is quite easier to work instead with joint probability mass function (pmf) for discrete random variables or joint probability density function (pdf) for continuous random variables.

6.3 The Joint Probability Mass Function of Two Random Variables

The notation for joint pmf is $p_{X,Y}(x, y)$, where we use the lowercase letter p – whose subscripts separated by a comma are the names of the random variables X and Y – as a function of possible values of the two random variables, denoted by lowercase letters x and y. The *joint pmf of two discrete random variables* X and Y, which is a complete probability model, is defined as follows:

$$p_{X,Y}(x, y) = P(X = x, Y = y) \qquad (x, y) \in S_{X,Y} \qquad (6.5)$$

The joint pmf has the following properties:

(i) $p_{X,Y}(x, y) \geq 0 \quad \forall(x, y)$

(ii) $\displaystyle\sum_{(x,y) \in S_{X,Y}} \sum p_{X,Y}(x, y) = 1$

(iii) $\displaystyle P((X, Y) \in B) = \sum_{(x,y) \in B \subset S_{X,Y}} \sum p_{X,Y}(x, y)$ \qquad (6.6)

Note that above summations have a finite or an infinite number of terms, depending on whether the range is finite or not. All these three properties are consequences of the axioms of probability. The joint cdf of a pair of discrete random variables (X, Y) is thus given by

$$F_{X,Y}(x, y) = \sum_{X \leq x} \sum_{Y \leq y} p_{X,Y}(x, y) \qquad (6.7)$$

6.3.1 Marginal Probability Mass Function

We can find the *marginal pmf* by removing the constraint on the variable that is of no interest. For discrete random variables, the marginal pmfs of X and Y are, respectively, defined as follows:

$$p_X(x) = \sum_{y \in S_Y} p_{X,Y}(x, y) \quad x \in S_X$$

$$p_Y(y) = \sum_{x \in S_X} p_{X,Y}(x, y) \quad y \in S_Y \qquad (6.8)$$

To find $p_X(x)$, in the sum, y is a constant and each term corresponds to a value of $x \in S_X$ and to find $p_Y(y)$, in the sum, x is a constant and each term corresponds to a value of $y \in S_Y$. Note that the marginal pmfs are probability models for the individual random variables X and Y, but they do not provide a complete probability model for the pair of random variables (X, Y). In other words, it is always possible to obtain marginal pmfs from the joint pmf, but in general, the converse is not true.

Example 6.2

Suppose the joint pmf of the two random variables X and Y is as follows:

$$p_{X,Y}(x,y) = \begin{cases} k\,xy & x = 1, 2, 3 \text{ and } y = 1, 2, 3, 4 \\ 0 & \text{elsewhere} \end{cases}$$

where k is a nonrandom constant. Determine the value of k, and find the marginal pmfs of X and Y.

Solution

Since the sum of all probabilities must be 1, we have

$$\sum_{x=1}^{3} \sum_{y=1}^{4} k\,xy = 1 \quad \rightarrow \quad k(1 + 2 + 3 + 2 + 4 + 6 + 3 + 6 + 9 + 4 + 8 + 12) = 1$$

$$\rightarrow \quad k = \frac{1}{60}$$

The marginal pmf is thus as follows:

$$p_X(x) = \sum_{y=1}^{4} p_{X,Y}(x,y) = \sum_{y=1}^{4} \frac{xy}{60} = \frac{x}{60} \sum_{y=1}^{4} y = \frac{x}{60}(1 + 2 + 3 + 4)$$

$$= \frac{x}{6} \quad x = 1, 2, 3$$

and

$$p_Y(y) = \sum_{x=1}^{3} p_{X,Y}(x,y) = \sum_{x=1}^{3} \frac{xy}{60} = \frac{y}{60} \sum_{x=1}^{3} x = \frac{y}{60}(1 + 2 + 3)$$

$$= \frac{y}{10} \quad y = 1, 2, 3, 4$$

Note that, as expected, we have

$$\sum_{x=1}^{3} p_X(x) = \sum_{y=1}^{4} p_Y(y) = 1$$

∎

6.4 The Joint Probability Density Function of Two Random Variables

The notation for joint pdf is $f_{X,Y}(x, y)$, where we use the lowercase letter f – whose subscripts separated by a comma are the names of the random variables X and Y – as a function of possible values of the two random variables, denoted by lowercase letters x and y. The **joint pdf of two continuous random variables** X and Y, which is a complete probability model, is defined as follows:

$$f_{X,Y}(x,y) = \frac{\partial^2 F_{X,Y}(x,y)}{\partial x\, \partial y} \tag{6.9}$$

In other words, if X and Y are jointly continuous random variables, then the joint pdf can be obtained from the joint cdf by differentiation. Note that if X and Y are not jointly continuous, it is then possible that the above partial derivative does not exist. If the joint cdf or its partial derivatives are discontinuous, then the joint pdf does not exist. The joint pdf properties are as follows:

(i) $f_{X,Y}(x,y) \geq 0 \quad \forall (x,y)$

(ii) $\displaystyle\int_{-\infty}^{\infty} \int_{-\infty}^{\infty} f_{X,Y}(x,y)\, dx\, dy = 1$ $\tag{6.10}$

(iii) $P((X,Y) \in B) = \displaystyle\int_{(x,y)\in B \subset S_{X,Y}} f_{X,Y}(x,y) dx\, dy$

All these three properties are consequences of the axioms of probability. Note that the joint cdf of a pair of continuous random variables (X, Y) is given by

$$F_{X,Y}(x,y) = \int_{-\infty}^{x} \int_{-\infty}^{y} f_{X,Y}(w,z)\, dw\, dz \tag{6.11}$$

The probability of a rectangular region in the x-y plane can be obtained as follows:

$$P(a < X \leq b, c < Y \leq d) = \int_{c}^{d} \int_{a}^{b} f_{X,Y}(x,y)\, dx\, dy \tag{6.12}$$

Note that for jointly continuous random variables X and Y, the probability of any event is zero if the region of the event in the x-y plane has zero area, for the volume over that area then becomes obviously zero. To this effect, a point, a straightline, a closed curve, and an open curve in the x-y plane are all events for which the probabilities are zero.

6.4.1 Marginal Probability Density Function

We can find the marginal pdf by removing the constraint on the variable that is of no interest. For continuous random variables, the **marginal pdfs** of X and Y are defined as follows:

$$f_X(x) = \int_{-\infty}^{\infty} f_{X,Y}(x,y)\, dy \quad x \in S_X$$

$$f_Y(y) = \int_{-\infty}^{\infty} f_{X,Y}(x,y)\, dx \quad y \in S_Y \tag{6.13}$$

A marginal pdf is thus obtained by integrating over the variable that is not of interest. The joint pdf in general cannot be obtained from the marginal pdfs, as the marginal pdfs can only provide partial information about the joint pdf.

Example 6.3

Suppose the joint pdf of the random variables X and Y is as follows:

$$f_{X,Y}(x, y) = \begin{cases} 2(1 - x) & 0 \leq x \leq 1 \text{ and } 0 \leq y \leq 1 \\ 0 & \text{elsewhere} \end{cases}$$

Calculate the probability of the event $\{x \leq 0.5, \, 0.4 \leq y \leq 0.7\}$ and find the marginal pdfs of X and Y.

Solution

The probability of interest is calculated as follows:

$$P(0 \leq X \leq 0.5, 0.4 \leq Y \leq 0.7) = \int_{0.4}^{0.7} \int_{0}^{0.5} 2(1 - x)\, dx\, dy = 0.225$$

The marginal pdfs are as follows:

$$f_X(x) = \int_{-\infty}^{\infty} f_{X,Y}(x, y)\, dy = \int_{0}^{1} 2(1 - x)\, dy = 2(1 - x) \qquad 0 \leq x \leq 1$$

and

$$f_Y(y) = \int_{-\infty}^{\infty} f_{X,Y}(x, y)\, dx = \int_{0}^{1} 2(1 - x)\, dx = 1 \qquad 0 \leq y \leq 1$$

Note that, as expected, we have

$$\int_{-\infty}^{\infty} f_X(x)\, dx = \int_{0}^{1} 2(1 - x)\, dx = 1$$

and

$$\int_{-\infty}^{\infty} f_Y(y)\, dy = \int_{0}^{1} dy = 1 \qquad \blacksquare$$

6.5 Expected Values of Functions of Two Random Variables

Suppose $g(X, Y)$ is a function of two random variables X and Y. The function $g(X, Y)$ is thus a random variable. The *expected value of a function of two random variables* is as follows:

$$E[g(X, Y)] = \begin{cases} \displaystyle\int_{-\infty}^{\infty} \int_{-\infty}^{\infty} g(x, y) f_{X,Y}(x, y) \, dx \, dy & \text{Continuous} \\ \displaystyle\sum_{x \in S_X} \sum_{y \in S_Y} g(x, y) p_{X,Y}(x, y) & \text{Discrete} \end{cases}$$ (6.14)

Note that in order to determine the expected value of the function $g(X, Y)$, the distribution of the random variable $g(X, Y)$ is not needed.

6.5.1 Joint Moments

The joint moments of two random variables X and Y can provide insights into their joint behavior. Assuming m and n are both nonnegative integers, the **joint moment** of X and Y, denoted by $E[X^m Y^n]$, and the **joint central moment** of X and Y, denoted by $E[(X - E[X])^m (Y - E[Y])^n]$, are respectively defined as follows:

$$E[X^m Y^n] = \begin{cases} \displaystyle\int_{-\infty}^{\infty} \int_{-\infty}^{\infty} x^m y^n f_{X,Y}(x, y) \, dx \, dy & \text{Continuous} \\ \displaystyle\sum_{x \in S_X} \sum_{y \in S_Y} x^m y^n p_{X,Y}(x, y) & \text{Discrete} \end{cases}$$

$$E[(X - E[X])^m (Y - E[Y])^n]$$

$$= \begin{cases} \displaystyle\int_{-\infty}^{\infty} \int_{-\infty}^{\infty} (x - E[X])^m (y - E[Y])^n f_{X,Y}(x, y) \, dx \, dy & \text{Continuous} \\ \displaystyle\sum_{x \in S_X} \sum_{y \in S_Y} (x - E[X])^m (y - E[Y])^n p_{X,Y}(x, y) & \text{Discrete} \end{cases}$$

(6.15)

Note that the sum $m + n$ is called the **order of the moments**. From the above definitions, we can determine the following measures for each random variable:

$$\begin{cases} m = 1, n = 0 \rightarrow E[X] = \mu_X \\ m = 2, n = 0 \rightarrow E[X^2] \end{cases} \rightarrow \sigma_X^2 = E[X^2] - (E[X])^2$$

$$\begin{cases} m = 0, n = 1 \rightarrow E[Y] = \mu_Y \\ m = 0, n = 2 \rightarrow E[Y^2] \end{cases} \rightarrow \sigma_Y^2 = E[Y^2] - (E[Y])^2$$ (6.16)

Example 6.4
Determine the mean and variance of the sum of the two random variables X and Y whose joint pdf is $f_{X,Y}(x, y)$.

Solution

In order to determine the expected value of $Z = X + Y$, we do not need to have the pdf of the random variable Z, as reflected in the following equation:

$$\begin{aligned}
E[Z] &= \int_{-\infty}^{\infty} \int_{-\infty}^{\infty} (x+y) f_{X,Y}(x,y) \, dx \, dy \\
&= \int_{-\infty}^{\infty} \int_{-\infty}^{\infty} x f_{X,Y}(x,y) \, dx \, dy + \int_{-\infty}^{\infty} \int_{-\infty}^{\infty} y f_{X,Y}(x,y) \, dx \, dy \\
&= \int_{-\infty}^{\infty} x \left(\int_{-\infty}^{\infty} f_{X,Y}(x,y) \, dy \right) dx + \int_{-\infty}^{\infty} y \left(\int_{-\infty}^{\infty} f_{X,Y}(x,y) \, dx \right) dy \\
&= \int_{-\infty}^{\infty} x f_X(x) \, dx + \int_{-\infty}^{\infty} y f_Y(y) \, dy = E[X] + E[Y]
\end{aligned}$$

Thus the expected value of a sum of two random variables is always equal to the sum of the individual expected values. The expected value of Z^2 is as follows:

$$\begin{aligned}
E[Z^2] &= \int_{-\infty}^{\infty} \int_{-\infty}^{\infty} (x+y)^2 f_{X,Y}(x,y) \, dx \, dy \\
&= \int_{-\infty}^{\infty} \int_{-\infty}^{\infty} (x^2 + y^2 + 2xy) f_{X,Y}(x,y) \, dx \, dy \\
&= \int_{-\infty}^{\infty} \int_{-\infty}^{\infty} x^2 f_{X,Y}(x,y) \, dx \, dy + \int_{-\infty}^{\infty} \int_{-\infty}^{\infty} y^2 f_{X,Y}(x,y) \, dx \, dy \\
&\quad + \int_{-\infty}^{\infty} \int_{-\infty}^{\infty} 2xy f_{X,Y}(x,y) \, dx \, dy \\
&= E[X^2] + E[Y^2] + 2E[XY]
\end{aligned}$$

The variance of Z is thus as follows:

$$\begin{aligned}
\sigma_Z^2 &= E[Z^2] - (E[Z])^2 = E[X^2] + E[Y^2] + 2E[XY] - (E[X] + E[Y])^2 \\
&= \sigma_X^2 + \sigma_Y^2 + 2(E[XY] - E[X]E[Y])
\end{aligned}$$

Unlike the expected value, the variance of a sum of two random variables is not, in general, equal to the sum of the variances, unless for the special case when we have $E[XY] = E[X]E[Y]$. ∎

6.6 Independence of Two Random Variables

The concept of statistical independence of two random variables is simple, but very important. When two random variables are independent, knowledge of one random variable gives no information about the value of the other. Generally, random variables that arise from different physical sources are quite often statistically independent, so are they if they come from the same source but at greatly different times. The random variables X and Y are **statistically independent** if and only if their joint cdf is equal to the product of their marginal cdfs that is we have

$$F_{X,Y}(x,y) = F_X(x)F_Y(y) \qquad \forall(x,y) \tag{6.17}$$

Equivalently, two continuous random variables X and Y are independent if and only if their joint pdf is equal to the product of their marginal pdfs, and two discrete random variables X and Y are independent if and only if their joint pmf is equal to the product of their marginal pmfs. In other words, we have

$$f_{X,Y}(x,y) = f_X(x)f_Y(y) \qquad \forall(x,y) \text{ Continuous}$$
$$p_{X,Y}(x,y) = p_X(x)\,p_Y(y) \qquad \forall(x,y) \text{ Discrete} \tag{6.18}$$

It is also important to note that if X and Y are independent random variables, then the random variables defined by any pair of deterministic functions $h(X)$ and $k(Y)$ are also independent.

Example 6.5

The joint pdf for the random variables X and Y is given as follows:

$$f_{X,Y}(x,y) = x+y \quad 0 \le x \le 1 \quad \text{and} \quad 0 \le y \le 1$$

Are X and Y statistically independent?

Solution

The marginal pdfs are as follows:

$$f_X(x) = \int_{-\infty}^{\infty} f_{X,Y}(x,y)\,dy = \int_0^1 (x+y)\,dy = x+0.5 \quad 0 \le x \le 1$$

and

$$f_Y(y) = \int_{-\infty}^{\infty} f_{X,Y}(x,y)\,dx = \int_0^1 (x+y)\,dx = y+0.5 \quad 0 \le y \le 1$$

We therefore have

$$(x+y) \ne (x+0.5)(y+0.5) \rightarrow f_{X,Y}(x,y) \ne f_X(x)f_Y(y)$$

which means the random variables X and Y are not statistically independent. ∎

Example 6.6

Suppose the random variables X and Y are statistically independent, and we have $g(X, Y) = h(X)\,k(Y)$. Show that $E[g(X, Y)] = E[h(X)]\,E[k(Y)]$.

Solution

Noting that we have $f_{X,Y}(x, y) = f_X(x)f_Y(y)$, the expected value is as follows:

$$E[g(X, Y)] = \int_{-\infty}^{\infty} \int_{-\infty}^{\infty} h(x)k(y)f_{X,Y}(x, y) \, dx \, dy$$

$$= \int_{-\infty}^{\infty} \int_{-\infty}^{\infty} h(x)k(y)f_X(x)f_Y(y) \, dx \, dy$$

$$= \left(\int_{-\infty}^{\infty} h(x)f_X(x) \, dx \right) \left(\int_{-\infty}^{\infty} k(y)f_Y(y) \, dy \right) = E[h(X)] \, E[k(Y)]$$

∎

It is easy to show that if the random variables X and Y are assumed to be statistically independent, and we also have $h(X) = X^j$ and $k(Y) = Y^k$, for any real j and k, we then have $E[X^j Y^k] = E[X^j]E[Y^k]$. Note that the converse is also true.

Example 6.7

Assuming the random variables X and Y are statistically independent, determine the mean and variance of the random variable $Z = aX + bY$, where a and b are both nonzero, nonrandom constants.

Solution

We have the following:

$$E[Z] = \int_{-\infty}^{\infty} \int_{-\infty}^{\infty} (ax + by)f_{X,Y}(x, y) \, dx \, dy$$

$$= \int_{-\infty}^{\infty} \int_{-\infty}^{\infty} ax f_{X,Y}(x, y) \, dx \, dy + \int_{-\infty}^{\infty} \int_{-\infty}^{\infty} by f_{X,Y}(x, y) \, dx \, dy$$

$$= \int_{-\infty}^{\infty} ax \left(\int_{-\infty}^{\infty} f_{X,Y}(x, y) \, dy \right) dx + \int_{-\infty}^{\infty} by \left(\int_{-\infty}^{\infty} f_{X,Y}(x, y) \, dx \right) dy$$

$$= \int_{-\infty}^{\infty} ax f_X(x) \, dx + \int_{-\infty}^{\infty} by f_Y(y) \, dy = aE[X] + bE[Y]$$

Thus the expected value of a linear combination of two random variables, whether they are independent or not, is always equal to the same linear combination of the individual expected values. We also have the following:

$$E[Z^2] = \int_{-\infty}^{\infty} \int_{-\infty}^{\infty} (ax + by)^2 f_{X,Y}(x, y) \, dx \, dy$$

$$= \int_{-\infty}^{\infty} \int_{-\infty}^{\infty} (a^2x^2 + b^2y^2 + 2abxy)f_X(x)f_Y(y) \, dx \, dy$$

$$= \int_{-\infty}^{\infty} \int_{-\infty}^{\infty} a^2x^2 f_X(x)f_Y(y) \, dx \, dy + \int_{-\infty}^{\infty} \int_{-\infty}^{\infty} b^2y^2 f_X(x)f_Y(y) \, dx \, dy$$

$$+ \int_{-\infty}^{\infty} \int_{-\infty}^{\infty} 2abxy f_X(x)f_Y(y) \, dx \, dy$$

$$= \int_{-\infty}^{\infty} a^2 x^2 f_X(x) dx \int_{-\infty}^{\infty} f_Y(y) \, dy + \int_{-\infty}^{\infty} b^2 y^2 f_Y(y) dy \int_{-\infty}^{\infty} f_X(x) \, dx$$

$$+ 2ab \int_{-\infty}^{\infty} x f_X(x) \, dx \int_{-\infty}^{\infty} y f_Y(y) \, dy$$

$$= a^2 E[X^2] + b^2 E[Y^2] + 2ab E[X] E[Y]$$

The variance of Z is thus as follows:

$$\sigma_Z^2 = E[Z^2] - (E[Z])^2 = a^2 E[X^2] + b^2 E[Y^2] + 2ab E[X] E[Y]$$
$$- (aE[X] + bE[Y])^2 = a^2 \sigma_X^2 + b^2 \sigma_Y^2$$

Note that for $a = b = 1$, we can conclude that the variance of a sum of two independent random variables is equal to the sum of the individual variances. ∎

6.7   Correlation between Two Random Variables

Suppose X and Y denote two random variables that represent two outcomes in a random experiment. A scatter diagram can show the locations of the observed points (x, y) on a rectangular coordinate system, i.e. the x–y plane. Should there be a linear correlation between X and Y, the location of the points $(x - E[X], y - E[Y])$ in a scatter diagram tend to cluster along a line.

In a random experiment with two random variables X and Y, the covariance of X and Y needs to be determined, as it is a meaningful measure to provide insight about the dependence between X and Y. The covariance measures the correlation, i.e. the degree of similarity and dependence, between two random variables through their deviations from their respective mean values. To this effect, in a practical application of interest, many trials of the random experiment are conducted and the values of the two random outcomes X and Y for each trial are recorded. By computing the product of $(x - E[X])$ and $(y - E[Y])$ for every trial, adding these products, and dividing it by the number of trials, the covariance of the random experiment is then obtained.

In order to define the correlation between two variables, we define the ***covariance*** of the two random variables X and Y as follows:

$$\text{COV}(X, Y) = E[(X - E[X])(Y - E[Y])] = E[XY] - E[X] E[Y] \quad (6.19)$$

In order to be able to compare the correlation between a pair of random variables with the correlation between another pair of random variables, the ***correlation coefficient***, which is the normalized version of the covariance, is defined as follows:

$$\rho_{X,Y} = \frac{\text{COV}(X, Y)}{\sigma_X \sigma_Y} = \frac{E[XY] - E[X] E[Y]}{\sigma_X \sigma_Y} \quad (6.20)$$

The correlation coefficient provides a measure of the linear dependence, i.e. the strength and direction of a linear relationship, between the two random variables X and Y, and can be at most 1 in magnitude, i.e. $0 \leq |\rho_{X,Y}| \leq 1$. The inter-relatedness of the two random variables X and Y, measured by their correlation, as shown in Figure 6.3, generally falls into three distinct categories:

(i) Suppose the random variables X and Y represent the height and weight of a child in a country, respectively. Many children are randomly chosen, and their heights and weights are measured. A kid with an above-average height is likely to have an above-average weight and a kid with a below-average height is likely to have a below-average weight. It is likely that the random variables $(X - E[X])$ and $(Y - E[Y])$ have the same sign, and their product is thus positive for most kids. The points $(X - E[X], Y - E[Y])$ mostly cluster along a line with a positive slope in the scatter diagram, thus giving rise to a ***positive covariance*** and reflecting a ***positive correlation*** or ***direct correlation*** between the two random variables X and Y.

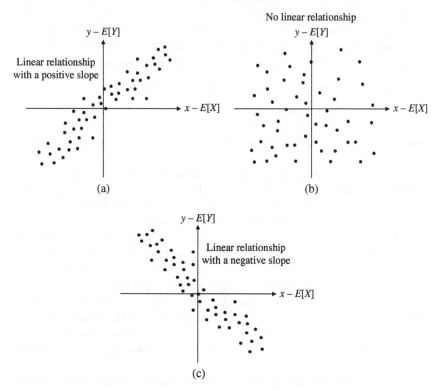

Figure 6.3 Scatter diagrams for different pairs of random variables, each with 50 observations: (a) positively correlated; (b) uncorrelated; (c) negatively correlated.

(ii) Suppose the random variables X and Y represent the amount of snowfall and the number of people driving on a winter day, respectively. The amount of snowfall and the number of people driving are measured on a number of different days during a winter. On a winter day, an above-average amount of snowfall is likely to bring about a below-average number of people driving and a below-average amount of snowfall is likely to bring about an above-average number of people driving. It is likely that the random variables $(X - E[X])$ and $(Y - E[Y])$ have opposite signs, and their product is thus negative on most winter days. The points $(X - E[X], Y - E[Y])$ mostly cluster along a line with a negative slope in the scatter diagram, thus giving rise to a ***negative covariance*** and reflecting a ***negative correlation*** or ***inverse correlation*** between the two random variables X and Y.

(iii) Suppose the random variables X and Y represent the birth weight of a new-born baby and the number of cousins the baby has at birth, respectively. This information is obtained from a random sample of population. The signs of the random variables $(X - E[X])$ and $(Y - E[Y])$ are sometimes the same and sometimes different. The points $(X - E[X], Y - E[Y])$ are scattered randomly in the x-y plane, and are not confined to a completely bounded region. This in turn gives rise to ***zero covariance***, thus reflecting the fact that the two random variables X and Y are ***uncorrelated***.

It is insightful to summarize the following distinct cases:

- $\rho_{X,Y} = +1 \to$ It indicates that Y can be predicted by a linear function of X whose slope is positive, an example is the temperature of a component measured in degrees Fahrenheit and its temperature measured in degrees Celsius.
- $0 < \rho_{X,Y} < 1 \to$ A ***positive correlation coefficient*** implies X and Y both increase together or both decrease together, examples are a growing baby's height and weight, and an old person's age and health problems.
- $\rho_{X,Y} = 0 \to X$ and Y have ***zero correlation coefficient***, an example is the number of years of university education a person has and the number of plants she keeps in her house. It is very important to note that a value of 0 for $\rho_{X,Y}$ does not mean that there is no correlation, but there could be a nonlinear correlation, as confounding variables might also be involved.
- $-1 < \rho_{X,Y} < 0 \to$ A ***negative correlation coefficient*** implies when X increases, Y decreases, and vice versa, examples are the distance from a transmitter to its receiver and the received signal power, and the time a student spends on social media during the academic year and his grade point average.
- $\rho_{X,Y} = -1 \to$ It indicates that Y can be predicted by a linear function of X whose slope is negative, an example is the gain and attenuation measured in a system both in decibels.

Example 6.8

Suppose the random variable Y is specified by the following deterministic function of the random variable X:

$$Y = aX^2 + c$$

where X has a uniform distribution over the interval $[-1, 1]$, and $a \neq 0$ and c are known nonrandom constants. Determine the correlation coefficient between X and Y. Comment on the results.

Solution

In order to find the correlation coefficient between X and Y, we first need to find the covariance of the random variables X and Y:

$$E[XY] - E[X]E[Y] = E[X(aX^2 + c)] - E[X]E[aX^2 + c]$$
$$= aE[X^3] + cE[X] - aE[X]E[X^2] - cE[X] = a(E[X^3] - E[X]E[X^2])$$

Since X has a uniform distribution over the interval $[-1, 1]$, its nth-moment is as follows:

$$E[X^n] = \frac{1}{2} \int_{-1}^{1} x^n \, dx = \frac{1 - (-1)^{n+1}}{2(n+1)} \rightarrow \begin{cases} E[X] = 0 \\ E[X^2] = \dfrac{1}{3} \\ E[X^3] = 0 \end{cases}$$

As a result, we have

$$E[XY] - E[X]E[Y] = a(E[X^3] - E[X]E[X^2]) = 0$$

Since the covariance is 0, the correlation coefficient is 0. The random variables X and Y are thus uncorrelated, even though the random variable Y is deterministically dependent on the random variable X, the reason lies in the fact that the correlation coefficient is a measure of how well Y is predicted by a linear function of X, not by a nonlinear one. ∎

The random variables X and Y are **orthogonal** if and only if $E[XY] = 0$. The random variables X and Y are **uncorrelated** if and only if their covariance is 0, i.e. $E[XY] = E[X]E[Y]$ or equivalently the random variables $(X - E[X])$ and $(Y - E[Y])$ are orthogonal. If the random variables X and Y are **orthogonal and uncorrelated,** then at least one of the two means is 0.

Example 6.9

Suppose the random variable Y is linearly related to the random variable X by $Y = mX + b$, where $m \neq 0$ and b are nonrandom constants. Determine the correlation coefficient in terms of m and b. Are X and Y orthogonal?

Solution

We first find the mean and standard deviation of Y in terms of those of X, and then the correlation of the random variables X and Y:

$$E[Y] = E[mX + b] = m\,E[X] + b$$
$$\sigma_Y^2 = E[(mX + b)^2] - (E[mX + b])^2 = m^2(E[X^2] - (E[X])^2)$$
$$= m^2\sigma_Y^2 \rightarrow \sigma_Y = |m|\sigma_X$$
$$E[XY] = E[X(mX + b)] = mE[X^2] + bE[X]$$

We are now in a position to determine the correlation coefficient:

$$\rho_{X,Y} = \frac{E[XY] - E[X]E[Y]}{\sigma_X\sigma_Y} = \frac{mE[X^2] + bE[X] - E[X](mE[X] + b)}{\sigma_X(|m|\sigma_X)}$$

$$= \frac{m\sigma_X^2}{|m|\sigma_X^2} = \begin{cases} 1 & m > 0 \\ -1 & m < 0 \end{cases}$$

We can thus conclude that if X and Y are linearly related, not only are they correlated, but the correlation coefficient between them is either +1 or −1. The random variables X and Y are orthogonal if we have the following:

$$E[XY] = mE[X^2] + bE[X] = 0 \rightarrow b = -m\frac{E[X^2]}{E[X]}$$

whenever $E[X] \neq 0$. ∎

Note that the mean value of a weighted sum of random variables always equals the weighted sum of mean values. However, the variance of a weighted sum of random variables equals the weighted sum of the variances of the random variables, if the random variables are uncorrelated.

As shown earlier, if X and Y are independent, we have $E[X^jY^k] = E[X^j]E[Y^k]$, including when $j = k = 1$. It is then obvious that when X and Y are independent, they are uncorrelated, i.e. we have $E[XY] = E[X]E[Y]$. In other words, independence indicates absence of correlation. However, the converse, in general, is not true, that is it is possible for X and Y to be uncorrelated, but not independent, as shown in Figure 6.4. There is only one special case, however, for which independence and uncorrelatedness are equivalent, and that is when the random variables X and Y are jointly Gaussian, as discussed extensively in the next chapter.

Example 6.10

Suppose the joint pmf $p_{X,Y}(i, j)$ and marginal pmfs $p_X(i)$ and $p_Y(j)$ for a pair of discrete random variables are reflected in the following table:

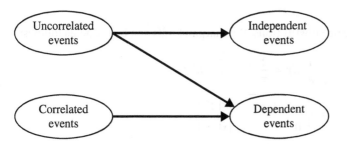

Figure 6.4 Relationship between uncorrelatedness and independence.

		$j = -1$	$j = 0$	$j = 1$	$p_X(i)$
$i = -1$	$\rightarrow$	0	0.25	0	0.25
$i = 0$	$\rightarrow$	0.25	0	0.25	0.5
$i = 1$	$\rightarrow$	0	0.25	0	0.25
$p_Y(j)$	$\rightarrow$	0.25	0.5	0.25	

Are X and Y uncorrelated? Are they independent?

Solution
We have

$$\text{COV}(X, Y) = E[XY] - E[X]E[Y] = \sum_{i=-1}^{1} \sum_{j=-1}^{1} ij\, p_{X,Y}(i,j)$$

$$- \left(\sum_{i=-1}^{1} i\, p_X(i) \right) \left(\sum_{j=-1}^{1} j\, p_Y(j) \right) = 0$$

Therefore, X and Y are uncorrelated. The random variables X and Y are independent if we have $p_X(i)\, p_Y(j) = p_{X,Y}(i,j)$ for all values of i and j. However, we have $p_X(1)\, p_Y(1) \neq p_{X,Y}(1,1)$, we can thus conclude that the random variables X and Y are not statistically independent. ∎

Example 6.11
Suppose we select one point at random from within a unit circle, where the center of the circle is the origin of the x–y plane and the coordinates of the point chosen are defined by the random variables X and Y. We know that (X, Y) is a pair of uniform random variables with joint pdf given by

$$f_{X,Y}(x,y) = \begin{cases} \dfrac{1}{\pi} & x^2 + y^2 \leq 1 \\ 0 & x^2 + y^2 > 1 \end{cases}$$

Are they independent? Are X and Y uncorrelated?

Solution

Using trigonometric substitutions, the marginal pdfs can be found as follows:

$$f_X(x) = \int_{-\sqrt{1-x^2}}^{\sqrt{1-x^2}} \frac{1}{\pi} \, dy = \frac{2}{\pi}\sqrt{1-x^2}, \quad x^2 \le 1$$

$$\rightarrow f_X(x) = \begin{cases} \frac{2}{\pi}\sqrt{1-x^2} & |x| \le 1 \\ 0 & |x| > 1 \end{cases}$$

and

$$f_Y(y) = \int_{-\sqrt{1-y^2}}^{\sqrt{1-y^2}} \frac{1}{\pi} \, dx = \frac{2}{\pi}\sqrt{1-y^2}, \quad y^2 \le 1$$

$$\rightarrow f_y(y) = \begin{cases} \frac{2}{\pi}\sqrt{1-y^2} & |y| \le 1 \\ 0 & |y| > 1 \end{cases}$$

Since we have

$$f_X(x) \times f_Y(y) \ne f_{X,Y}(x,y)$$

the random variables X and Y are not independent. Using trigonometric substitutions, the means of the random variables X, Y, and XY are as follows:

$$E[X] = \frac{2}{\pi} \int_{-1}^{1} x\sqrt{1-x^2} \, dx = 0$$

$$E[Y] = \frac{2}{\pi} \int_{-1}^{1} y\sqrt{1-y^2} \, dy = 0$$

and

$$E[XY] = \frac{1}{\pi} \iint_{x^2+y^2 \le 1} xy \, dx = 0$$

Since we have $E[XY] = E[X]E[Y]$, the random variables X and Y are uncorrelated. ∎

It is imperative to highlight the fact that correlation between two random variables does not indicate a causal relationship between them. In other words, we cannot conclude that if two random quantities are correlated, then certainly one causes the other. For instance, when there is an increase in drowning deaths, there is an increase in ice cream sales. This clearly indicates there is a correlation between them. However, neither of them causes the other one, as the correlation is solely due to the fact that it is summer with many hot and sunny days. This indicates there is a correlation between them, but clearly no causation. There is a difference between **causation** (cause and effect) on the one hand and **correlation** (dependence and association) on the other hand. Note that no causal or physical relationship need to exist for correlation, and correlation does not necessarily imply a direct causal interdependence of the two random variables.

It is worth highlighting the fact that the existence of correlation between a group of particular people on the one hand and the occurrences of certain events on the other is no proof in any way to conclude causation, and therefore must never lead to stereotypes, generalizations, and eventually discrimination.

6.8 Conditional Distributions

In many applications, the random variables of interest are not independent. In such situations, we may need to find the conditional distribution of one variable, given that the other variable takes on a particular value, as having a value for one of the two random variables provides partial information about the other.

Assuming X and Y are two dependent continuous random variables, the **conditional pdf** of Y given $X = x$, and the conditional pdf of X given $Y = y$ are defined as follows:

$$
\begin{cases}
f_{Y|X}(y \mid x) = \dfrac{f_{X,Y}(x,y)}{f_X(x)} & f_X(x) > 0 \\[2ex]
f_{X|Y}(x \mid y) = \dfrac{f_{X,Y}(x,y)}{f_Y(y)} & f_Y(y) > 0
\end{cases}
\rightarrow
$$

$$
f_{X,Y}(x,y) = f_{Y|X}(y \mid x)f_X(x) = f_{X|Y}(x \mid y)f_Y(y) \tag{6.21}
$$

Assuming X and Y are two dependent discrete random variables, the **conditional pmf** of Y given $X = x$, and the conditional pmf of X given $Y = y$ are defined as follows:

$$
\begin{cases}
p_{Y|X}(y \mid x) = \dfrac{p(x,y)}{p_X(x)} & p_X(x) > 0 \\[2ex]
p_{X|Y}(x \mid y) = \dfrac{p_{X,Y}(x,y)}{p_Y(y)} & p_Y(y) > 0
\end{cases}
\rightarrow
$$

$$
p_{X,Y}(x,y) = p_{Y|X}(y \mid x)p_X(x) = p_{X|Y}(x \mid y)p_Y(y) \tag{6.22}
$$

Note that if the random variables X and Y are statistically independent, then their conditional cdfs (equivalently pmfs or pdfs) are equal to their corresponding marginal cdfs (equivalently pmfs or pdfs).

Example 6.12

Suppose we have

$$f_{X,Y}(x,y) = \begin{cases} \dfrac{1}{2} & 0 \le x \le y \text{ and } 0 \le y \le 2 \\ 0 & \text{elsewhere} \end{cases}$$

Determine the conditional pdf of X, given $Y = y$, and evaluate the probability that $X < \frac{1}{2}$, given $Y = 1$.

Solution

We first need to find the marginal pdf of Y:

$$f_Y(y) = \int_{-\infty}^{\infty} f_{X,Y}(x,y)dx = \int_0^y \left(\frac{1}{2}\right) dx = \frac{y}{2}$$

The conditional pdf of interest is thus as follows:

$$f_{X|Y}(x \mid y) = \frac{f_{X,Y}(x,y)}{f_y(y)} = \frac{\dfrac{1}{2}}{\dfrac{y}{2}} = \frac{1}{y} \quad 0 \le y \le 2$$

We can now determine the conditional probability of interest:

$$P\left(X \le \frac{1}{2} \mid Y = 1\right) = \int_0^{1/2} f_{X|Y}(x \mid y = 1)dx = \int_0^{1/2} \left(\frac{1}{1}\right) dx = \frac{1}{2} \qquad \blacksquare$$

6.8.1 Conditional Expectations

The *conditional expected value of a function* $g(X, Y)$ given a random variable for continuous random variables and discrete random variables are defined as follows:

$$\begin{cases} E[g(X,Y) \mid Y = y] = \displaystyle\int_{-\infty}^{\infty} g(x,y)f_{X|Y}(x \mid y)\,dx \\[2mm] E[g(X,Y) \mid X = x] = \displaystyle\int_{-\infty}^{\infty} g(x,y)f_{Y|X}(y \mid x)\,dy \end{cases} \quad \text{Continuous}$$

$$\begin{cases} E[g(X,Y) \mid Y = y] = \displaystyle\sum_{x \in S_X} g(x,y)p_{X|Y}(x \mid y) \quad y \in S_Y \\[2mm] E[g(X,Y) \mid X = x] = \displaystyle\sum_{y \in S_Y} g(x,y)p_{Y|X}(y \mid x) \quad x \in S_X \end{cases} \quad \text{Discrete} \quad (6.23)$$

Therefore, the conditional expectation of X given y, which is a function of y and hence a random variable, and the conditional expectation of Y given x, which is a function of x and hence a random variable, can be obtained as follows:

$$
\begin{cases}
E[X \mid Y = y] = \displaystyle\int_{-\infty}^{\infty} x f_{X|Y}(x \mid y)\, dx \\[2mm]
\qquad\qquad\qquad\qquad\qquad\qquad \text{Continuous} \\[2mm]
E[Y \mid X = x] = \displaystyle\int_{-\infty}^{\infty} y f_{Y|X}(y \mid x)\, dy
\end{cases}
$$

$$
\begin{cases}
E[X \mid Y = y] = \displaystyle\sum_{x \in S_X} x\, p_{X|Y}(x \mid y) \quad y \in S_Y \\[2mm]
\qquad\qquad\qquad\qquad\qquad\qquad \text{Discrete} \qquad (6.24) \\[2mm]
E[Y \mid X = x] = \displaystyle\sum_{y \in S_Y} y\, p_{Y|X}(y \mid x) \quad x \in S_X
\end{cases}
$$

In general, the conditional expected value $E[X \mid Y = y]$ is a function of y only and the conditional expected value $E[Y \mid X = x]$ is a function of x only. However, when X and Y are independent, the observation $Y = y$ provides no information about X, nor does the observation $X = x$ provide information about Y. Therefore, when X and Y are independent, the conditional expected values are the same as the unconditional expected values.

Example 6.13
The joint pdf of a pair of random variables (X, Y) is given by

$$
f_{X,Y}(x, y) = \begin{cases} 2 & 0 < y \le x < 1 \\ 0 & \text{elsewhere} \end{cases}
$$

Determine the marginal pdfs, the means of the two random variables, the conditional pdfs, the conditional means, and the expected values of the conditional means.

Solution
The marginal pdfs of the random variables as well as their means are as follows:

$$
f_X(x) = \int_0^x 2\, dy = 2x \quad 0 < x < 1 \to E[X] = \int_0^1 2x^2\, dx = \frac{2}{3}
$$

and

$$
f_Y(y) = \int_y^1 2\, dx = 2(1 - y) \quad 0 < y < 1 \to E[Y] = \int_0^1 2y(1 - y)\, dy = \frac{1}{3}
$$

The conditional pdfs are as follows:

$$
f_{Y|X}(y \mid x) = \frac{f_{X,Y}(x, y)}{f_X(x)} = \frac{2}{2x} = \frac{1}{x} \quad 0 < y \le x < 1
$$

and

$$f_{X|Y}(x \mid y) = \frac{f_{X,Y}(x,y)}{f_Y(y)} = \frac{2}{2(1-y)} = \frac{1}{1-y} \quad 0 < y \le x < 1$$

The conditional means are as follows:

$$E[Y \mid X = x] = \int_{-\infty}^{\infty} y f_{Y|X}(y \mid x) dy = \int_0^x y \frac{1}{x} dy = \frac{x}{2} \quad 0 < x < 1$$

and

$$E[X \mid Y = y] = \int_{-\infty}^{\infty} x f_{X|Y}(x \mid y) dx = \int_y^1 x \frac{1}{1-y} dx = \frac{1+y}{2} \quad 0 < y < 1$$

The expected values of the conditional means are interestingly as follows:

$$E[E[Y \mid X]] = \int_{-\infty}^{\infty} E[Y \mid X = x] f_X(x) dx = \int_0^1 \frac{x}{2} \times 2x \, dx = \frac{1}{3} \rightarrow$$
$$E[E[Y \mid X]] = E[Y]$$

and

$$E[E[X \mid Y]] = \int_{-\infty}^{\infty} E[X \mid Y = y] f_Y(y) \, dx = \int_0^1 \frac{1+y}{2} \times 2(1-y) \, dy = \frac{2}{3} \rightarrow$$
$$E[E[X \mid Y]] = E[X]$$

∎

6.9 Distributions of Functions of Two Random Variables

It is often required to determine the distribution of a function of two random variables, such as the sum of two random variables or the ratio of two random variables. The objective is thus to determine the distribution of the random variable $W = g(X, Y)$, where we have the joint distribution of the random variables X and Y.

When X and Y are both discrete random variables, the range of W is a countable set corresponding to all possible values of $g(X, Y)$. For the discrete random variable W, the pmf $p_W(w)$ can be obtained by adding the values of $p_{X,Y}(x, y)$ corresponding to the (x, y) pairs for which $g(x, y) = w$. In other words, we have

$$p_W(w) = \sum_{(x,y) \in g(x,y)=w} p_{X,Y}(x,y) \tag{6.25}$$

Example 6.14
A person sends 1, 2, or 3 text messages an hour, and when she sends it, a text message has 40, 80, or 120 characters. The following table provides the corresponding probabilities for all nine possible cases:

		1 Message	2 Messages	3 Messages
40 Characters	→	0.05	0.10	0.05
80 Characters	→	0.10	0.15	0.15
120 Characters	→	0.15	0.20	0.05

Let the random variable X represent the number of messages sent and the random variable Y represent the length of a message, measured in characters. Let $W = g(X, Y) = XY$ be the random variable representing the number of characters sent by her during an hour, determine the pmf $p_W(w)$.

Solution
The following table provides the range of W:

		$X = 1$	$X = 2$	$X = 3$
$Y = 40$	→	40	80	120
$Y = 80$	→	80	160	240
$Y = 120$	→	120	240	360

The range of W is thus as follows: {40, 80, 120, 160, 240, 360}. Using the table providing the probabilities, the following shows the pmf $p_W(w)$:

w	$p_W(w)$
40	0.05
80	0.20 (=0.10 + 0.10)
120	0.20 (=0.15 + 0.05)
160	0.15
240	0.35 (=0.15 + 0.20)
360	0.05

∎

When X and Y are both continuous random variables, the range of W is an uncountable set corresponding to all possible values of $g(X, Y)$. For the continuous random variable W, the pdf $f_W(w)$ can be obtained by first deriving the cdf $F_W(w)$ and then differentiating it with respect to w, as reflected below:

$$F_W(w) = P(W \le w) = \iint_{g(x,y) \le w} f_{X,Y}(x, y)dx\, dy \to f_W(w) = \frac{dF_W(w)}{dw}$$

$$(6.26)$$

Note that for most functions, performing the integration to find the cdf can be a complex process.

Example 6.15

Determine the pdf of the sum of two continuous random variables X and Y whose joint pdf is $f_{X,Y}(x, y)$.

Solution

Noting $W = X + Y$, we first determine the cdf of W:

$$F_W(w) = P(X + Y \leq w) = \begin{cases} \int_{-\infty}^{\infty} \left(\int_{-\infty}^{w-x} f_{X,Y}(x, y) dy \right) dx \\ \text{or} \\ \int_{-\infty}^{\infty} \left(\int_{-\infty}^{w-y} f_{X,Y}(x, y) dx \right) dy \end{cases}$$

Taking the derivative of the cdf to find the pdf, we have

$$f_W(w) = \frac{dF_W(w)}{dw} = \begin{cases} \int_{-\infty}^{\infty} \left(\frac{d}{dw} \left(\int_{-\infty}^{w-x} f_{X,Y}(x, y) dy \right) \right) dx \\ = \int_{-\infty}^{\infty} f_{X,Y}(x, w - x) \, dx \\ \text{or} \\ \int_{-\infty}^{\infty} \left(\frac{d}{dw} \left(\int_{-\infty}^{w-y} f_{X,Y}(x, y) dx \right) \right) dy \\ = \int_{-\infty}^{\infty} f_{X,Y}(w - y, y) \, dy \end{cases}$$

When X and Y are independent random variables, the pdf of the sum is thus as follows:

$$f_W(w) = \begin{cases} \int_{-\infty}^{\infty} f_X(w - y) f_Y(y) dy \\ \text{or} \\ \int_{-\infty}^{\infty} f_Y(w - x) f_X(x) dx \end{cases}$$

Note that the above operation is known as convolution, denoted by the symbol *, that is we have

$$f_W(w) = f_X(x) * f_Y(y)$$

∎

6.9.1 Joint Distribution of Two Functions of Two Random Variables

It is sometimes required to determine the joint distribution of two functions of two random variables. The objective is thus to determine the joint distribution of the random variable $Z = g(X, Y)$ and $W = h(X, Y)$, where we have the joint distribution of the continuous random variables X and Y. Assuming these functions are one-to-one, we have the following inverse functions:

$$\begin{cases} Z = g(X, Y) \\ W = h(X, Y) \end{cases} \rightarrow \begin{cases} X = k(Z, W) \\ Y = l(Z, W) \end{cases} \tag{6.27}$$

We also define the following determinants, respectively known as the *Jacobian of the transformation* and the *Jacobian of the inverse transformation*:

$$J(x, y) = \begin{vmatrix} \dfrac{\partial z}{\partial x} & \dfrac{\partial z}{\partial y} \\ \dfrac{\partial w}{\partial x} & \dfrac{\partial w}{\partial y} \end{vmatrix}$$

$$J(z, w) = \begin{vmatrix} \dfrac{\partial x}{\partial z} & \dfrac{\partial x}{\partial w} \\ \dfrac{\partial y}{\partial z} & \dfrac{\partial y}{\partial w} \end{vmatrix} \tag{6.28}$$

where $|J(x, y)|$ and $|J(z, w)|$ represent the absolute values of the determinants $J(x, y)$ and $J(z, w)$ respectively, and $|J(x, y)| \times |J(z, w)| = 1$. It can then be shown that the joint pdf of Z and W, in terms of the joint pdf of X and Y, is as follows:

$$f_{Z,W}(z, w) = |J(z, w)| f_{X,Y}(k(z, w), l(z, w)) = \frac{1}{|J(x, y)|} \times f_{X,Y}(k(z, w), l(z, w)) \tag{6.29}$$

Example 6.16
Assuming X and Y are two independent random variables, each uniformly distributed over the interval $[0, 1]$, determine the pdf of $Z = XY$.

Solution
We first define the arbitrary function $W = X$. With the functions $Z = XY$ and $W = X$, and the inverse functions $X = W$ and $Y = \frac{Z}{W}$, we find the joint pdf of Z and W as follows:

$$f_{Z,W}(z, w) = \begin{Vmatrix} \dfrac{\partial x}{\partial z} & \dfrac{\partial x}{\partial w} \\ \dfrac{\partial y}{\partial z} & \dfrac{\partial y}{\partial w} \end{Vmatrix} f_{X,Y}\left(w, \frac{z}{w}\right) = \begin{Vmatrix} \dfrac{\partial w}{\partial z} & \dfrac{\partial w}{\partial w} \\ \partial \dfrac{z}{w} & \partial \dfrac{z}{w} \\ \dfrac{\partial}{\partial z} & \dfrac{\partial}{\partial w} \end{Vmatrix} f_{X,Y}\left(w, \frac{z}{w}\right)$$

$$= \begin{Vmatrix} 0 & 1 \\ \dfrac{1}{w} & -\dfrac{z}{w^2} \end{Vmatrix} f_{X,Y}\left(w, \frac{z}{w}\right) = \left|-\frac{1}{w}\right| f_{X,Y}\left(w, \frac{z}{w}\right) = \frac{1}{w} f_{X,Y}\left(w, \frac{z}{w}\right)$$

The marginal pdf of Z is thus as follows:

$$f_Z(z) = \int_{-\infty}^{\infty} \frac{1}{w} f_{X,Y}\left(w, \frac{z}{w}\right) dw$$

Since X and Y are independent, we have

$$f_Z(z) = \int_{-\infty}^{\infty} \frac{1}{w} f_X(w) f_Y\left(\frac{z}{w}\right) dw$$

As X and Y are uniformly distributed over the interval $[0, 1]$, we have

$$f_X(x) = \begin{cases} 1 & 0 < x < 1 \\ 0 & \text{elsewhere} \end{cases} \rightarrow f_X(w) = \begin{cases} 1 & 0 < w < 1 \\ 0 & \text{elsewhere} \end{cases}$$

and

$$f_Y(y) = \begin{cases} 1 & 0 < y < 1 \\ 0 & \text{elsewhere} \end{cases} \rightarrow f_Y\left(\frac{z}{w}\right) = \begin{cases} 1 & 0 < z < w \\ 0 & \text{elsewhere} \end{cases}$$

From $f_X(w)$ and $f_Y\left(\frac{z}{w}\right)$, we have $0 < z < w < 1$. We thus obtain the pdf of Z as follows:

$$f_Z(z) = \int_{-\infty}^{\infty} \frac{1}{w} f_X(w) f_Y\left(\frac{z}{w}\right) dw = \int_z^1 \frac{1}{w} \times 1 \times 1 \, dw = -\ln z \quad 0 < z < 1$$

∎

6.10 Random Vectors

We now proceed to briefly discuss when there are more than two random variables. The probability model of an experiment that produces n random variables, $X_1, X_2, \ldots, X_n$, can be represented as an n-dimensional cdf, where n is a nonnegative integer. Noting that the superscript T denotes the transpose operation, let the boldface uppercase letter $X = [X_1, X_2, \ldots, X_n]^T$ be an n-dimensional ***random vector*** and the boldface lowercase letter $x = [x_1, x_2, \ldots, x_n]^T$ be an n-dimensional ***sample value of a random vector***, i.e. representing n sample values (observations). The joint cdf of $X_1, X_2, \ldots, X_n$ or the ***cdf of the jointly random vector*** X is as follows:

$$F_X(x) = F_{X_1, \ldots, X_n}(x_1, \ldots, x_n) = P(X_1 \leq x_1, \ldots, X_n \leq x_n)$$
$$-\infty < x_1, \ldots, x_n < \infty \tag{6.30}$$

Suppose the random variables $X_1, X_2, \ldots, X_n$ are all continuous, such as the Gaussian random vector which is discussed in Chapter 7. For a random vector, the marginal cdfs are obtained by setting the appropriate entries to ∞ in the joint cdf.

The **pdf of the jointly random vector** X is then defined as the nth partial derivative of the cdf with respect to all n variables, if the derivative exists. We thus have

$$f_X(x) = f_{X_1,X_2,...,X_n}(x_1, x_2, ..., x_n) = \frac{\partial^n F_{X_1,X_2,...,X_n}(x_1, x_2, ..., x_n)}{\partial x_1 \partial x_2 ... \partial x_n} \quad (6.31)$$

For a random vector consisting of continuous random variables, the marginal pdfs are obtained by integrating the other variables out. Suppose all of the random variables are discrete, such as the multinomial distribution which will be discussed later in this section. The **pmf of the jointly random vector** X is then defined as follows:

$$p_X(x) = p_{X_1,X_2,...,X_n}(x_1, x_2, ..., x_n) = P(X_1 = x_1, X_2 = x_2, ..., X_n = x_n) \quad (6.32)$$

For a random vector consisting of discrete random variables, marginal pmfs are obtained by adding the joint pmfs over all random variables other than those interested. Moreover, the cdf, pdf, and pmf of the jointly random vector have properties that are generalization of the three axioms of probability.

Example 6.17
Let X, Y, and Z be three jointly continuous random variables whose joint pdf is as follows:

$$f_{X,Y,Z}(x, y, z) = \begin{cases} \frac{2}{3}(x + y + z) & 0 \le x \le 1, \ 0 \le y \le 1, \ 0 \le z \le 1 \\ 0 & \text{otherwise} \end{cases}$$

Determine the marginal pdf $f_X(x)$ and the joint pdf $f_{X,Y}(x, y)$.

Solution
The marginal pdf of the random variable X is as follows:

$$f_X(x) = \int_0^1 \int_0^1 \frac{2}{3}(x + y + z) dy\, dz = \frac{2}{3}(x + 1)$$

The joint pdf of the random variables X and Y is as follows:

$$f_{X,Y}(x, y) = \int_0^1 \frac{2}{3}(x + y + z) dz = \frac{2}{3}\left(x + y + \frac{1}{2}\right)$$ ∎

If the random variables X_1, X_2, ..., X_n are **independent**, then we have

$$F_{X_1,X_2,...,X_n}(x_1, x_2, ..., x_n) = F_{X_1}(x_1) \times F_{X_2}(x_2) \times \cdots \times F_{X_n}(x_n)$$
$$f_{X_1,X_2,...,X_n}(x_1, x_2, ..., x_n) = f_{X_1}(x_1) \times f_{X_2}(x_2) \times \cdots \times f_{X_n}(x_n) \quad \text{Continuous}$$
$$p_{X_1,X_2,...,X_n}(x_1, x_2, ..., x_n) = p_{X_1}(x_1) \times p_{X_2}(x_2) \times \cdots \times p_{X_n}(x_n) \quad \text{Discrete}$$
$$(6.33)$$

For n statistically independent random variables, any group of them is independent of any other group, and a function of any group is independent of any function of any other group of the random variables.

Random variables $X_1, X_2, \ldots, X_n$ are said to be ***independent and identically distributed*** (iid), if they are independent and have the same marginal distribution. As a result, for iid random variables, we have

$$F_{X_1, X_2, \ldots, X_n}(x_1, x_2, \ldots, x_n) = (F_X(x))^n$$

$$f_{X_1, X_2, \ldots, X_n}(x_1, x_2, \ldots, x_n) = (f_X(x))^n \quad \text{Continuous}$$

$$p_{X_1, X_2, \ldots, X_n}(x_1, x_2, \ldots, x_n) = (p_X(x))^n \quad \text{Discrete} \quad (6.34)$$

The ***expected value (mean) vector*** $\boldsymbol{\mu_X}$ and the ***covariance matrix*** $\boldsymbol{K_X}$, which can provide valuable information about a random vector, are defined as follows:

$$\mu_X = E[X] = [E[X_1], E[X_2], \ldots, E[X_n]]^T$$

$$K_X = \begin{bmatrix} \text{COV}(X_1, X_1) & \cdots & \text{COV}(X_1, X_n) \\ \vdots & \ddots & \vdots \\ \text{COV}(X_n, X_1) & \cdots & \text{COV}(X_n, X_n) \end{bmatrix}$$

$$\text{COV}(X_i, X_j) = E[X_i X_j] - E[X_i]E[X_j] \quad i \text{ and } j = 1, \ldots, n \quad (6.35)$$

Note that K_X is an $n \times n$ symmetric matrix, as we have $\text{COV}(X_i, X_j) = \text{COV}(X_j, X_i)$. The matrix K_X has the variances along the main diagonal and its determinant is non-negative, i.e. $|K_X| > 0$. If the random variables $X_1, X_2, \ldots, X_n$ are independent or even uncorrelated, then the covariance matrix is a diagonal matrix.

Example 6.18

The joint pdf of the random variables X and Y is as follows:

$$f_{X,Y}(x, y) = \begin{cases} \dfrac{2}{3}\left(x + y + \dfrac{1}{2}\right) & 0 \le x \le 1 \text{ and } 0 \le y \le 1 \\ 0 & \text{otherwise} \end{cases}$$

Determine the covariance matrix.

Solution

The marginal pdfs are as follows:

$$f_X(x) = \int_0^1 \frac{2}{3}\left(x + y + \frac{1}{2}\right) dy = \frac{2}{3}(x + 1)$$

and

$$f_Y(y) = \int_0^1 \frac{2}{3}\left(x + y + \frac{1}{2}\right) dx = \frac{2}{3}(y + 1)$$

The means and variances of the random variables X and Y are as follows:

$$E[X] = \frac{5}{9} \rightarrow \sigma_X^2 = \frac{13}{162}$$

and

$$E[Y] = \frac{5}{9} \rightarrow \sigma_Y^2 = \frac{13}{162}$$

The correlation and covariance are as follows:

$$E[XY] = \frac{11}{36} \rightarrow COV(X, Y) = E[XY] - E[X]E[Y] = \frac{11}{36} - \frac{5}{9} \times \frac{5}{9} = -\frac{1}{324}$$

We thus have

$$K_X = \begin{bmatrix} \frac{13}{162} & -\frac{1}{324} \\ -\frac{1}{324} & \frac{13}{162} \end{bmatrix} = \frac{1}{324} \begin{bmatrix} 26 & -1 \\ -1 & 26 \end{bmatrix}$$

∎

It is important to emphasize that all concepts introduced, measures defined, probabilities calculated, and conclusions drawn for one or two random variables can be extended to a random vector, including, but not limited to:

- The joint cdf, pdf, or pmf of a random vector can fully characterize the statistical behavior of one or some or all random variables in the random vector.
- Using the joint cdf, pdf, or pmf of a random vector, all marginal cdfs, pdfs, or pmfs, all joint cdfs, pdfs, or pmfs for subsets of the n random variables, and all conditional cdfs, pdfs, or pmfs conditioned on subsets of the n random variables can be obtained.
- Determining the joint moments and joint central moments involves an n-fold integration for jointly continuous random vector or summation for jointly discrete random vector.
- Transformations of a random vector generates another random vector, and the joint distribution of the new random vector can be determined.

The most widely used continuous random vector is the jointly Gaussian random vector, which will be discussed in the next chapter, but here we briefly discuss the multinomial distribution, a well-known discrete random vector.

The multinomial distribution is a multidimensional generalization of the binomial distribution. The multinomial distribution occurs in counting experiments, including natural language processing, multicomponent system reliability, and card games, to name just few examples. The multinomial distribution, as an example, can model the probability of counts for rolling a k-sided die n times, where $k \geq 2$ and $n \geq 1$ are integers. For instance, when $n = 1$ and $k = 2$, the multinomial distribution becomes the Bernoulli distribution or when $n > 1$ and $k = 2$, the multinomial distribution becomes the binomial distribution.

Consider a sequence of n independent trials where each individual trial can have $k \geq 2$ mutually exclusive outcomes that can occur with constant probabilities $p_1, p_2, \ldots, p_k$, where the probabilities sum to one, i.e. $p_1 + p_2 + \cdots + p_k = 1$. Suppose the discrete random variable X_i counts the number of trials that the outcome $i \in \{1, 2, \ldots, k\}$ is observed over n trials, and has a constant probability p_i. While the trials are independent, their outcomes, i.e. the random variables $X_1, X_2, \ldots, X_k$, are dependent because they must be summed to n, i.e. $x_1 + x_2 + \cdots + x_k = n$.

The random variables $X_1, X_2, \ldots, X_k$ that count the number of occurrences of each outcome are said to have a ***multinomial distribution***, and their joint pmf is as follows:

$$P(X_1 = x_1, \ldots, X_k = x_k) = n! \prod_{i=1}^{k} \frac{(p_i)^{x_i}}{x_i!} \quad x_i \in \{0, 1, \ldots, n\}$$

$$i \in \{1, 2, \ldots, k\} \quad x_1 + \cdots + x_k = n \tag{6.36}$$

The marginal distribution of X_i is the binomial distribution with parameters n and p_i. Noting that the random variables $X_1, X_2, \ldots, X_k$ are not independent, they have expectations, variances, and covariances respectively given by

$$\mu_{X_i} = np_i \quad i \in \{1, 2, \ldots, k\}$$
$$\sigma_{X_i}^2 = np_i(1 - p_i) \quad i \in \{1, 2, \ldots, k\}$$
$$\mathrm{COV}(X_i, X_j) = -np_i p_j \quad i \in \{1, 2, \ldots, k\}, \ j \in \{1, 2, \ldots, k\} \text{ and } i \neq j$$
$$\tag{6.37}$$

Note that in the covariance matrix, each diagonal entry is the variance of a binomially distributed random variable, and the off-diagonal entries are the covariances. The covariance is a measure of how one variable varies with changes to the other variable. All covariances are thus negative because for fixed n, an increase in one component of a multinomial vector requires a decrease in another component.

Example 6.19

Consider a random experiment that constitutes rolling a fair six-sided cube-shaped die and coming to rest on a flat surface, where the face of the die that is uppermost yields the outcome. Suppose the die is rolled $n = 6m$ times, where m is a positive integer, and the random variables $X_1, X_2, \ldots, X_6$ count the number of occurrences of outcomes 1, 2, ..., 6, respectively. We now define event A where the occurrence of each of the six outcomes 1, 2, ..., 6 is m. Determine the probability of event A, i.e. $P(A)$. Comment on the results.

Solution

Using the multinomial distribution, and noting that we have $n = 6m$, $k = 6$, $p_1 = p_2 = p_3 = p_4 = p_5 = p_6 = \frac{1}{6}$ and $x_1 = x_2 = x_3 = x_4 = x_5 = x_6 = m$, the probability of interest is then as follows:

$$P(A) = P(X_1 = m, X_2 = m, \ldots, X_6 = m) = (6m)! \prod_{i=1}^{6} \frac{\left(\frac{1}{6}\right)^m}{m!} = \frac{(6m)!}{(m!6^m)^6}$$

n	m	P(A)
6	1	0.015 432
12	2	0.003 438
18	3	0.001 351
24	4	0.000 685
30	5	0.000 402

As n increases, m increases, thus the corresponding number of possibilities increases; hence, the probability of interest diminishes. ∎

Example 6.20

A particular new drug is being experimented by a pharmaceutical company on a group of 20 independent (i.e. unrelated) patients, who have a certain disease. Reactions to the drug, which are mutually exclusive, can occur. Drug allergy symptoms include 30% skin rash, 25% fever, 20% swelling, and 15% nausea, while with 10% probability, there may be no allergy. Determine the probability that 50% of patients have no reactions, 20% experience nausea, 15% swelling, 10% fever, and 5% skin rash.

Solution

Suppose X_1, X_2, X_3, X_4, and X_5 are the random variables exhibiting the skin rash, fever, swelling, nausea, and no allergy, respectively. Noting that $X_1 = 20 \times 5\% = 1, X_2 = 20 \times 10\% = 2, X_3 = 20 \times 15\% = 3, X_4 = 20 \times 20\% = 4$, and $X_5 = 20 \times 50\% = 10$ and using the multinomial distribution with $n = 20$ and $k = 5$, the joint pmf is thus as follows:

$$P(X_1 = 1, X_2 = 2, X_3 = 3, X_4 = 4, X_5 = 10)$$
$$= \frac{20!}{1!2!3!4!10!}(0.3)^1(0.25)^2(0.2)^3(0.15)^4(0.1)^{10} \cong 17.7 \times 10^{-9}$$

∎

6.11 Summary

In this chapter, we first introduced the joint statistical behavior of multiple random variables, and then the focus turned toward the marginal and conditional cdfs, pdfs, and pmfs. In the context of a pair of random variables, the important concepts of independence and correlation were discussed. After highlighting conditional distributions and distributions of functions of two random variables, random vectors were briefly discussed.

Problems

6.1 Let the random variable Θ be uniformly distributed in the interval $[0, 2\pi]$. Suppose we have $X = \cos(\Theta)$ and $Y = \sin(\Theta)$. Are the random variables X and Y independent? Are they uncorrelated? Are they orthogonal?

6.2 Two friends, who are very busy, are going to have lunch between 1 p.m. and 2 p.m., with the understanding that each will wait no longer than $0 \le z \le 1$ hours for the other, where z is a nonrandom constant. What is the probability that they will have lunch together?

6.3 The two random variables X and Y are independent, where each has a uniform probability density function in the interval $[0, 1]$. Determine the probability that the sum of their squares is no greater than one.

6.4 The joint pdf of the random variables X and Y is as follows:

$$f_{X,Y}(x,y) = \begin{cases} \dfrac{1}{(x_2 - x_1)(y_2 - y_1)} & x_1 < x < x_2, \ y_1 < y < y_2 \\ 0 & \text{elsewhere} \end{cases}$$

Are X and Y independent?

6.5 Suppose the random variables X and Y, whose means are μ_X and μ_Y and standard deviations are σ_X and σ_Y, respectively, are correlated with the correlation coefficient $\rho_{X,Y}$. The random variables W and Z are related to the random variables X and Y by the following coordinate rotation:

$$\begin{cases} W = X \cos c + Y \sin c \\ Z = -X \sin c + Y \cos c \end{cases}$$

Determine the value of the nonrandom constant c such that the random variables W and Z are uncorrelated.

6.6 Suppose the pdf of the random variable X is an even function and we also have the random variable $Y = X^4$. Are X and Y correlated? Are X and Y statistically independent?

6.7 Let X and Y be a pair of two random variables. Show that $(E[XY])^2 \leq E[X^2]E[Y^2]$. Note that this is known as the Cauchy–Schwarz inequality.

6.8 The joint cdf of a pair of random variables (X, Y) is as follows:

$$F_{X,Y}(X, Y) = \begin{cases} (1 - e^{-x})(1 - e^{-y}) & x \geq 0, \ y \geq 0 \\ 0 & \text{elsewhere} \end{cases}$$

Find the marginal cdfs of X and Y. Find $P(X \geq x, Y \geq y)$.

6.9 The joint pdf of a pair of random variables (X, Y) is as follows:

$$f_{X,Y}(x, y) = \begin{cases} 0.125 (x + y) & 0 < x \leq 2, \ 0 < y \leq 2 \\ 0 & \text{elsewhere} \end{cases}$$

Determine the conditional pdfs $f_{Y|X}(y \mid x)$ and $f_{X|Y}(x \mid y)$ as well as the conditional probability $P(0 < Y \leq 0.5 \mid X = 1)$.

6.10 Consider the joint pdf of a pair of random variables (X, Y) given by

$$f_{X,Y}(x, y) = \begin{cases} x + y & 0 < x \leq 1, \ 0 < y \leq 1 \\ 0 & \text{elsewhere} \end{cases}$$

Are the random variables X and Y independent?

6.11 Consider the joint pdf of a pair of random variables (X, Y) given by

$$f_{X,Y}(x, y) = \begin{cases} 1 & 0 < x \leq 1, \ 0 < y \leq 1 \\ 0 & \text{elsewhere} \end{cases}$$

Are the random variables X and Y independent?

6.12 Let X and Y be independent random variables, whose pdfs are as follows:

$$f_X(x) = \begin{cases} 1 & 0 < x \leq 1 \\ 0 & \text{elsewhere} \end{cases}$$

$$f_Y(y) = \begin{cases} e^{-y} & y \geq 0 \\ 0 & \text{elsewhere} \end{cases}$$

Assuming $Z = X + Y$, determine the pdf of the random variable Z.

6.13 Let Φ be a uniformly distributed random variable in the interval $\left[0, \frac{\pi}{2}\right]$, and let the random variables X and Y be defined by $X = \cos \Phi$ and $Y = \sin \Phi$. Are X and Y uncorrelated?

6.14 The joint pdf of the random variables X and Y is as follows:

$$f_{X,Y}(x,y) = \begin{cases} 2\exp(-x-y) & x \geq y \geq 0 \\ 0 & \text{otherwise} \end{cases}$$

Are X and Y independent?

6.15 Let Z be a uniform random variable in the interval $[-1, 1]$. Assuming $X = Z$ and $Y = Z^2$, are X and Y correlated?

6.16 Let Y be the uniform random variable over the interval $[0, 1]$. Given $Y = y$, X is the uniform random variable over the interval $[0, y]$. Determine the conditional pdf of Y given X.

6.17 The joint pdf of the random variables X and Y is given by

$$f_{X,Y}(x,y) = \frac{1}{y}\exp\left(-\left(\frac{x+y^2}{y}\right)\right)u(x)$$

where $u(x)$ is the unit step function. Determine $P(X > 1 \mid Y = 2)$.

6.18 Suppose the joint pdf of the random variables X, Y, and Z is as follows:

$$f_{X,Y,Z}(x,y,z) = \begin{cases} (abc)\exp(-(ax+by+cz)) & x > 0, \quad y > 0, \quad z > 0 \\ 0 & \text{otherwise} \end{cases}$$

where $a > 0$, $b > 0$, and $c > 0$. Are X, Y, and Z independent?

6.19 Let X and Y be two random variables and $g(X)$ and $h(Y)$ be two functions. Prove the following:

$$E[g(X)h(Y) \mid X] = g(X)E[h(Y) \mid X]$$

6.20 Let X be a continuous random variable whose pdf is $f_X(x) = 2x\,(u(x) - u(x-1))$, where $u(x)$ is the unit step function. For $X = x$, the random variable Y is uniformly distributed over the interval $[-x, x]$. Determine the joint pdf of X and Y and the marginal pdf of Y.

7

The Gaussian Distribution

The Gaussian distribution was named after Johann Carl Friedrich Gauss, a German mathematician. The Gaussian distribution, also known as the normal distribution, is by far the most important of all probability distributions, as it can be used to describe a very large variety of random phenomena in a multitude of areas in science, engineering, social sciences, business, and medicine. The Gaussian distribution holds a unique role in probability and statistics and provides a crucial bridge between these two areas of great importance. This chapter discusses various aspects of the Gaussian distribution and explores the impact of the Gaussian distribution on other probability distributions, as they can be constructed on the Gaussian distribution or connected by means of approximations and limits to the Gaussian distribution. In addition, the unique role of the Gaussian distribution on long-term averages describing the outcomes of a large number of repetitions of a single experiment is briefly discussed.

7.1 The Gaussian Random Variable

Since the Gaussian random variable can have any real value, it has a continuous distribution. The **probability density function of the Gaussian random variable** X is defined as follows:

$$f_X(x) = \left(\frac{1}{\sqrt{2\pi\sigma_X^2}} \right) \exp\left(-\frac{(x - \mu_X)^2}{2\sigma_X^2} \right) \quad -\infty < x < \infty \tag{7.1}$$

where the parameter μ_X, which is its mean, can have any finite real value, i.e. $-\infty < \mu_X < \infty$, and the parameter σ_X^2, which is its variance, can have any finite positive value, i.e. $0 < \sigma_X^2 < \infty$. The values of the mean and variance of a Gaussian random variable, which are independent of one another, specify the pdf of a Gaussian random variable. The mean μ_X is often referred to as a *location parameter* and the standard deviation σ_X is often referred to as

Probability, Random Variables, Statistics, and Random Processes: Fundamentals & Applications,
First Edition. Ali Grami.
© 2020 John Wiley & Sons, Inc. Published 2020 by John Wiley & Sons, Inc.
Companion Website: www.wiley.com/go/grami/PRVSRP

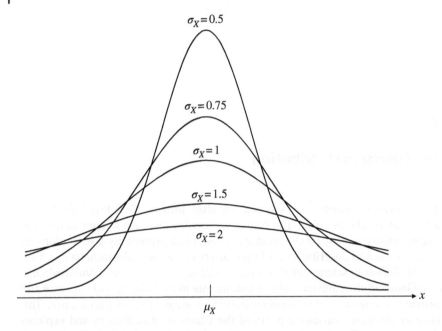

Figure 7.1 pdfs of the Gaussian distribution with mean μ_X and variance σ_X^2.

a **shape parameter**. Note that the notation $X \sim N(\mu_X, \sigma_X^2)$ denotes that the random variable X has a Gaussian distribution with the mean μ_X and variance σ_X^2. The Gaussian random variable is sometimes referred to as being normally distributed.

As shown in Figure 7.1, the Gaussian pdf is a bell-shaped curve symmetric about the mean μ_X. The inflection points of the Gaussian pdf are at $x = \mu_X \pm \sigma_X$, where $\sigma_X > 0$ represents the standard deviation. Thus, distances from the mean μ_X to points of inflection provide a good visual sense of the magnitude of the standard deviation σ_X. The area under the Gaussian pdf between the two inflection points, i.e. $P(\mu_X - \sigma_X \leq X \leq \mu_X + \sigma_X)$, is about 68.3%. The Gaussian pdf is increasing for $x < \mu_X$ and decreasing for $x > \mu_X$. The Gaussian pdf is concave down between the two inflection points. In the limit, when σ_X tends to 0, the Gaussian pdf eventually tends to 0 at any $x \neq \mu_X$, but grows without limit at $x = \mu_X$, while the area under the pdf remains equal to 1.

The maximum value of the Gaussian pdf, also known as mode, occurs at the mean μ_X, and is inversely proportional to σ_X. The median, the value x for which $P(X \leq x) = P(X \geq x) = 0.5$, also occurs at the mean μ_X. If σ_X is large, the bell is wide with a low peak, thus resulting in a long, flat bell-shaped curve, and if σ_X is small, the bell is narrow with a high peak, thus yielding a thin, sharp bell-shaped curve. The Gaussian pdf, which is very smooth, goes to near 0 quite rapidly and approaches 0 as $x \to \pm \infty$. In other words, the Gaussian pdf curve approaches the horizontal axis asymptomatically as x moves in either direction away from the mean.

The **cumulative distribution function of the Gaussian random variable,** which is a continuous increasing function, is obtained as follows:

$$F_X(x) = \int_{-\infty}^{x} \left(\frac{1}{\sqrt{2\pi\sigma_X^2}} \right) \exp\left(-\frac{(t-\mu_X)^2}{2\sigma_X^2} \right) dt \quad -\infty < x < \infty \quad (7.2)$$

The above integration cannot be done analytically; there is thus no closed-form solution for the cdf of a Gaussian distribution. Since a simple mathematical expression cannot be written down for this definite integral, a numerical solution must always be sought.

Example 7.1

Verify the mean and variance of the Gaussian random variable, whose pdf is presented in (7.1), are μ_X and σ_X^2, respectively.

Solution:

The moment generating function of the Gaussian random variable of interest is as follows:

$$M_X(v) = E[e^{vX}] = \int_{-\infty}^{\infty} \left(\frac{1}{\sqrt{2\pi\sigma_X^2}} \right) \exp\left(-\frac{(x-\mu_X)^2}{2\sigma_X^2} \right) \exp(vx)dx$$

By the change of variable $t = \frac{x-\mu_X}{\sigma_X}$, we then have

$$M_X(v) = E[e^{vX}] = \int_{-\infty}^{\infty} \left(\frac{1}{\sqrt{2\pi}} \right) \exp\left(-\frac{t^2}{2} \right) \exp(v\sigma_X t + v\mu_X)dt$$

$$= \exp(v\mu_X) \int_{-\infty}^{\infty} \left(\frac{1}{\sqrt{2\pi}} \right) \exp\left(-\frac{t^2}{2} \right) \exp(v\sigma_X t)dt$$

By combining the exponents and completing the square, i.e. by replacing $-\frac{t^2}{2} + v\sigma_X t$ with $-\frac{(t-v\sigma_X)^2}{2} + \frac{(v\sigma_X)^2}{2}$, we obtain the following:

$$M_X(v) = \exp(v\mu_X) \int_{-\infty}^{\infty} \left(\frac{1}{\sqrt{2\pi}} \right) \exp\left(-\frac{(t-v\sigma_X)^2}{2} + \frac{(v\sigma_X)^2}{2} \right) dt$$

$$= \exp\left(\frac{\sigma_X^2 v^2}{2} + v\mu_X \right) \int_{-\infty}^{\infty} \left(\frac{1}{\sqrt{2\pi}} \right) \exp\left(-\frac{(t-v\sigma_X)^2}{2} \right) dt$$

$$= \exp\left(\frac{\sigma_X^2 v^2}{2} + v\mu_X \right)$$

Note that due to the fact that the last integrand is the pdf of $N(v\sigma_X, 1)$, the area under it over the entire range is 1. We then find the first two derivatives of $M_X(v)$ and subsequently the first two moments:

$$M'_X(v) = (\mu_X + v\sigma_X^2)\exp\left(\frac{\sigma_X^2 v^2}{2} + v\mu_X\right) \rightarrow E[X] = M'_X(0) = \mu_X$$

and

$$M''_X(v) = (\sigma_X^2 + (\mu_X + v\sigma_X^2)^2)\exp\left(\frac{\sigma_X^2 v^2}{2} + v\mu_X\right) \rightarrow$$

$$E[X^2] = M''_X(0) = \sigma_X^2 + \mu_X^2$$

Therefore, we have

$$E[X^2] - (E[X])^2 = \sigma_X^2 + \mu_X^2 - \mu_X^2 = \sigma_X^2 \qquad \blacksquare$$

7.2 The Standard Gaussian Distribution

It is important to be able to calculate the probability of an event for any Gaussian distribution regardless of the values of its mean μ_X and variance σ_X^2. The usual approach to calculating the integral in (7.2) is to reduce a general Gaussian distribution to the special Gaussian distribution, called the standard Gaussian distribution. To this end, we make a change of variable $z = \frac{t-\mu_X}{\sigma_X}$, (7.2), therefore, becomes as follows:

$$F_X(x) = \int_{-\infty}^{\frac{x-\mu_X}{\sigma_X}} \left(\frac{1}{\sqrt{2\pi}}\right)\exp\left(-\frac{z^2}{2}\right)dz \quad -\infty < x < \infty \qquad (7.3)$$

The integrand in (7.3), as reflected by

$$f_Z(z) = \left(\frac{1}{\sqrt{2\pi}}\right)\exp\left(-\frac{z^2}{2}\right) \quad -\infty < z < \infty \qquad (7.4)$$

is known as the zero-mean ($\mu_Z = 0$), unit-variance ($\sigma_Z^2 = 1$) Gaussian pdf, and is widely referred to as the **standard Gaussian distribution**, i.e. $Z \sim N(0, 1)$. The letter Z is generally used to indicate the standard Gaussian variable. To determine (7.2), we can use the **Q-function** or the **Φ-function**, as shown in Figure 7.2, which are defined as follows:

$$Q(x) \triangleq \int_x^\infty \left(\frac{1}{\sqrt{2\pi}}\right)\exp\left(-\frac{z^2}{2}\right)dz \quad -\infty < x < \infty$$

$$\Phi(x) \triangleq \int_{-\infty}^x \left(\frac{1}{\sqrt{2\pi}}\right)\exp\left(-\frac{z^2}{2}\right)dz \quad -\infty < x < \infty \qquad (7.5)$$

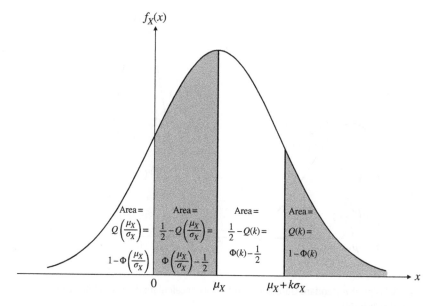

Figure 7.2 Area interpretation of $Q(x)$ and $\Phi(x)$ for the Gaussian distribution $N(\mu_X, \sigma_X^2)$.

Based on the definitions of $Q(x)$ and $\Phi(x)$ and due to symmetry of the standard Gaussian distribution around $x = 0$, we have the following properties:

$$Q(x) + \Phi(x) = 1$$
$$Q(-x) + \Phi(-x) = 1$$
$$\Phi(0) = Q(0) = 0.5$$
$$Q(-x) = 1 - Q(x)$$
$$\Phi(-x) = 1 - \Phi(x) \tag{7.6}$$

Using (7.5), (7.3) can be thus written as follows:

$$F_X(x) = 1 - Q\left(\frac{x - \mu_X}{\sigma_X}\right) = \Phi\left(\frac{x - \mu_X}{\sigma_X}\right) \tag{7.7}$$

Obviously, (7.7) indicates that either $Q(x)$ or $\Phi(x)$ is sufficient to calculate all Gaussian probabilities. If X is a Gaussian random variable whose mean and variance are μ_X and σ_X^2 respectively, then the probability that the random variable X lying in the interval $[a, b]$ is as follows:

$$P(a \leq X \leq b) = F_X(b) - F_X(a) = Q\left(\frac{a - \mu_X}{\sigma_X}\right) - Q\left(\frac{b - \mu_X}{\sigma_X}\right)$$
$$= \Phi\left(\frac{b - \mu_X}{\sigma_X}\right) - \Phi\left(\frac{a - \mu_X}{\sigma_X}\right) \tag{7.8}$$

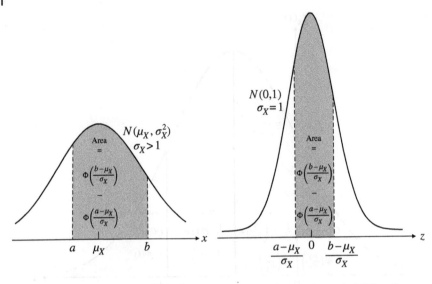

Figure 7.3 Using the standard Gaussian distribution to calculate the probabilities of a Gaussian distribution.

As shown in Figure 7.3, the area under the Gaussian distribution with mean μ_X and variance σ_X^2 between a and b is equal to the area under the standard Gaussian distribution between $\frac{a-\mu_X}{\sigma_X}$ and $\frac{b-\mu_X}{\sigma_X}$. The normalized values $\left(\frac{a-\mu_X}{\sigma_X}\right)$ and $\left(\frac{b-\mu_X}{\sigma_X}\right)$ are known as *z-scores* of a and b, where the z-score of a number represents the difference from that number to the mean, in units of the standard deviation.

A common misconception is that the use of z-scores – by subtracting the mean from the random variable X and then dividing by the standard deviation to introduce the new random variable $Z = \frac{X-\mu_X}{\sigma_X}$ – will automatically create a scale that has a Gaussian distribution, this is only true if X has a Gaussian distribution.

Note that if X has a Gaussian distribution with the mean μ_X and variance σ_X^2 and Z has a Gaussian distribution with zero mean and unit variance, we then have the following:

$$P(\mu_X - c\sigma_X \leq X \leq \mu_X + c\sigma_X) = P(-c \leq Z \leq c) \quad c > 0 \tag{7.9}$$

Table 7.1 presents the values of the Q-function for an array of x values. When the probability of interest is very small, such as the bit error rate of interest in digital transmission systems, it is more practical to compute $Q(x)$. The widely used Q-function can be evaluated by using look-up tables or approximate expressions that provide a high degree of accuracy for $Q(x)$ when x is large,

Table 7.1 $Q(x)$ represents area to the right of x in the standard Gaussian distribution.

x	Q(x)	x	Q(x)	x	Q(x)
0.00	5.000E−01	0.40	3.446E−01	2.4	8.171E−03
0.01	4.960E−01	0.45	3.264E−01	2.5	6.191E−03
0.02	4.920E−01	0.50	3.085E−01	2.6	4.649E−03
0.03	4.880E−01	0.55	2.912E−01	2.7	3.458E−03
0.04	4.841E−01	0.60	2.743E−01	2.8	2.549E−03
0.05	4.801E−01	0.65	2.579E−01	2.9	1.862E−03
0.06	4.761E−01	0.70	2.420E−01	3.0	1.347E−03
0.07	4.721E−01	0.75	2.266E−01	3.1	9.659E−04
0.08	4.681E−01	0.80	2.119E−01	3.2	6.860E−04
0.09	4.641E−01	0.85	1.977E−01	3.3	4.827E−04
0.10	4.602E−01	0.90	1.841E−01	3.4	3.365E−04
0.11	4.562E−01	0.95	1.711E−01	3.5	2.323E−04
0.12	4.522E−01	1.00	1.587E−01	3.6	1.589E−04
0.13	4.483E−01	1.05	1.469E−01	3.7	1.077E−04
0.14	4.443E−01	1.10	1.357E−01	3.8	7.227E−05
0.15	4.404E−01	1.15	1.251E−01	3.9	4.805E−05
0.16	4.364E−01	1.20	1.151E−01	4.0	3.164E−05
0.17	4.325E−01	1.25	1.057E−01	4.1	2.064E−05
0.18	4.286E−01	1.30	9.680E−02	4.2	1.334E−05
0.19	4.247E−01	1.35	8.851E−02	4.3	8.534E−06
0.20	4.207E−01	1.40	8.076E−02	4.4	5.409E−06
0.21	4.168E−01	1.45	7.353E−02	4.5	3.396E−06
0.22	4.129E−01	1.50	6.681E−02	4.6	2.111E−06
0.23	4.091E−01	1.55	6.057E−02	4.7	1.300E−06
0.24	4.052E−01	1.60	5.480E−02	4.8	7.929E−07
0.25	4.013E−01	1.65	4.947E−02	4.9	4.790E−07
0.26	3.974E−01	1.70	4.457E−02	5.0	2.865E−07
0.27	3.936E−01	1.75	4.006E−02	5.1	1.698E−07
0.28	3.897E−01	1.80	3.593E−02	5.2	9.961E−08
0.29	3.895E−01	1.85	3.216E−02	5.3	5.788E−08
0.30	3.821E−01	1.90	2.820E−02	5.4	3.331E−08
0.31	3.783E−01	1.95	2.559E−02	5.5	1.898E−08
0.32	3.745E−01	2.00	2.275E−02	5.6	1.071E−08
0.33	3.707E−01	2.05	2.009E−02	5.7	5.989E−09
0.34	3.669E−01	2.10	1.779E−02	5.8	3.315E−09
0.35	3.632E−01	2.15	1.571E−02	5.9	1.817E−09
0.36	3.594E−01	2.20	1.385E−02	6.0	9.863E−10
0.37	3.557E−01	2.25	1.218E−02	6.1	5.302E−10
0.38	3.520E−01	2.30	1.069E−02	6.2	2.823E−10
0.39	3.483E−01	2.35	9.355E−03	6.3	1.488E−10

such as the following:

$$Q(x) \approx \left(\frac{\pi}{(\pi - 1)x + \sqrt{(x^2 + 2\pi)}} \right) \left(\frac{1}{\sqrt{2\pi}} \exp\left(-\frac{x^2}{2} \right) \right) \qquad (7.10)$$

In fact, if we have $x \geq 5$, then the above approximation and $Q(x)$ both give almost the same value.

Example 7.2
In a digital transmission system, the noise level is modeled as a Gaussian random variable with $\mu_X = 0$ and $\sigma_X^2 = 0.0001$. Using the Q-function, determine $P(X > 0.01)$ and $P\left(-\frac{0.04}{3} \leq X \leq \frac{0.05}{3} \right)$.

Solution:
We have

$$P(X > 0.01) = Q\left(\frac{0.01 - 0}{\sqrt{0.0001}} \right) = Q(1) \cong 0.159$$

and

$$P\left(-\frac{0.04}{3} \leq X \leq \frac{0.05}{3} \right) = P\left(-\frac{0.04}{3} \leq X \right) - P\left(\frac{0.05}{3} \leq X \right)$$

$$= Q\left(\frac{-\frac{0.04}{3} - 0}{0.01} \right) - Q\left(\frac{\frac{0.05}{3} - 0}{0.01} \right)$$

$$= Q\left(-\frac{4}{3} \right) - Q\left(\frac{5}{3} \right)$$

$$= 1 - Q\left(\frac{4}{3} \right) - Q\left(\frac{5}{3} \right)$$

$$\cong 0.858.$$

∎

The **percentiles of the standard Gaussian distribution** are used so frequently that they have their own notation. For $\alpha < 0.5$, the $(1 - \alpha) \times 100$th percentile of the distribution is denoted by z_α, so that

$$P(Z \leq z_\alpha) = \Phi(z_\alpha) = 1 - Q(z_\alpha) = 1 - \alpha \qquad (7.11)$$

where $1 - \alpha$ is called the **confidence level**. As shown in Figure 7.4, the symmetry of the standard Gaussian distribution reflects the fact that we have the following:

$$P(|Z| \leq z_{\alpha/2}) = P(-z_{\alpha/2} \leq Z \leq z_{\alpha/2}) = \Phi(z_{\alpha/2}) - \Phi(-z_{\alpha/2})$$

$$= 1 - Q(z_{\alpha/2}) - Q(z_{\alpha/2}) = 1 - \frac{\alpha}{2} - \frac{\alpha}{2} = 1 - \alpha \qquad (7.12)$$

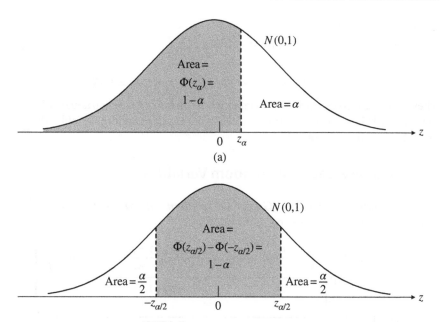

Figure 7.4 The critical points of the standard Gaussian distribution. (a) z_α, (b) $z_{\alpha/2}$.

The percentiles z_α and $z_{\alpha/2}$ are often referred to as the ***critical points*** of the standard Gaussian distribution. The percentiles of a $N(\mu_X, \sigma_X^2)$ distribution are related to the percentiles of the standard Gaussian distribution $N(0, 1)$ through the following relationships:

$$P(X \leq \mu_X + \sigma_X z_\alpha) = P(Z \leq z_\alpha)$$
$$P(\mu_X - \sigma_X z_{\alpha/2} \leq X \leq \mu_X + \sigma_X z_{\alpha/2}) = P(-z_{\alpha/2} \leq Z \leq z_{/2}) \qquad (7.13)$$

Example 7.3
Determine the critical points of z_α and $z_{\alpha/2}$ for the 95th percentile of a Gaussian random variable X whose mean is -5 and variance is 16.

Solution:
With $\mu_X = -5$, $\sigma_X = \sqrt{16} = 4$ and $1 - \alpha = 95\%$, we thus have

$$\alpha = 5\% \rightarrow Q^{-1}(5\%) = z_{0.05} \cong 1.65 \rightarrow$$
$$\mu_X + \sigma_X z_{0.05} = -5 + 4 \times 1.65 = 1.6 \rightarrow X \leq 1.6$$

and

$$\frac{\alpha}{2} = 2.5\% \rightarrow Q^{-1}(2.5\%) = z_{0.025} \cong 1.95 \rightarrow$$

$$\begin{cases} \mu_X + \sigma_X z_{0.025} = 2.8 \\ \mu_X - \sigma_X z_{0.025} = -12.8 \end{cases} \rightarrow -12.8 \leq X \leq 2.8$$

We therefore have

$$P(Z \leq 1.65) = 95\% \rightarrow P(X \leq 1.6) = 95\%$$

and

$$P(-1.95 \leq Z \leq 1.95) = 95\% \rightarrow P(-12.8 \leq X \leq 2.8) = 95\%$$

where Z is a Gaussian random variable whose mean is 0 and variance is 1, i.e. the standard Gaussian distribution, and X is a Gaussian random variable whose mean is -5 and variance is 16. ∎

7.3 Bivariate Gaussian Random Variables

The *joint pdf of bivariate Gaussian random variables* X and Y, for $-\infty < x < \infty$ and $-\infty < y < \infty$, is as follows:

$$f_{X,Y}(x,y) = \frac{\exp\left[-\frac{\left(\frac{x-\mu_X}{\sigma_X}\right)^2 - 2\rho_{X,Y}\left(\frac{x-\mu_X}{\sigma_X}\right)\left(\frac{y-\mu_Y}{\sigma_Y}\right) + \left(\frac{y-\mu_Y}{\sigma_Y}\right)^2}{2(1-\rho_{X,Y}^2)}\right]}{2\pi\sigma_X\sigma_Y\sqrt{1-\rho_{X,Y}^2}}$$

(7.14)

where μ_X and μ_Y are the mean values, and σ_X and σ_Y are the standard deviations of the random variables X and Y, respectively, and $\rho_{X,Y}$ is the correlation coefficient between the random variables X and Y. To determine the probability, we need to find the volume under the function $f_{X,Y}(x, y)$, i.e. to perform double integration, for the particular region of interest in the x–y plane. To this effect, the *joint cdf of bivariate Gaussian random variables* X and Y, for $-\infty < x < \infty$ and $-\infty < y < \infty$, is as follows:

$$F_{X,Y}(x,y) = P(X \leq x, Y \leq y) = \int_{-\infty}^{x} \int_{-\infty}^{y} f_{X,Y}(u,v)dv\,du$$

(7.15)

Obviously, the integration in (7.15) cannot be done analytically; a numerical solution is thus required. The important properties of jointly Gaussian random variables are as follows:

- The marginal pdf of X, i.e. $f_X(x)$, and the marginal pdf of Y, i.e. $f_Y(y)$, both have Gaussian distributions.
- The conditional pdf of X given $Y = y$, i.e. $f_{X|Y}(x\,|\,y)$, and the conditional pdf of Y given $X = x$, i.e. $f_{Y|X}(y\,|\,x)$, both have Gaussian distributions.
- If $\rho_{X,Y} = 0$, then the random variables X and Y are not only uncorrelated, but they are also statistically independent.
- If $\sigma_X = \sigma_Y$ and $\rho_{X,Y} = 0$, then the equal-pdf contour is a circle, otherwise it is an ellipse.

Example 7.4

Suppose X and Y are jointly Gaussian random variables. Show that if they are uncorrelated, i.e. $\rho_{X,Y} = 0$, then they are statistically independent.

Solution:

With $\rho_{X,Y} = 0$, the joint pdf of bivariate Gaussian random variables X and Y simplifies to the following:

$$f_{X,Y}(x,y) = \frac{1}{2\pi\sigma_X\sigma_Y}\exp\left(-\frac{1}{2}\left(\left(\frac{x-\mu_X}{\sigma_X}\right)^2 + \left(\frac{y-\mu_Y}{\sigma_Y}\right)^2\right)\right)$$

We can obtain the marginal pdfs $f_X(x)$ and $f_Y(y)$ from the above joint pdf $f_{X,Y}(x,y)$ as follows:

$$f_X(x) = \int_{-\infty}^{\infty} \frac{1}{2\pi\sigma_X\sigma_Y}\exp\left(-\frac{1}{2}\left(\left(\frac{x-\mu_X}{\sigma_X}\right)^2 + \left(\frac{y-\mu_Y}{\sigma_Y}\right)^2\right)\right)dy$$

$$= \frac{1}{\sqrt{2\pi\sigma_X^2}}\exp\left(-\frac{(x-\mu_X)^2}{2\sigma_X^2}\right)\int_{-\infty}^{\infty}\frac{1}{\sqrt{2\pi\sigma_Y^2}}$$

$$\times\exp\left(-\frac{1}{2}\left(\frac{y-\mu_Y}{\sigma_Y}\right)^2\right)dy$$

$$= \left(\frac{1}{\sqrt{2\pi\sigma_X^2}}\right)\exp\left(-\frac{(x-\mu_X)^2}{2\sigma_X^2}\right)$$

Note that the last integrand is a Gaussian pdf and the area under it over the entire range is thus one. Similarly, we can get the following:

$$f_Y(y) = \left(\frac{1}{\sqrt{2\pi\sigma_Y^2}}\right)\exp\left(-\frac{(y-\mu_Y)^2}{2\sigma_Y^2}\right)$$

Since we have $f_{X,Y}(x,y) = f_X(x)f_Y(y)$, we can conclude when the bivariate Gaussian random variables X and Y are uncorrelated, they are also statistically independent. ∎

Example 7.5

Suppose the random variables X and Y are bivariate Gaussian. Prove that the marginal pdfs $f_X(x)$ and $f_Y(y)$ and the conditional pdfs $f_{X|Y}(x|y)$ and $f_{Y|X}(y|x)$ all have Gaussian distributions.

Solution:

The joint pdf of bivariate Gaussian random variables X and Y can be written as the product of two Gaussian pdfs, one with parameters μ_X and σ_X and the other with parameters $\widetilde{\mu_Y} = \mu_Y + \rho_{X,Y}\left(\frac{\sigma_Y}{\sigma_X}\right)(x - \mu_X)$ and $\widetilde{\sigma_Y} = \sigma_Y\sqrt{1 - \rho_{X,Y}^2}$, we thus have

$$f_{X,Y}(x,y) = \left(\frac{1}{\sqrt{2\pi\sigma_X^2}}\exp\left(-\frac{(x-\mu_X)^2}{2\sigma_X^2}\right)\right)\left(\frac{1}{\sqrt{2\pi\widetilde{\sigma_Y^2}}}\exp\left(-\frac{(y-\widetilde{\mu_Y})^2}{2\widetilde{\sigma_Y^2}}\right)\right)$$

By integrating $f_{X,Y}(x,y)$ over the entire range, we obtain

$$f_X(x) = \int_{-\infty}^{\infty} f_{X,Y}(x,y)dy$$

$$= \int_{-\infty}^{\infty}\left(\frac{1}{\sqrt{2\pi\sigma_X^2}}\exp\left(-\frac{(x-\mu_X)^2}{2\sigma_X^2}\right)\right)$$

$$\times \left(\frac{1}{\sqrt{2\pi\widetilde{\sigma_Y^2}}}\exp\left(-\frac{(y-\widetilde{\mu_Y})^2}{2\widetilde{\sigma_Y^2}}\right)\right)dy$$

$$= \left(\frac{1}{\sqrt{2\pi\sigma_X^2}}\exp\left(-\frac{(x-\mu_X)^2}{2\sigma_X^2}\right)\right)$$

$$\times \int_{-\infty}^{\infty}\frac{1}{\sqrt{2\pi\widetilde{\sigma_Y^2}}}\exp\left(-\frac{(y-\widetilde{\mu_Y})^2}{2\widetilde{\sigma_Y^2}}\right)dy$$

$$= \frac{1}{\sqrt{2\pi\sigma_X^2}}\exp\left(-\frac{(x-\mu_X)^2}{2\sigma_X^2}\right)$$

Note that the last integral equals 1 because it is the integral of a Gaussian pdf over the entire range. We now determine the conditional pdf $f_{Y|X}(y|x)$ as follows:

$$f_{Y|X}(y \mid x) = \frac{f_{X,Y}(x,y)}{f_X(x)}$$

$$= \frac{\left(\frac{1}{\sqrt{2\pi\sigma_X^2}} \exp\left(-\frac{(x-\mu_X)^2}{2\sigma_X^2}\right)\right)\left(\frac{1}{\sqrt{2\pi\widetilde{\sigma_Y^2}}} \exp\left(-\frac{(y-\widetilde{\mu_Y})^2}{2\widetilde{\sigma_Y^2}}\right)\right)}{\frac{1}{\sqrt{2\pi\sigma_X^2}} \exp\left(-\frac{(x-\mu_X)^2}{2\sigma_X^2}\right)}$$

$$= \frac{1}{\sqrt{2\pi\widetilde{\sigma_Y^2}}} \exp\left(-\frac{(y-\widetilde{\mu_Y})^2}{2\widetilde{\sigma_Y^2}}\right)$$

$$= \frac{1}{\sqrt{2\pi\sigma_Y^2(1-\rho_{X,Y}^2)}}$$

$$\times \exp\left(-\frac{\left(y-\mu_Y-\rho_{X,Y}\left(\frac{\sigma_Y}{\sigma_X}\right)(x-\mu_X)\right)^2}{2\sigma_Y^2(1-\rho_{X,Y}^2)}\right)$$

This conditional pdf has a Gaussian distribution, where its mean is $\mu_Y + \rho_{X,Y}\left(\frac{\sigma_Y}{\sigma_X}\right)(x-\mu_X)$ and its variance is $\sigma_Y^2(1-\rho_{X,Y}^2)$. With the same line of reasoning and reversing the roles of X and Y, we could also obtain $f_Y(y)$ and $f_{X|Y}(x|y)$, respectively.

Note that when $\rho_{X,Y} = 0$, the conditional pdf of X given $Y = y$ equals the marginal pdf of X and the conditional pdf of Y given $X = x$ equals the marginal pdf of Y. This is consistent with the fact that when $\rho_{X,Y} = 0$, the random variables X and Y are independent. ∎

7.3.1 Linear Transformations of Bivariate Gaussian Random Variables

Suppose X and Y are bivariate Gaussian random variables, where μ_X and μ_Y are the mean values, and σ_X and σ_Y are the standard deviations of the random variables X and Y, respectively, and $\rho_{X,Y}$ is the correlation coefficient between the random variables X and Y. Also, suppose V and W are linear transformations of X and Y, where V and W are linearly independent in the context of linear algebra, i.e. V and W cannot be written in terms of one another. It can then be shown that V and W are also bivariate Gaussian random variables and their means, variances, and covariance can be obtained as follows:

$$\begin{cases} V = aX + bY \\ W = cX + dY \end{cases} \rightarrow \begin{cases} \mu_V = a\mu_X + b\mu_Y \\ \mu_W = c\mu_X + d\mu_Y \\ \sigma_V^2 = a^2\sigma_X^2 + b^2\sigma_Y^2 + 2ab\rho_{X,Y}\sigma_X\sigma_Y \\ \sigma_W^2 = c^2\sigma_X^2 + d^2\sigma_Y^2 + 2cd\rho_{X,Y}\sigma_X\sigma_Y \\ COV(V, W) = ac\sigma_X^2 + bd\sigma_Y^2 \\ \qquad\qquad\qquad + (ad + bc)\sigma_X\sigma_Y\rho_{X,Y} \end{cases} \tag{7.16}$$

where a, b, c, and d are nonrandom constants.

Note that as we have $\rho_{V,W} = \frac{ac\sigma_X}{\sigma_Y} + \frac{bd\sigma_Y}{\sigma_X} + (ad + bc)\rho_{X,Y}$, by the appropriate selection of the four parameters a, b, c, and d, $|\rho_{V,W}| \leq 1$ can be any value of interest, regardless of the given value of $\rho_{X,Y}$. For instance, if X and Y are correlated, i.e. $\rho_{X,Y} \neq 0$, then a, b, c, and d can be easily chosen in such a way that V and W become uncorrelated, i.e. $\rho_{V,W} = 0$.

Example 7.6

Suppose X and Y are bivariate Gaussian random variables, whose means are zero, variances are 4 and 9, respectively, and their correlation coefficient is 0.5. Assuming we have $V = X - 2Y$ and $W = 3X + 4Y$, determine the means and variances of the bivariate Gaussian random variables V and W. Are V and W uncorrelated?

Solution:

We have

$$\begin{cases} \mu_x = 0 \\ \mu_Y = 0 \\ \sigma_X^2 = 4 \\ \sigma_Y^2 = 9 \\ \rho_{X,Y} = 0.5 \end{cases} \rightarrow \begin{cases} \mu_V = \mu_x - 2\mu_Y = 0 \\ \mu_W = 3\mu_x + 4\mu_Y = 0 \\ \sigma_V^2 = 4 + 36 - 12 = 28 \\ \sigma_W^2 = 36 + 144 + 72 = 252 \\ COV(V, W) = 12 - 72 - 6 = -66 \end{cases}$$

Since the covariance of the random variables V and W is not zero, they are not uncorrelated. ∎

Example 7.7

Suppose X and Y are bivariate Gaussian random variables, where X and Y are both zero mean and unit variance, and their correlation coefficient is $\rho_{X,Y}$. If we have $V = \frac{X+Y}{\sqrt{2}}$ and $W = \frac{-X+Y}{\sqrt{2}}$, then determine the joint pdf of V and W.

Solution:

We have

$$
\begin{cases} V = \dfrac{X+Y}{\sqrt{2}} \\[3mm] W = \dfrac{-X+Y}{\sqrt{2}} \end{cases} \rightarrow \begin{cases} X = \dfrac{V-W}{\sqrt{2}} \\[3mm] Y = \dfrac{V+W}{\sqrt{2}} \end{cases}
$$

Using the linear transformation of jointly Gaussian random variables and noting that the Jacobian of transformation turns to be 1, we have the following:

$$
f_{V,W}(v,w) = f_{X,Y}\left(\frac{v-w}{\sqrt{2}}, \frac{v+w}{\sqrt{2}} \right)
$$

$$
= \frac{1}{2\pi\sqrt{1-\rho_{X,Y}^2}}
$$

$$
\times \exp\left[-\frac{\left(\dfrac{v-w}{\sqrt{2}}\right)^2 - 2\rho_{X,Y}\left(\dfrac{v-w}{\sqrt{2}}\right)\left(\dfrac{v+w}{\sqrt{2}}\right) + \left(\dfrac{v+w}{\sqrt{2}}\right)^2}{2(1-\rho_{X,Y}^2)} \right]
$$

$$
= \frac{1}{2\pi\sqrt{1-\rho_{X,Y}^2}} \exp\left(-\frac{v^2}{2(1+\rho_{X,Y})} - \frac{w^2}{2(1-\rho_{X,Y})} \right)
$$

$$
= \frac{1}{\sqrt{2\pi(1+\rho_{X,Y})}} \exp\left(-\frac{v^2}{2(1+\rho_{X,Y})} \right)
$$

$$
\times \frac{1}{\sqrt{2\pi(1-\rho_{X,Y})}} \exp\left(-\frac{w^2}{2(1-\rho_{X,Y})} \right)
$$

Therefore, the transformed variables V and W are independent zero-mean Gaussian random variables with variances $1+\rho_{X,Y}$ and $1-\rho_{X,Y}$, respectively. Note that if $\rho_{X,Y} = 0$, then V and W are also independent standard Gaussian random variables. ∎

7.4 Jointly Gaussian Random Vectors

We now turn our focus toward the multivariate Gaussian distribution. The multivariate Gaussian pdf is a probability model for a vector of $n \geq 1$ random variables in which the marginal pdf for each of the n random variables is Gaussian. The pdf of a **jointly Gaussian random vector** X is defined as follows:

$$f_X(x) = \frac{\exp(-0.5(x - \mu_X)^T K_X^{-1}(x - \mu_X))}{\sqrt{(2\pi)^n \mid K_X \mid}} \tag{7.17}$$

where the superscript T denotes transpose, $X = [X_1, X_2, ..., X_n]^T$ and $x = [x_1, x_2, ..., x_n]^T$ as well as μ_X represents the mean vector and K_X is the covariance matrix. It can be shown that all the marginal pdfs are also Gaussian, and they can be completely specified by the same set of means and variances. A Gaussian random vector X has independent random variables if and only if the covariance matrix K_X is a diagonal matrix, where the elements on the diagonal are the variances and the off-diagonal elements are all zero.

Example 7.8

In a digital transmission system, the level of noise is measured three times, where the measurements are all Gaussian random variables X_1, X_2, and X_3. Suppose the expected values are 1, 5, and 6, respectively, and the covariance matrix of the three measurements is as follows:

$$K_X = \begin{bmatrix} 4 & 3 & 2 \\ 3 & 4 & 3 \\ 2 & 3 & 4 \end{bmatrix}$$

Determine the pdf of X_1, the joint pdf of X_1 and X_2, and the joint pdf of X_1, X_2, and X_3.

Solution:

X_1 is a Gaussian random variable with mean 1 and variance 4. The pdf of X_1 is thus as follows:

$$f_{X_1}(x_1) = \left(\frac{1}{2\sqrt{2\pi}}\right) \exp\left(-\frac{(x_1 - 1)^2}{8}\right)$$

X_1 and X_2 are bivariate Gaussian random variables, where the expected values are 1 and 5, respectively, and their variances are both 4. We can thus determine the correlation coefficients as follows:

$$\rho_{X,Y} = \frac{\text{COV}(X_1, X_2)}{\sigma_1 \sigma_2} = \frac{3}{2 \times 2} = 0.75$$

The pdf of X_1 and X_2 is thus as follows:

$$f_{X_1,X_2}(x_1, x_2) = \frac{1}{2\pi\sqrt{7}}$$
$$\times \exp\left(-\frac{2(x_1 - 1)^2 - 3(x_1 - 1)(x_2 - 5) + 2(x_2 - 5)^2}{7}\right)$$

X_1, X_2, and X_3 are three Gaussian random variables. The determinant of the covariance matrix $|K_X|$ is 12 and its inverse is as follows:

$$K_X^{-1} = \frac{1}{12} \begin{bmatrix} 7 & -6 & 1 \\ -6 & 12 & -6 \\ 1 & -6 & 7 \end{bmatrix}$$

The pdf of X_1, X_2, and X_3 is thus as follows:

$$f_{X_1,X_2,X_3}(x_1, x_2, x_3) = f_X(x) = \frac{1}{4\pi\sqrt{6\pi}} \exp\left(-\frac{(x - \mu_X)^T K_X^{-1}(x - \mu_X)}{2}\right)$$

where

$$x = [x_1, x_2, x_3]^T$$

and

$$\mu_X = [1, 5, 6]^T$$

∎

7.5 Sums of Random Variables

Suppose we have a set of n random variables, $X_1, X_2, \ldots, X_n$, where n is a positive integer. We also assume the pdf or pmf of each of these random variables is known. Let S_n be their sum, i.e. we have

$$S_n = X_1 + X_2 + \cdots + X_n \tag{7.18}$$

The sum S_n is thus a random variable. It is possible to derive the probability model of S_n, i.e. to determine the pdf or pmf of S_n from those of the n random variables. However, analyzing a general n-dimensional probability model is not easy, nor is it generally needed for most applications, unless the random variables are independent.

7.5.1 Mean and Variance of Sum of Random Variables

Whether or not the n random variables are independent, the expected value of a sum of random variables is the sum of the individual expected values and the variance of a sum of random variables is the sum of all pairwise covariances, we thus have the following:

$$E[S_n] = E[X_1] + E[X_2] + \cdots + E[X_n]$$

$$\sigma_{S_n}^2 = \sum_{i=1}^{n} \sum_{j=1}^{n} \text{COV}(X_i, X_j) = \sum_{k=1}^{n} \sigma_{X_k}^2 + \sum_{\substack{i=1 \\ i \neq j}}^{n} \sum_{j=1}^{n} \text{COV}(X_i, X_j) \tag{7.19}$$

However, if the n random variables are uncorrelated, let alone independent, the variance of the sum of random variables is then equal to the sum of the individual variances, as all pairwise covariances are zero that is we have

$$\sigma_{S_n}^2 = \sigma_{X_1}^2 + \sigma_{X_2}^2 + \cdots + \sigma_{X_n}^2 \tag{7.20}$$

7.5.2 Mean and Variance of Sum of Independent, Identically Distributed Random Variables

If we have n independent, identically distributed (iid) random variables, $X_1, X_2,$ ..., X_n, each with mean μ_X and variance σ_X^2, the mean and variance of the sum are, respectively, defined as follows:

$$E[S_n] = E[X_1] + E[X_2] + \cdots + E[X_n] = n\mu_X$$
$$\sigma_{S_n}^2 = \sigma_{X_1}^2 + \sigma_{X_2}^2 + \cdots + \sigma_{X_n}^2 = n\sigma_X^2 \tag{7.21}$$

7.5.3 Distribution of Sum of Independent Random Variables

Suppose we have n independent random variables $X_1, X_2, ..., X_n$. Using transform methods, such as characteristic function and moment generating function, it can be shown that the pdf of a sum of continuous random variables or the pmf of the sum of discrete random variables, are respectively as follows:

$$f_{S_n}(x) = f_{X_1}(x) * f_{X_2}(x) * \cdots * f_{X_n}(x)$$
$$p_{S_n}(x) = p_{X_1}(x) * p_{X_2}(x) * \cdots * p_{X_n}(x) \tag{7.22}$$

where the symbol $*$ denotes the convolution operation. In view of this, we have the following:

– The sum of n iid Bernoulli random variables is a binomial random variable.
– The sum of n iid geometric random variables is a Pascal random variable.
– The sum of n iid Poisson random variables is a Poisson random variable.
– The sum of n iid exponential random variables is an Erlang random variable.
– The sum of n iid Gaussian random variables is a Gaussian random variable.

Example 7.9
Show that a linear combination of n independent Gaussian random variables is also a Gaussian random variable. In other words, if $X_1, X_2, ..., X_n$ are n independent Gaussian random variables with means $\mu_1, \mu_2, ..., \mu_n$ and variances $\sigma_1^2, \sigma_2^2, ..., \sigma_n^2$, then $S_n = a_1X_1 + a_2X_2 + \cdots + a_nX_n$, where $a_1, a_2, ..., a_n$ are all nonrandom, nonzero constants, is also a Gaussian random variable.

Solution:
We have S_n as the sum of n independent Gaussian random variables, whose respective means are $a_1\mu_1, a_2\mu_2, ..., a_n\mu_n$ and variances are $a_1^2\sigma_1^2, a_2^2\sigma_2^2, ..., a_n^2\sigma_n^2$. Noting that the moment generating function of a Gaussian random variable

X with mean μ_X and variance σ_X^2 is $M_X(v) = e^{\mu_X v + \frac{\sigma_X^2 v^2}{2}}$, the moment generation function of a Gaussian random variable $a_i X_i$ with mean $a_i \mu_i$ and variance $a_i^2 \sigma_i^2$ is thus as follows:

$$M_{a_i X_i}(v) = e^{a_i \mu_i v + \frac{a_i^2 \sigma_i^2 v^2}{2}}$$

As the n random variables are independent, the moment generating function of the sum S_n is the product of the moment generating functions of individual terms in the sum. Therefore, we have

$$M_{S_n}(v) = \left(e^{a_1 \mu_1 v + \frac{a_1^2 \sigma_1^2 v^2}{2}} \right) \times \cdots \times \left(e^{a_n \mu_n v + \frac{a_n^2 \sigma_n^2 v^2}{2}} \right)$$

$$= e^{(a_1 \mu_1 + \cdots + a_n \mu_n)v + \frac{(a_1^2 \sigma_1^2 + \cdots + a_n^2 \sigma_n^2)v^2}{2}}$$

By finding the inverse moment generating function, we see that the pdf of the sum S_n has a Gaussian distribution whose mean is $a_1 \mu_1 + \cdots + a_n \mu_n$ and variance is $a_1^2 \sigma_1^2 + \cdots + a_n^2 \sigma_n^2$. ∎

A special case of a linear combination of n independent Gaussian random variables is the sum of n iid Gaussian random variables. In other words, if X_i for $i = 1, 2, \ldots, n$ are n iid Gaussian random variables with the same mean μ_X and variance σ_X^2, then the sum $S_n = X_1 + X_2 + \cdots + X_n$ is also a Gaussian random variable with the mean $n\mu_X$ and variance $n\sigma_X^2$.

7.5.4 Sum of a Random Number of Independent, Identically Distributed Random Variables

Let S_N be the sum of N iid random variables $X_1, X_2, \ldots, X_N$ with finite mean μ_X and finite variance σ_X^2, where N representing the number of random variables is itself a random variable. The **random sum** is thus defined as follows:

$$S_N = X_1 + X_2 + \cdots + X_N \tag{7.23}$$

Suppose the random variable N, with mean μ_N and variance σ_N^2, is independent of each of the N random variables $X_1, X_2, \ldots, X_N$, i.e. the number of terms in the sum does not depend on the values of the terms in the sum. It can then be shown that the mean and variance of the random sum are, respectively, as follows:

$$E[S_N] = \mu_N \mu_X$$
$$\sigma_{S_N}^2 = \mu_N \sigma_X^2 + \sigma_N^2 \mu_X^2 \tag{7.24}$$

Suppose N is deterministic, say $N = n$, we then have $\mu_N = n$ and $\sigma_N^2 = 0$. In such a case, the mean and variance of the sum are $n\mu_X$ and $n\sigma_X^2$, respectively, as shown earlier. Suppose each of the N terms is a deterministic constant c, we then have $\mu_X = c$ and $\sigma_X^2 = 0$. In such a case, the mean and variance of the sum are $c\mu_N$ and $c^2 \sigma_N^2$, respectively.

Example 7.10

There are N people attending a local charity event, where N is a discrete uniform random variable over the interval [94, 106]. Suppose the amount of money an attendee may donate is X dollars, where X is a discrete uniform random variable over the interval [89, 111]. Noting that the total money collected is a random variable, determine its mean and standard deviation.

Solution:

As N and X both have uniform distributions, we have the following:

$$\mu_N = \frac{94 + 106}{2} = 100$$

$$\sigma_N^2 = \frac{(106 - 94 + 1)^2 - 1}{12} = 14$$

$$\mu_X = \frac{89 + 111}{2} = 100$$

$$\sigma_X^2 = \frac{(111 - 89 + 1)^2 - 1}{12} = 44$$

We thus get

$$E[S_N] = \mu_N \mu_X = 10\,000$$

$$\sigma_{S_N}^2 = \mu_N \sigma_X^2 + \sigma_N^2 \mu_X^2 = 144\,400 \rightarrow \sigma_{S_N} = 380$$

Note that most of the variance is contributed by the randomness in N. ∎

7.6 The Sample Mean

Suppose we have n iid random variables, $X_1, X_2, \ldots, X_n$, each with mean μ_X and variance σ_X^2. The **sample mean** is then defined as follows:

$$\overline{X}_n = \frac{S_n}{n} = \frac{X_1 + X_2 + \cdots + X_n}{n} \tag{7.25}$$

The sample mean $\overline{X}_n$, which is a function of the random variables $X_1, X_2, \ldots, X_n$, is thus a random variable itself. It is important to distinguish the sample mean $\overline{X}_n$ from the expected value μ_X, as $\overline{X}_n$ is a random variable, and μ_X is a nonrandom number. The expected value and variance of the sample mean $\overline{X}_n$ are, respectively, as follows:

$$E[\overline{X}_n] = E\left[\frac{S_n}{n}\right] = E\left[\frac{X_1 + X_2 + \cdots + X_n}{n}\right]$$

$$= \frac{1}{n}(E[X_1] + E[X_2] + \cdots + E[X_n]) = \frac{1}{n}(n\mu_X) = \mu_X$$

$$\sigma_{\overline{X}_n}^2 = \frac{1}{n^2}(\sigma_{X_1}^2 + \sigma_{X_2}^2 + \cdots + \sigma_{X_n}^2) = \frac{1}{n^2}(n\sigma_X^2) = \frac{\sigma_X^2}{n} \tag{7.26}$$

As n increases without bound, the variance of the sample mean goes to zero, that is the sample mean converges to the expected value. It is important to note that averaging reduces the variance from σ_X^2 to $\frac{\sigma_X^2}{n}$, while maintaining the same mean and by increasing the sample size n, the level of randomness, which is manifested by variance, is reduced. Note that the average has a tendency to be closer to the mean μ_X than the individual random variables do, and by increasing n, this tendency increases.

Example 7.11
Compare the probability that a Gaussian random variable X taking a value within m standard deviations of its mean to the probability that the sample mean $\overline{X}_n$ based on n of these random variables taking a value within the same interval, for $m = 1$ and $m = 2$ as well as $n = 9$ and $n = 16$. Comment on the results.

Solution:
The probability for the random variable X, whose mean is μ_X and variance is σ_X^2, to take a value within m standard deviations of its mean is as follows:

$$P(\mu_X - m\sigma_X \leq X \leq \mu_X + m\sigma_X) = \Phi(m) - \Phi(-m) = 2\Phi(m) - 1$$

and the probability for the sample mean $\overline{X}_n$, whose mean is μ_X and variance is $\frac{\sigma_X^2}{n}$, to take a value within the same interval is as follows:

$$P(\mu_X - m\sigma_X \leq \overline{X}_n \leq \mu_X + m\sigma_X) = \Phi\left(\frac{(\mu_X + m\sigma_X) - \mu_X}{\frac{\sigma_X}{\sqrt{n}}}\right)$$
$$- \Phi\left(\frac{(\mu_X - m\sigma_X) - \mu_X}{\frac{\sigma_X}{\sqrt{n}}}\right)$$
$$= \Phi(m\sqrt{n}) - \Phi(-m\sqrt{n})$$
$$= 2\Phi(m\sqrt{n}) - 1$$

We thus have the following sets of probabilities:

m	n	$P(\mu_X - m\sigma_X \leq X \leq \mu_X + m\sigma_X)$	$P(\mu_X - \sigma_X m \leq \overline{X}_n \leq \mu_X + m\sigma_X)$
1	9	0.682 68	$1 - 3 \times 10^{-3}$
1	16	0.682 68	$1 - 7 \times 10^{-5}$
2	9	0.954 50	$1 - 2 \times 10^{-9}$
2	16	0.954 50	$1 - 1 \times 10^{-15}$

For a given m, while there is only a rather small chance that a $N(\mu_X, \sigma_X^2)$ random variable lies within the interval $[\mu_X - m\sigma_X, \mu_X + m\sigma_X]$, the mean of n independent random variables of this kind has a very significantly higher chance of taking a value within the same interval. For instance, there is only about 68.27% chance that a $N(\mu_X, \sigma_X^2)$ random variable X lies within the interval $[\mu_X - \sigma_X, \mu_X + \sigma_X]$, whereas the average of 9 independent random variables of this kind has more than 99.7% chance of taking a value within this interval. Also, an increase in n brings about a significant increase in the chance for the sample mean to take a value within an interval. For instance, the 99.7% chance can be further increased to 99.993%, if 16 samples rather than 9 samples are employed. Moreover, for a given n, say $n = 9$, an increase in m, say from 1 to 2, makes the difference between the probability that a random variable takes a value within an interval and the probability that the sample mean takes a value within the same interval much more pronounced. ∎

7.6.1 Laws of Large Numbers

The sample mean $\overline{X}_n$ converges to the expected value μ_X and the relative frequency of an event approaches to the probability of that event as n representing the number of trials of an experiment increases without bound, i.e. as $n \to \infty$. The law of large numbers, as the name implies, guarantees a stable long-term result for the average of a random event when a large number of observations is considered. The laws of large numbers consist of the weak law of large numbers and the strong law of large numbers.

The *weak law of large numbers*, which is an interesting consequence of the Chebyshev inequality, states that if $X_1, X_2, \ldots, X_n$ is a sequence of iid random variables with finite expected value μ_X, then for $\varepsilon > 0$, we have

$$\lim_{n \to \infty} P(|\overline{X}_n - \mu_X| < \varepsilon) = 1 \tag{7.27}$$

The weak law essentially highlights the fact that for any nonzero margin ε specified, no matter how small, with a sufficiently large fixed value of n, there will be a very high probability that the average of the observations, i.e. the sample mean $\overline{X}_n$, will be close to the expected value μ_X that is within the margin. The weak law of large numbers does not address what happens to the sample mean $\overline{X}_n$ if additional samples (measurements) are made.

The *strong law of large numbers* states that if $X_1, X_2, \ldots, X_n$ is a sequence of iid random variables with finite expected value μ_X and finite variance, then we have

$$P(\lim_{n \to \infty} \overline{X}_n = \mu_X) = 1 \tag{7.28}$$

The strong law of large numbers highlights the fact that the sequence of the sample means $\overline{X}_1, \overline{X}_2, ..., \overline{X}_n$, where $\overline{X}_i$ is the sample mean computed using X_1 through X_i, converges to μ_X with probability 1. Note that the terms convergence almost surely, convergence almost always, and convergence almost everywhere are synonymous for convergence with probability 1.

The strong law of large numbers implies the weak law of large numbers but not vice versa, when the strong law conditions hold, the variable converges both strongly (almost surely) and weakly (in probability). The difference between the strong law and the weak law of large numbers is not much and rarely arises in practical applications. By the law of large numbers, the probability distribution of the sample mean random variable decreases in width until all the probability is concentrated about the mean. However, it does not much say about the distribution itself. This task is left to the central limit theorem (CLT), as discussed in the next section.

7.7 Approximating Distributions with the Gaussian Distribution

It is sometimes quite easier to provide a very good approximation of a probability using the continuous Gaussian distribution rather than to provide an accurate, but unwieldy, evaluation of the probability of interest using the discrete binomial or Poisson distributions. Note that in order to treat discrete data as continuous, an adjustment of 0.5, which is called **continuity correction**, is made to the value of the random variable of interest. The shifting of end points by 0.5 allows a discrete probability distribution, such as the binomial and Poisson distributions, to be approximated by a continuous probability distribution, such as a Gaussian distribution. Note that there are statistical software packages that can calculate the exact probabilities of the binomial or Poisson distributions, however, if n is very large, say $n > 100$, then the statistical software packages actually use the Gaussian approximation to calculate the binomial or Poisson probabilities.

7.7.1 Relation between the Gaussian and Binomial Distributions

The probability mass function of a binomial distribution has a bell-shaped curve similar to the probability density function of a Gaussian distribution. In fact, if in a binomial distribution, the number of trials n is large and neither the probability of success p nor the probability of failure $1 - p$ is too close to 0, the binomial distribution can be closely approximated by a Gaussian distribution whose mean is $\mu_X = np$ and variance is $\sigma_X^2 = np(1 - p)$. If X is a binomial random variable with mean $\mu_X = np$ and variance $\sigma_X^2 = np(1-p)$, we can then have the following Gaussian approximation to the binomial distribution:

$$P(X \le x) = \sum_{j=0}^{x} \binom{n}{j} p^j (1-p)^{n-j} \cong \Phi\left(\frac{x+0.5-np}{\sqrt{np(1-p)}}\right)$$

$$P(X \ge x) = \sum_{j=x}^{n} \binom{n}{j} p^j (1-p)^{n-j} \cong 1 - \Phi\left(\frac{x-0.5-np}{\sqrt{np(1-p)}}\right) \qquad (7.29)$$

In practice, the approximation is very good if we have $np \ge 5$ and $n \ge 10$. The approximation becomes better with increasing n and is exact in the limiting case. The binomial distribution asymptotically approaches the Gaussian distribution as reflected below:

$$\lim_{n\to\infty} P\left(a \le \frac{X-np}{\sqrt{np(1-p)}} \le b\right) = \int_a^b \left(\frac{1}{\sqrt{2\pi}}\right) \exp\left(-\frac{x^2}{2}\right) dx \qquad (7.30)$$

In other words, as $n \to \infty$, the limiting form of the distribution of $\frac{X-np}{\sqrt{np(1-p)}}$ is the standard Gaussian distribution.

Example 7.12
Suppose that a fair coin is tossed 10 times. Determine the probability of getting between 3 and 6 heads by using the binomial distribution and the Gaussian approximation to the binomial distribution.

Solution:
We have $n = 10$ and $p = \frac{1}{2}$. By using the binomial distribution, we obtain:

$$P(3 \le X \le 6) = \sum_{j=3}^{6} \binom{10}{j} \left(\frac{1}{2}\right)^j \left(\frac{1}{2}\right)^{10-j} \cong 0.7734$$

By using the Gaussian distribution, we get

$$\Phi\left(\frac{6+0.5-5}{\sqrt{2.5}}\right) - \Phi\left(\frac{3-0.5-5}{\sqrt{2.5}}\right)$$

$$= \Phi\left(\frac{1.5}{\sqrt{2.5}}\right) - \left(1 - \Phi\left(\frac{2.5}{\sqrt{2.5}}\right)\right)$$

$$= 1 - Q\left(\frac{1.5}{\sqrt{2.5}}\right) - Q\left(\frac{2.5}{\sqrt{2.5}}\right) \cong 0.7718$$

This approximation obviously compares very well with the true value. Note that a larger value of n can bring about even a better approximation. ∎

7.7.2 Relation between the Gaussian and Poisson Distributions

The probability mass function of the Poisson distribution has a bell-shaped curve similar to the probability density function of a Gaussian distribution. In fact, if in a Poisson distribution, the mean and variance λ is sufficiently large, the Poisson distribution can be closely approximated by a Gaussian distribution whose mean is $\mu_X = \lambda$ and variance is $\sigma_X^2 = \lambda$. If X is a Poisson random variable with mean $\mu_X = \lambda$ and variance $\sigma_X^2 = \lambda$, we can then have the following Gaussian approximation to the Poisson distribution:

$$P(X \leq x) = \sum_{j=0}^{x} \frac{\lambda^j e^{-\lambda}}{j!} \cong \Phi\left(\frac{x + 0.5 - \lambda}{\sqrt{\lambda}}\right)$$

$$P(X \geq x) = \sum_{j=x}^{n} \frac{\lambda^j e^{-\lambda}}{j!} \cong 1 - \Phi\left(\frac{x - 0.5 - \lambda}{\sqrt{\lambda}}\right) \qquad (7.31)$$

In practice, the approximation is very good if we have $\lambda \geq 10$. The approximation becomes better with increasing λ and is exact in the limiting case. The Poisson distribution asymptotically approaches the Gaussian distribution as reflected below:

$$\lim_{\lambda \to \infty} P\left(a \leq \frac{X - \lambda}{\sqrt{\lambda}} \leq b\right) = \int_a^b \left(\frac{1}{\sqrt{2\pi}}\right) \exp\left(-\frac{x^2}{2}\right) dx \qquad (7.32)$$

In other words, as $\lambda \to \infty$, the limiting form of the distribution of $\frac{X - \lambda}{\sqrt{\lambda}}$ is the standard Gaussian distribution.

Example 7.13
Suppose that the number of phone calls made by a mobile user is a Poisson random variable with an average rate of 100 calls per week. Determine the probability of making between 97 and 102 calls in a week.

Solution:
By using the Poisson distribution with $\lambda = 100$, we have

$$P(97 \leq X \leq 102) = \sum_{j=97}^{102} \frac{\lambda^j e^{-\lambda}}{j!} = \sum_{j=97}^{102} \frac{100^j e^{-100}}{j!} \cong 0.2376$$

By the Gaussian distribution, we get

$$\Phi\left(\frac{102 + 0.5 - 100}{\sqrt{100}}\right) - \Phi\left(\frac{97 - 0.5 - 100}{\sqrt{100}}\right)$$

$$= \Phi\left(\frac{2.5}{10}\right) - \left(1 - \Phi\left(\frac{3.5}{10}\right)\right)$$

$$= 1 - Q(0.25) - Q(0.35) \cong 0.2355$$

This approximation clearly compares very well with the true value. ∎

7.7.3 The Central Limit Theorem

The CLT is a remarkable result in probability theory. The importance of the CLT is hard to overstate, as it is the basis of many statistical procedures. One of the main reasons for the great importance of the Gaussian distribution is the CLT, through which many practical phenomena whose values depend on the sum of a large number of small random contributions can be described and analyzed. Application categories may include noise characterization, polling process, stock market, medical treatment, and population growth, to name just a few. Specific examples include when noise in a communication system is well approximated by a Gaussian distribution as it is the sum of a wide variety of noise components in the transmitter, channel, and receiver, and when many physiologic measures are determined in part by a combination of several genetic and environmental risk factors and can often be well approximated by a Gaussian distribution. In summary, data which can be influenced by many small and unrelated random effects are approximately normally distributed.

Noting that if S_n is the sum of n iid random variables $X_1, \ldots, X_n$ with a finite mean μ_X and a finite variance σ_X^2, then the mean of S_n is $n\mu_X$ and its variance is $n\sigma_X^2$. Let Z_n be the normalized sum with zero-mean and unit-variance that is we have

$$Z_n = \frac{S_n - n\mu_X}{\sqrt{n\sigma_X^2}} = \frac{\frac{S_n}{n} - \mu_X}{\frac{\sigma_X}{\sqrt{n}}} = \frac{\overline{X}_n - \mu_X}{\frac{\sigma_X}{\sqrt{n}}} = \sqrt{n}\left(\frac{\overline{X}_n - \mu_X}{\sigma_X}\right) \qquad (7.33)$$

where $\overline{X}_n$, as discussed earlier, is the average value of the random variable, known as the sample mean. The *central limit theorem* states as $n \to \infty$, the pdf of Z_n, regardless of the underlying distribution of the n random variables, can be approximated by a Gaussian random variable with zero mean and unit variance, i.e. $N(0, 1)$, that is the pdf of Z_n will converge to the standard Gaussian distribution. In other words, we have

$$\lim_{n \to \infty} P(Z_n \leq z) = \int_{-\infty}^{z} \left(\frac{1}{\sqrt{2\pi}} \right) \exp\left(-\frac{t^2}{2} \right) dt = \Phi(z) \qquad (7.34)$$

There are no restrictions on the type of the random variable, as they can be continuous, discrete, or mixed. The CLT basically highlights a major property of convolutions, in that the convolution of a large number of positive, identical functions is approximately a Gaussian function.

The accuracy of the approximation improves, i.e. the distribution of the sum more and more resembles a Gaussian distribution, as the number of random variables n increases. The practical value of the CLT is not so much in the exact shape of the Gaussian distribution for $n \to \infty$, as the variance of the sum then becomes infinite. Instead, the value derives from the fact that the sum for any finite n may have a distribution that is closely approximated as Gaussian. The approximation, however, can be quite inaccurate in the tails regions away from the mean, even when n is very large. In practice, values of n close to 30 can usually be adequate for most applications, and for very smooth functions, values of n can be as low as 6.

Example 7.14
A random variable is defined as the sum of four iid random variables. Assuming the pdf of each has a uniform distribution in the interval $[-0.5, 0.5]$, determine the pdf of the sum.

Solution:
Noting that the symbol * denotes the convolution operation, the convolution of the two functions $f_{X_1}(x)$ and $f_{X_2}(x)$ is defined as follows:

$$f_{X_1}(x) * f_{X_2}(x) = \int_{-\infty}^{\infty} f_{X_1}(t) f_{X_2}(x - t) dt$$

We first find the pdf of the sum of two uniformly distributed random variables. The pdf of $b = a_1 + a_2$ is the convolution of the pdfs of a_1 and a_2. The convolution of two identical square-shape functions results in a triangular-shape function, and is as follows:

$$f_B(b) = \begin{cases} 1 - |b| & |b| \leq 1 \\ 0 & |b| > 1 \end{cases}$$

We then find the pdf of the sum of three random variables. The pdf of $c = a_1 + a_2 + a_3 = b + a_3$ is the convolution of the pdfs of b and a_3, and is thus as follows:

$$f_C(c) = \begin{cases} 0.75 - c^2 & |c| \leq 0.5 \\ 0.5(1.5 - |c|)^2 & 0.5 \leq |c| \leq 1.5 \\ 0 & 1.5 \leq |c| \end{cases}$$

We finally find the pdf of the sum of four random variables. The pdf of $d = a_1 + a_2 + a_3 + a_4 = c + a_4$ is the convolution of the pdfs of c and a_4, and is thus as follows:

$$f_D(d) = \begin{cases} \frac{|d|^3}{2} - d^2 + \frac{2}{3} & 0 \leq |d| < 1 \\ -\frac{|d|^3}{6} + d^2 - 2|d| + \frac{4}{3} & 1 \leq |d| < 2 \\ 0 & 2 \leq |d| \end{cases}$$

Figure 7.5 shows the pdfs of one single variable, the sum of two variables, the sum of three random variables, and the sum of four random variables. The total interval of the convolution is the sum of the intervals of the functions being convoluted. To this effect, by adding more random variables, the pdf of the sum becomes wider and closer to that of a Gaussian random variable. ∎

Example 7.15

In a binary communication channel, each data packet has 10 000 bits. Each bit can be in error independently with probability of 10^{-3}. Using the CLT, determine the approximate probability that more than 0.4% of the bits are in error.

Solution:

There are 10 000 iid random variables. Each random variable is assumed to be 1 if the bit is in error and 0 if not. We therefore have the mean and variance for each random variable as follows:

$$\mu_X = 10^{-3}(1) + (1 - 10^{-3})(0) = 10^{-3}$$

and

$$\sigma_X^2 = (10^{-3}(1)^2 + (1 - 10^{-3})(0)^2) - (10^{-3})^2 = 0.000\,999$$

Based on the CLT, S_n is assumed to be the number of errors in the packet, and has a Gaussian distribution with the following mean and variance, respectively:

$$E[S_n] = n\mu_X = 10\,000 \times 0.001 = 10$$

and

$$\sigma_{S_n}^2 = n\sigma_X^2 = 10\,000 \times 0.000\,999 = 9.99$$

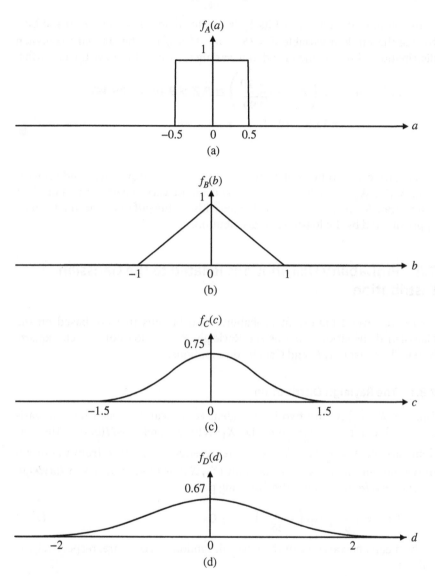

Figure 7.5 (a) pdf of one variable; (b) pdf of sum of two variables; (c) pdf of sum of three variables; (d) pdf of sum of four variables.

The number of erroneous bits S_n is greater than 40 (=10 000 × 0.4%) bits. Noting the random variable $Z = (S_n - n\mu_X)/\sigma_X\sqrt{n}$ is the standard Gaussian distribution with zero-mean and unit-variance, we can thus have the following:

$$P(S_n > 40) = P\left(Z > \frac{40 - 10}{\sqrt{9.99}}\right) \cong P(Z > 9.49) = Q(9.49)$$

$$\cong 1.05 \times 10^{-21}$$

∎

The CLT can also be applied to the product of n iid positive random variables $X_1, \ldots, X_n$ with a finite mean μ_X and a finite variance σ_X^2. Since a product can be transformed into a sum using logarithms, the pdf of the product can be approximated by the log-normal distribution.

7.8 Probability Distributions Related to the Gaussian Distribution

There are several important probability distributions that are based on the Gaussian distribution, such as the Rayleigh, Ricean, log-normal, chi-square, Maxwell, Student's t, F, and Cauchy distributions.

7.8.1 The Rayleigh Distribution

Suppose X_1 and X_2 are two independent zero-mean Gaussian random variables with common variance σ^2, i.e. $X_1 \sim N(0, \sigma^2)$ and $X_2 \sim N(0, \sigma^2)$. The random variable $X = \sqrt{X_1^2 + X_2^2}$, which represents the distance from the origin in two-dimensional space to the point (X_1, X_2), is then known as a **Rayleigh random variable**, and has the following pdf:

$$f_X(x) = \left(\frac{x}{\sigma^2}\right)\exp\left(-\frac{x^2}{2\sigma^2}\right) \quad x \geq 0 \tag{7.35}$$

The mean and variance of the Rayleigh random variable are, respectively, as follows:

$$\mu_X = \sigma\sqrt{0.5\pi} \quad \sigma > 0$$
$$\sigma_X^2 = \sigma^2(2 - 0.5\pi) \tag{7.36}$$

Unlike the Gaussian random variable, the mean and variance of a Rayleigh random variable depend on a single parameter σ and cannot thus be set independently. In mobile communication systems, in the absence of the transmitted direct line-of-sight signal, the signal amplitude value of a randomly received signal usually can be modeled as a Rayleigh distribution. It also arises in connection with the errors associated with the aiming of projectiles, if the two errors in the rectangular coordinates have independent zero-mean Gaussian distributions.

7.8.2 The Ricean Distribution

Suppose X_1 and X_2 are two independent Gaussian random variables with means μ_1 and μ_2, respectively, but with common variance σ^2, i.e. $X_1 \sim N(\mu_1, \sigma^2)$ and $X_2 \sim N(\mu_2, \sigma^2)$. The random variable $X = \sqrt{X_1^2 + X_2^2}$ is then known as a *Ricean random variable* and has the following pdf:

$$f_X(x) = \left(\frac{x}{\sigma^2}\right) I_0\left(\frac{sx}{\sigma^2}\right)\left(\exp\left(-\frac{(x^2 + s^2)}{2\sigma^2}\right)\right) \quad x > 0 \qquad (7.37)$$

where $s = \sqrt{\mu_1^2 + \mu_2^2}$ and $I_0(\beta)$ is known as the modified Bessel function of the first kind with order zero. The mean and variance of the Ricean random variable, which can be independently specified, are as follows:

$$\mu_X = \sigma\sqrt{\frac{\pi}{2}} \exp\left(-\frac{s^2}{4\sigma^2}\right)\left(\left(1 + \frac{s^2}{2\sigma^2}\right) I_0\left(\frac{s^2}{4\sigma^2}\right) + \frac{s^2}{2\sigma^2} I_1\left(\frac{s^2}{4\sigma^2}\right)\right)$$
$$\sigma_X^2 = 2\sigma^2 + s^2 - (\mu_X)^2 \qquad (7.38)$$

where $I_1(\beta)$ is known as the modified Bessel function of the first kind with order one. Note that for $s = 0$, the Ricean random variable reduces to the Rayleigh random variable, and for large $\frac{s}{\sigma}$, the Ricean random variable can be well approximated by a Gaussian random variable. The Ricean distribution is often used in fading multipath situations where the mean reflects the line of sight signal and the zero-mean Gaussian random variable part could be due to random multipath components adding up incoherently. The envelope of such a signal is said to be Ricean.

7.8.3 The Log-Normal Distribution

Suppose Y is a Gaussian random variable with mean μ_Y and variance σ_Y^2. The random variable $X = \exp(Y)$ is then known as a *log-normal random variable* and has the following pdf:

$$f_X(x) = \left(\frac{1}{x\sqrt{2\pi\sigma_Y^2}}\right)\exp\left(-\frac{(\ln x - \mu_Y)^2}{2\sigma_Y^2}\right) \quad x > 0 \qquad (7.39)$$

Note that $Y = \ln X$. The log-normal distribution is statistical realization of multiplying many independent random variables, as justified by the CLT in the log domain. The mean and variance of the log-normal random variable, which can be independently specified, are respectively as follows:

$$\mu_X = \exp(\mu_Y + 0.5\sigma_Y^2)$$
$$\sigma_X^2 = (\exp(\sigma_Y^2) - 1)\exp(2\mu_Y + \sigma_Y^2) \tag{7.40}$$

As the log-normal distribution has a long gradually decreasing tail, its median is always smaller than its mean. Examples with the log-normal distribution may include distribution of wealth, machine downtimes, and duration of time off due to sickness. In mobile communications, the log-normal distribution is used to model the effect of shadowing the signal due to large obstructions, such as tall buildings.

7.8.4 The Chi-Square Distribution

Suppose X_1, X_2, ..., X_n are all independent standard Gaussian random variables. The random variable $X = X_1^2 + X_2^2 + \cdots + X_n^2$ is then known as a **chi-square random variable** with n degrees of freedom, and has the following pdf:

$$f_X(x) = \frac{x^{\frac{n-2}{2}}}{2^{\frac{n}{2}}\,\Gamma\left(\frac{n}{2}\right)}\, e^{-\left(\frac{x}{2}\right)} \quad x > 0 \tag{7.41}$$

where $\Gamma(.)$ is the gamma function. The mean and variance of the chi-square random variable, which cannot be independently specified, are as follows:

$$\mu_X = n$$
$$\sigma_X^2 = 2n \tag{7.42}$$

Since a chi-square distribution with n degrees of freedom is generated as the sum of n iid random variables, the CLT implies that for large values of n a chi-square distribution can be approximated by $N(n, 2n)$ distribution. For $n = 2$, a chi-square distribution becomes an exponential distribution. In statistics, the chi-square distribution is used in hypothesis testing and construction of confidence intervals, and can be used to test the goodness of fit between the observed data points and the values predicted by the model, subject to differences being normally distributed.

7.8.5 The Maxwell–Boltzmann Distribution

Suppose X_1, X_2, X_3 are three independent zero-mean Gaussian random variables with common variance σ^2, i.e. we have $X_1 \sim N(0, \sigma^2)$, $X_2 \sim N(0, \sigma^2)$, and $X_3 \sim N(0, \sigma^2)$. The random variable $X = \sqrt{X_1^2 + X_2^2 + X_3^2}$, which is the magnitude of a three-dimensional random vector, is known as a *Maxwell–Boltzmann random variable*, and has the following pdf:

$$f_X(x) = \sqrt{\frac{2}{\pi}} \frac{x^2}{\sigma^3} e^{-\left(\frac{x^2}{2\sigma^2}\right)} \quad x > 0 \tag{7.43}$$

The Maxwell–Boltzmann distribution is the square root of a chi-square random variable with three degrees of freedom. The mean and variance of the Maxwell–Boltzmann random variable, which are related to one another, are as follows:

$$\mu_X = 2\sqrt{\frac{2}{\pi}}\sigma$$

$$\sigma_X^2 = \left(\frac{3\pi - 8}{\pi}\right)\sigma^2 \tag{7.44}$$

The Maxwell–Boltzmann distribution has a number of applications in settings where magnitudes of Gaussian random variables are important, such as the determination of the velocity of a molecule in thermodynamics.

7.8.6 The Student's *t*-Distribution

The Student's *t*-distribution is also referred to as the *t*-distribution. Suppose Y and Z are independent random variables, where Y is the standard Gaussian random variable and Z has a chi-square distribution with n degrees of freedom. The random variable $X = \dfrac{Y}{\sqrt{\frac{Z}{n}}}$ then has the Student's *t* distribution. A ***Student's t random variable*** has the following pdf:

$$f_X(x) = \frac{\Gamma\left(\dfrac{n+1}{2}\right)}{\sqrt{n\pi}\,\Gamma\left(\dfrac{n}{2}\right)} \left(1 + \frac{x^2}{n}\right)^{-\left(\frac{n+1}{2}\right)} \quad -\infty < x < \infty \tag{7.45}$$

If n is large, say $n \geq 30$, the pdf of the *t*-distribution, has a symmetric bell-shaped curve centered at zero, but it is actually a little flatter, and it approximates the standard Gaussian distribution. The mean and variance of the Student's *t*-distribution, which cannot be independently specified, are as follows:

$$\mu_X = 0 \quad n > 1$$

$$\sigma_X^2 = \frac{n}{n-2} \quad n > 2 \tag{7.46}$$

When a Gaussian distribution describes a population, a *t*-distribution describes samples drawn from the population. The *t*-distribution for each sample size is different, and the larger the sample, the more the distribution resembles a Gaussian distribution. The standard Gaussian distribution is in fact the limiting value of the *t*-distribution as $n \to \infty$. The Student's *t*-distribution is employed in estimating the mean of a population with Gaussian distribution whose standard deviation is unknown, while the sample size is small.

7.8.7 The F Distribution

Suppose Y and Z are independent random variables, where Y has a chi-square distribution with n_1 degrees of freedom and Z has a chi-square distribution with n_2 degrees of freedom. Then the random variable $X = \frac{\frac{Y}{n_1}}{\frac{Z}{n_2}}$ has the F distribution with n_1 and n_2 degrees of freedom. An **F random variable** has the following pdf:

$$f_X(x) = \frac{\Gamma\left(\dfrac{n_1 + n_2}{2}\right)}{\Gamma\left(\dfrac{n_1}{2}\right)\Gamma\left(\dfrac{n_2}{2}\right)} \sqrt{\frac{n_1^{\,n_1} n_2^{\,n_2} x^{n_1 - 2}}{(n_2 + n_1 x)^{n_1 + n_2}}} \quad x > 0 \tag{7.47}$$

The mean and variance of the F random variable, which can be independently specified, are as follows:

$$\mu_X = \frac{n_2}{n_2 - 2} \quad n_2 > 2$$

$$\sigma_X^2 = \frac{2(n_2)^2(n_1 + n_2 - 2)}{n_1(n_2 - 4)(n_2 - 2)^2} \quad n_2 > 4 \tag{7.48}$$

The F distribution, similar to the Student's t-distribution, plays key roles in statistical tests of significance, notably in the analysis of variance.

7.8.8 The Cauchy Distribution

The Cauchy distribution arises as the pdf of the ratio of two independent zero-mean Gaussian random variables. A **Cauchy random variable** has the following pdf:

$$f_X(x) = \frac{1}{\pi(1 + x^2)} \tag{7.49}$$

It is symmetric about $x = 0$, but its mean and variance are undefined. The Cauchy distribution has many applications in physics, including resonance behavior.

7.9 Summary

All major aspects of the Gaussian distribution were extensively discussed in this chapter. We then highlighted that the Gaussian distribution is also the asymptotic form of the sum of random variables. In addition, some well-known distributions related to the Gaussian distribution were introduced.

Problems

7.1 The amplitude of a radar signal is Rayleigh distributed. Determine the parameter σ, if measurements show that 0.1% of the time the amplitude exceeds 2.

7.2 Assuming X is a Rayleigh random variable with parameter $\sigma > 0$, determine its mean and variance.

7.3 The random variable X is uniformly distributed over the interval $[0, 1]$ and the random variable Y is uniformly distributed over the interval $[0, 2]$. Assuming the random variables X and Y are independent, determine the pdf of the random variable representing their sum $Z = X + Y$.

7.4 Determine the median of the Rayleigh distribution.

7.5 A data packet consists of 10 000 bits, where each bit is a 0 or a 1 with equal probability. Estimate the probability of having at least 5020 ones.

7.6 Determine the variance of the Gaussian random variable by not using transform methods.

7.7 Suppose the mean and variance of the Gaussian random variable X are 4 and 9, respectively. Using the Q-function, determine $P(0 < X \le 9)$.

7.8 The pdf of a Gaussian random variable X is given by $f_X(x) = K e^{-\frac{(x-4)^2}{18}}$. Find the value of K and determine the integral expression for the probability of X to be negative.

7.9 In a binary transmission system, 0s and 1s are equally likely to be transmitted, and the symbol 0 is represented by -1 and the symbol 1 is represented by $+1$. The channel introduces additive zero-mean unit-variance Gaussian noise. As a result, there are erroneous bits at the receiver. Determine (i.e. write the integral expression for) the average probability of bit error (bit error rate).

7.10 A random variable has a zero-mean, unit-variance Gaussian distribution. A constant 1 is added to this random variable. Determine the probability that the average of two independent measurements of this composite signal is negative.

7.11 Suppose X is a random variable whose mean and variance are $E[X]$ and σ_X^2, respectively. Determine the Chernoff bound on $P(X \geq kE[X])$, where k is a positive integer, and X is a Gaussian random variable.

7.12 Let $Y = \exp X$. Find the pdf of the random variable Y if X is a Gaussian random variable whose mean and variance are μ_X and σ_X^2, respectively.

7.13 Prove that the area under the pdf of a Gaussian random variable with mean μ_X and variance σ_X^2 is equal to one, i.e. show that we have

$$\int_{-\infty}^{\infty} \left(\frac{1}{\sqrt{2\pi\sigma_X^2}} \right) \exp\left(-\frac{(x-\mu_X)^2}{2\sigma_X^2} \right) dx = 1$$

7.14 Suppose the random variable X is a zero-mean unit-variance Gaussian random variable. Determine the variance of the random variable, if we know it is positive, i.e. $\text{Var}(X \mid X > 0)$.

7.15 Suppose we have $Y = X_1 X_2 \cdots X_n$, where the random variables X_1, X_2, ..., X_n are all independent and uniformly distributed in the interval $[0, 1]$. Determine the distribution of the random variable Y.

7.16 The noise at the input of a noise canceller is a Gaussian random variable n_{in} with a mean value of 1 and a standard deviation of 4. The noise canceller makes a noise that is a Gaussian random variable n_0 with a mean value of -1 and a standard deviation of 3. Determine the distribution of the noise n_{out} at the output of the noise canceller, where $n_{\text{out}} = n_{\text{in}} - n_0$, assuming n_{in}, n_0 are independent.

7.17 Calculate and estimate the width or spread of a Gaussian random variable using the Q-function and the Chebyshev inequality, respectively.

7.18 Determine the joint probability density function of $R = \sqrt{X^2 + Y^2}$ and $\Theta = \tan^{-1}\frac{Y}{X}$, if the joint probability density function of the random variable X and Y is $f_{X,Y}(x,y)$. Assuming X and Y are zero-mean independent Gaussian random variables with common variance σ^2, determine the pdfs of R and Θ.

7.19 A source of noise in a communication system is modeled as a Gaussian random variable with zero-mean unit-variance. Determine the probability that the value of the noise exceeds 1. Given that the value of the noise is positive, determine the probability that it exceeds 1.

7.20 The input X to a communication channel takes values 1 or -1 with equal probabilities. The output of the channel is given by $Y = X + N$, where N is a zero-mean, unit-variance Gaussian random variable. Determine $P(X = 1 \mid Y > 0)$.

7.21 Let X be a Gaussian random variable whose mean and variance are 0 and σ_X^2, respectively. Determine the expected value and the variance of X given that $X > 0$.

7.22 Assuming that X is a Gaussian random variable whose mean is 3 and variance is 4, Y is a Gaussian random variable whose mean is -1 and variance is 2, and X and Y are independent, determine the covariance of the two random variables $Z = X - Y$ and $W = 2X + 3Y$. Are the random variables Z and W uncorrelated?

7.23 Let X be a Gaussian random variable whose mean is μ_X and variance is σ_X^2. Determine the expected value of X given that $X < b$, i.e. $E[X \mid X < b]$, where b is a known nonrandom constant.

7.24 Suppose the random variable Y is defined as $Y = a X + b$, where X is a Gaussian random variable whose mean and variance are μ_X and σ_X^2 respectively, and a and b are both nonzero, nonrandom constants. Noting m is a nonrandom constant, write the integral expression for the probability of $Y > m$, in terms of a, b, μ_X, σ_X^2, and m.

7.25 There is a probability of 0.6 that a flawless item can be produced. Determine the number of items that needs to be produced in order to be 99% confident of having at least 1000 flawless items.

7.26 Suppose the brain reaction time can be modeled by a log-normal distribution with parameters $\mu_X = -0.35$ and $\sigma_X = 0.2$. Determine a reaction time that is less than 0.6 seconds as well as the fifth percentile of the reaction times.

7.27 Let S_N be the sum of N independent Gaussian random variables $X_1, X_2,$ ..., X_n with mean 100 and variance 100, where the number of random variables N is itself a Poisson random variable with parameter $\lambda = 1$. Determine the mean and variance of the random sum.

7.28 Let X be a Poisson random variable with parameter λ. Using the central limit theorem, derive the following approximation formula:

$$P(X \le x) \approx \Phi\left(\frac{x - \lambda}{\sqrt{\lambda}}\right)$$

7.29 Suppose the mean and variance of the random variable X are 4 and 3, respectively. Determine $P(2 < X \leq 5)$, if X has a uniform distribution and if it has a Gaussian distribution.

7.30 Suppose that orders at a restaurant are independent identically distributed random variables with mean of \$8 and standard deviation of \$2. Estimate the probability that the first 100 customers spend between \$780 and \$820.

7.31 Let X_1, X_2, and X_3 be independent standard Gaussian random variables. Let

$$Y_1 = X_1 + X_2 + X_3$$
$$Y_2 = X_1 - X_2$$
$$Y_3 = X_2 - X_3$$

Determine the joint pdf of the random variables Y_1, Y_2, and Y_3.

Part III
Statistics

8

Descriptive Statistics

Statistics is a field that bridges probability theory to actual data from the real world. *Statistics* is concerned with the collecting, arranging, summarizing, and analyzing data and with drawing conclusions or inferences from data based on the results of such analysis. Modern statistics has an essential role in a diverse array of major applications, including manufacturing and quality control, climate change and weather forecasts, opinion polling and marketing forecasts, public health, safety and security, economic and stock market forecasts, testing the effectiveness of drugs and treatments, plant breeding and crop production, risk analysis and gambling, and human behavior in psychology, to name a few.

8.1 Overview of Statistics

The raw material of statistics is *data*, generally defined as numbers that result from making a measurement or from the process of counting, and *data set* refers to the set of all observations made for a given purpose. A characteristic is known as a *quantitative variable* when observations made convey information regarding amount, such as patients' age in a hospital. On the other hand, a characteristic is referred to as a *qualitative variable* when observations taken provide information on attribute, such as the ethnic designations of citizens of a county. Data can be gathered from one or more sources, such as records, surveys, experiments, and already published data.

In collecting data, the totality of observations in which we are interested at a particular time is called a *population*. The number of observations in the population is defined to be the *population size*. In a *finite population*, the total number of observations is a finite number. For instance, the set of observations made by measuring the heights of people in a town is an example of a population with finite size. In an *infinite population*, the total number of observations is either an infinite number or so large that it is not possible to count. For instance,

Probability, Random Variables, Statistics, and Random Processes: Fundamentals & Applications,
First Edition. Ali Grami.
© 2020 John Wiley & Sons, Inc. Published 2020 by John Wiley & Sons, Inc.
Companion Website: www.wiley.com/go/grami/PRVSRP

the observations obtained by measuring the temperature at every single point on the Earth's land surface or by gauging the lengths of fish in all oceans are examples of populations whose sizes are infinite. Therefore, it is sometimes impossible or often totally impractical to observe the entire population. So instead, a small segment (subset) of a population, called a *sample*, is randomly chosen to be observed in the course of an investigation. The process of obtaining samples to estimate characteristics of the whole population is called *sampling*. Sampling where each member of the population may be chosen more than once is called *sampling with replacement*, while if each member cannot be chosen more than once is called *sampling without replacement*.

The objective of using samples is to gain insight into the population. In order to eliminate any bias in the sampling procedure and thus to be able to make valid statistical inferences, a sample must be representative of the population from which it is drawn. In the *simple random sampling*, every object in the population must have an equal probability of being selected for the sample. In simple random sampling, samples of the same size have the same chance of being selected.

A characteristic with a numerical value, i.e. a quantity that describes a population, is referred to as a *population parameter*. Any quantity obtained from a sample for the purpose of estimating a population parameter is called a *sample statistic* or briefly a *statistic*. A sample statistic is a random variable, as its value highly depends on which particular sample is obtained. The probability distribution of a sample statistic is called the *sampling distribution of the statistic*. The standard deviation of the sampling distribution of a statistic is known as the *standard error of the statistic*. A sample statistic can help to estimate a population parameter. A conclusion drawn about a population parameter based on a sample statistic is called an *inference*, as shown in Figure 8.1.

As an example, suppose we want to determine the mean height of tens of millions of female adults in a country, a population parameter. We first choose a random sample of a couple of thousand female adults of various ages and races who live in different parts of the country to represent the population of interest. We then determine that their mean height is 165 cm, a sample statistic. Noting concepts in probability help quantify the confidence in our conclusion, we conclude that the population mean height is likely to be close to 165 cm as well. Therefore, the random sample, along with the inferential statistics using probability theory, can allow us to draw conclusions about the population.

The study of statistics can be categorized into two branches: descriptive statistics and inferential statistics, where probability theory provides the transition between them.

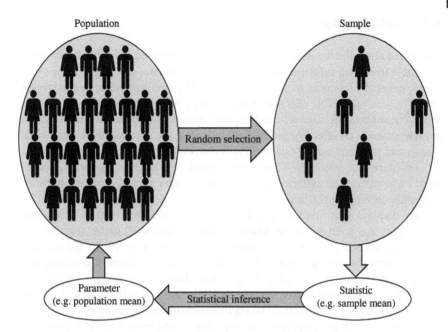

Figure 8.1 Relationship between population parameter and sample statistic.

Descriptive statistics or ***deductive statistics*** is concerned with collecting, organizing, analyzing, summarizing, and presenting the data set in concise and meaningful ways to convey the necessary and relevant information, such as trends and patterns, as clearly as possible. Descriptive statistics is simply a way to describe data rather to draw conclusions. Descriptive statistics is briefly discussed in this chapter, addresses various data displays, and assesses data distributions using measures of central tendency (location), measures of spread (dispersion), and measures of shape.

Inferential statistics or *inductive statistics*, on the other hand, focusses on examining a sample so as to make generalizations about the population from which the sample was drawn, using probability theory. It is, therefore, important that the sample accurately represents the population. Inferential statistics uses sample statistics to infer what the unknown population parameters are. Inferential statistics arises due to sampling error as a sample cannot perfectly represent the population. The inferential statistics is a broad field. However, our primary focus in this book is on two key areas of inferential statistics, namely, estimation and hypothesis testing, that will be briefly discussed in the next two chapters.

8.2 Data Displays

Raw data are collected data that have not been organized numerically. In order to be able to extract information from the collected data, data needs to be organized, generally in order of magnitude from the smallest value to the largest value. In general, a table of numbers is not very revealing, whereas a graphical representation of the data set can be quite informative. *Data display* or *data presentation* is an effective method of data analysis, through which principal trends and patterns can be observed and inferences may be made using methods of inferential statistics. To achieve a good data display, the primary goal must be to ensure the material is as self-contained as possible and understandable without reading the text.

Data displays may include histograms and cumulative relative frequency charts that are used to graph *quantitative (numerical) data*, such as the heights of school children in a town, the weights of newborn babies in a hospital, and the annual income taxes paid by income earners in a country. On the other hand, data displays may include bar, Pareto, and pie charts that are used to graph *qualitative (categorical) data*, such as the political parties with which people are affiliated, levels of education of the people who vote in a general election, the number of sunny, cloudy and rainy days in a year, and annual inflation rates in various countries.

The *frequency* of a data value is the number of times that the data value occurs in the data set. A *frequency distribution* is a table consisting of all possible data values and their corresponding frequencies. The data values can be presented in a grouped or ungrouped manner. A *histogram* is a graphical presentation of frequency distribution, in which vertical bars are drawn above different data values or groups of data values. The height of a bar represents the frequency of data values in the frequency distribution, and the widths of bars are of equal length with no spacing between them.

The *relative frequency* is the ratio of frequency of a data value to the total number of data values in the data set, and the *cumulative relative frequency distribution* gives the proportion of data values less than or equal to a specified value. A *cumulative relative frequency chart* is constructed by plotting cumulative relative frequencies against data values. Such charts are always nondecreasing graphs and are very similar to the graphs of cumulative distribution functions.

A *bar chart* is created by plotting all the categories in the data on one axis and the frequency of occurrence of each category in the data on the other axis. The bar height (if vertical) or length (if horizontal) shows the frequency of each category, where there is no intrinsic ordering to the categories and gaps are included between the bars.

A **Pareto chart** is a bar chart where categories are arranged in order of decreasing frequency. The Pareto chart is used in quality control and process improvement with the objective to sort out the few main causes of most problems, and solve them first.

A **pie chart** is created by dividing a circle (pie) into sectors (pie slices), where the size of a sector reflects the relative frequency of a category in percentage. In other words, the size of each slice is proportional to the probability of the category that the slice represents.

Example 8.1

In a class of 50 students, the distribution of final marks, represented by the values of x, is as follows:

Marks	No. of students	Marks	No. of students
$0\% \leq x < 5\%$	1	$50\% \leq x < 55\%$	3
$5\% \leq x < 10\%$	0	$55\% \leq x < 60\%$	3
$10\% \leq x < 15\%$	0	$60\% \leq x < 65\%$	4
$15\% \leq x < 20\%$	0	$65\% \leq x < 70\%$	8
$20\% \leq x < 25\%$	1	$70\% \leq x < 75\%$	8
$25\% \leq x < 30\%$	0	$75\% \leq x < 80\%$	6
$30\% \leq x < 35\%$	0	$80\% \leq x < 85\%$	5
$35\% \leq x < 40\%$	2	$85\% \leq x < 90\%$	3
$40\% \leq x < 45\%$	1	$90\% \leq x < 95\%$	2
$45\% \leq x < 50\%$	1	$95\% \leq x \leq 100\%$	2

Plot the histogram and the cumulative relative frequency chart for class marks. Assuming we can use the following information to assign grades, plot the bar, Pareto, and pie charts for class grades:

Marks	Grades
$0\% \leq x < 50\%$	F
$50\% \leq x < 60\%$	D
$60\% \leq x < 70\%$	C
$70\% \leq x < 85\%$	B
$85\% \leq x \leq 100\%$	A

Solution

The following table presents various distributions for class marks:

Class marks	Frequency distribution	Relative frequency distribution	Cumulative relative frequency distribution
$0\% \leq x < 5\%$	1	2%	2%
$5\% \leq x < 10\%$	0	0%	2%
$10\% \leq x < 15\%$	0	0%	2%
$15\% \leq x < 20\%$	0	0%	2%
$20\% \leq x < 25\%$	1	2%	4%
$25\% \leq x < 30\%$	0	0%	4%
$30\% \leq x < 35\%$	0	0%	4%
$35\% \leq x < 40\%$	2	4%	8%
$40\% \leq x < 45\%$	1	2%	10%
$45\% \leq x < 50\%$	1	2%	12%
$50\% \leq x < 55\%$	3	6%	18%
$55\% \leq x < 60\%$	3	6%	24%
$60\% \leq x < 65\%$	4	8%	32%
$65\% \leq x < 70\%$	8	16%	48%
$70\% \leq x < 75\%$	8	16%	64%
$75\% \leq x < 80\%$	6	12%	76%
$80\% \leq x < 85\%$	5	10%	86%
$85\% \leq x < 90\%$	3	6%	92%
$90\% \leq x < 95\%$	2	4%	96%
$95\% \leq x \leq 100\%$	2	4%	100%

Figure 8.2a,b shows the histogram and the cumulative relative frequency chart for class marks. Using the mark percentage to grade conversion, we have the following:

Grades	No. of students	Percentage of students
F	6	12%
D	6	12%
C	12	24%
B	19	38%
A	7	14%

Figure 8.3a–c shows the bar, Pareto, and pie charts for class grades. ∎

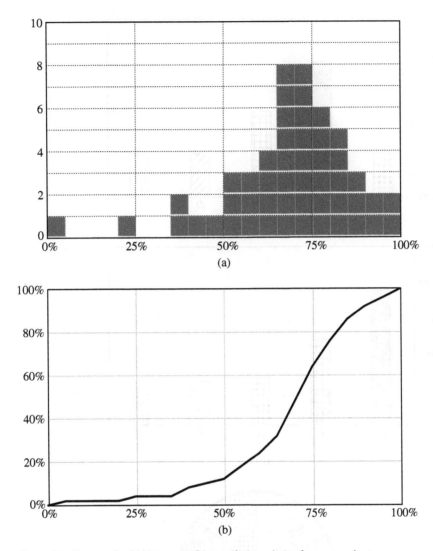

Figure 8.2 Class marks: (a) histogram; (b) cumulative relative frequency chart.

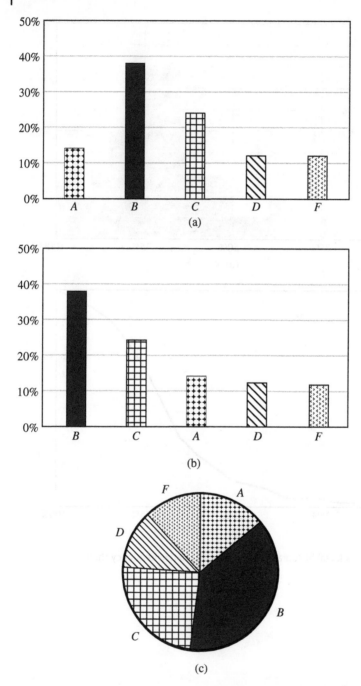

Figure 8.3 Class grades: (a) bar chart; (b) Pareto chart; (c) pie chart.

8.3   Measures of Location

The *measures of location*, also known as the *measures of central tendency*, are ways of describing the central position of distribution for a given set of data.

An important measure of central tendency for a set of data is the *sample mean*, also known as the *arithmetic mean*, or commonly called the *average*. The sample mean summarizes the set of data values into one approximate value. The sample mean, however, may not be a data value in the set. The sample mean takes all of the data values into account, as all of the data values are added and then divided by the number of data values in the set. Assuming the data set consists of $n \geq 1$ data values, $X_1, X_2, \ldots, X_n$, the sample mean is defined as follows:

$$\overline{X}_n = \left(\frac{1}{n}\right) \sum_{i=1}^{n} X_i \tag{8.1}$$

Note that the sum of the deviations of the individual observations of a sample about the sample mean is always zero. It is important to note that $\overline{X}_n$ is a function of the data set, therefore when the data set changes, the value of the sample mean likely changes. The sample mean is thus a random variable, has a probability density function, and as such, has an expected value and a variance.

The sample mean is influenced by all data values in the data set, including *outliers*, i.e. data values that appear remote from all or most of the other data values. Note that an outlier might indicate something went wrong with the data collection process or something important caused an unusual value. Outliers are not necessarily mistakes, and once they are detected, should not be removed from the data set without further investigation.

The median, another measure of central tendency, is resilient to outliers, and may be a better descriptor of a typical value. For instance, housing prices or household incomes are often characterized by the median because of outliers. To determine the median, all data values are arranged in increasing or decreasing order of magnitude. The *sample median* of the set of n sorted data values is defined as the middle value, if n is odd, and it is the average of the two middle values, if n is even. The difference between the mean and the median becomes proportionally larger if the data distribution is not very symmetrical or the sample size is rather small.

Another measure, which is not as sensitive to the tails of the data set as the overall sample mean, is sample trimmed mean. A *sample trimmed mean* is obtained by deleting some of the largest and some of the smallest data observations, and then by taking the mean of the remaining observations. For instance, a 10% trimmed mean removes the top 10% and the bottom 10% of the sorted data values.

The *sample mode*, which is also a measure of central tendency, is defined as the most frequently occurring data value in the data set. For instance, an election is generally determined by the mode, i.e. the candidate who receives the most votes is elected. The mode need not be unique, as a data set can have more than one mode, in which case it is called a ***multi-modal data set***.

An estimate is said to be ***robust*** if its value does not change much when spurious data occur in the sample. Median and mode are more robust than the mean. For instance, a very significant change to a data value in the data set can bring about a significant change to the mean, but results in modest changes to the median and mode. Although robustness is a desirable property, the mean, as it utilizes all of the data, is used most often.

Example 8.2
A student took 10 different courses during the last academic year. Her marks were as follows: 98 %, 86 %, 64 %, 88 %, 53 %, 71 %, 64 %, 69 %, 79 %, and 58%. Determine the mean, median, and mode of her marks.

Solution
Noting that the total number of data values n is 10, the sample mean is as follows:

$$\overline{X}_{10} = \frac{98\% + 86\% + 64\% + 88\% + 53\% + 71\% + 64\% + 69\% + 79\% + 58\%}{10}$$

$$= 73\%$$

As the number of marks is even, the sample median, which is the average of the middle two marks, is as follows:

$$\frac{(71\% + 69\%)}{2} = 70\%$$

Since her most frequent mark is 64%, the sample mode is 64%. ∎

8.4 Measures of Dispersion

The ***measures of spread***, also known as the ***measures of dispersion*** or the ***measures of variability***, are ways of describing the variation of distribution around the central tendency for a given set of data.

The term ***quantile*** refers to a family of measures that can provide information about the spread of data. More specifically, a quantile divides the sorted data into groups, each containing the same fraction of the data set. ***Percentiles*** are well-known quantiles, where 99 percentiles divide the sorted data into 100 equal parts. They are denoted by $P_1, P_2, ..., P_{99}$, while noting that P_k represents approximately k percent of the sorted data that are at or below it.

The most frequently used quantiles are the three ***quartiles*** Q_L, Q_M, and Q_H, which divide the sorted data into four equal parts. Note that the quartile Q_L, known as the ***first quartile*** or ***lower quartile***, is the middle value of the lower

Table 8.1 Various percentiles for the Gaussian distribution.

$F_X(\mu_X - 6\sigma_X)$	$\rightarrow$	0.000 000 01st percentile
$F_X(\mu_X - 5\sigma_X)$	$\rightarrow$	0.000 03rd percentile
$F_X(\mu_X - 4\sigma_X)$	$\rightarrow$	0.003rd percentile
$F_X(\mu_X - 3\sigma_X)$	$\rightarrow$	0.135th percentile
$F_X(\mu_X - 2.326\sigma_X)$	$\rightarrow$	1st percentile
$F_X(\mu_X - 2\sigma_X)$	$\rightarrow$	2.275th percentile
$F_X(\mu_X - \sigma_X)$	$\rightarrow$	15.866th percentile
$F_X(\mu_X - 0.675\sigma_X)$	$\rightarrow$ Q_L:	25th percentile
$F_X(\mu_X)$	$\rightarrow$ Q_M:	50th percentile
$F_X(\mu_X + 0.675\sigma_X)$	$\rightarrow$ Q_H:	75th percentile
$F_X(\mu_X + \sigma_X)$	$\rightarrow$	84.134th percentile
$F_X(\mu_X + 2\sigma_X)$	$\rightarrow$	97.725th percentile
$F_X(\mu_X + 2.326\sigma_X)$	$\rightarrow$	99th percentile
$F_X(\mu_X + 3\sigma_X)$	$\rightarrow$	99.865th percentile
$F_X(\mu_X + 4\sigma_X)$	$\rightarrow$	99.997th percentile
$F_X(\mu_X + 5\sigma_X)$	$\rightarrow$	99.999 97th percentile
$F_X(\mu_X + 6\sigma_X)$	$\rightarrow$	0.999 999 99th percentile

half of the sorted data, the quartile Q_M, known as the **second quartile** or **middle quartile**, is the middle value of the sorted data, and the quartile Q_H, known as the **third quartile** or **upper quartile**, is the middle value of the upper half of sorted data. Moreover, the **interquartile range (IQR)** is defined as the difference between P_{75} and P_{25}, or equivalently between Q_H and Q_L.

For a probability distribution, each specific value of the random variable represents a certain percentile. For a Gaussian (normal) distribution whose cumulative distribution function is represented by $F_X(x)$, and its mean and standard deviation are μ_X and σ_X, respectively, Table 8.1 presents various percentiles of importance.

Unlike the mean, variance, and standard deviation, quartiles and the interquartile range cannot be overly influenced by outliers. Since interquartile range is used when the relatively large and small data values need to be ignored, it is a relatively robust measure of dispersion. Usually, any data value that lies below $Q_L - 1.5 \times \text{IQR}$ or above $Q_H + 1.5 \times \text{IQR}$ may be considered as an outlier. For a Gaussian distribution whose mean and standard deviation are μ_X and σ_X, respectively, any data value that lies below $\mu_X - 2.7\sigma_X$ or above $\mu_X + 2.7\sigma_X$ may be considered as an outlier.

The **range** is the simplest measure of statistical variation or spread and is defined as the difference between the largest value in the data set and the

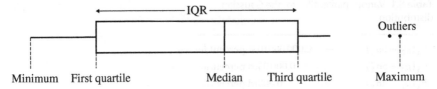

Figure 8.4 Boxplot of a data set.

smallest. This measure is useless, if either of the extreme values is infinite. Another disadvantage of the range is that it provides no information at all about the actual distribution of the data set.

Although histograms are excellent graphic displays for focusing attention on key aspects of the shape of a distribution of data (symmetry, skewness, clustering, gaps), but they are not good tools for making comparisons among multiple datasets. A simple graphical display that is ideal for making comparisons is the boxplot. A **boxplot**, also known as **box-and-whisker diagram**, is a schematic presentation summarizing the measures of position, including the smallest and largest data values, the lower and upper sample quartiles (25% and 75%), the median, and the outliers, as shown in Figure 8.4.

Example 8.3
A random sample of 100 students were asked how many text messages they have sent during the past 24 hours. A summary statistics of a preliminary analysis of data is as follows:

No. of data values	100
Mean	23
Medium	24
Standard deviation	20
Minimum	3
Maximum	81
Q_L	8
Q_H	33

Comment on the presence of outliers.

Solution
Using these summary statistics, we have

$$IQR = 33 - 8 = 25 \rightarrow \begin{cases} Q_L - 1.5 \times IQR = 8 - 1.5 \times 25 = -29.5 \\ Q_H + 1.5 \times IQR = 33 + 1.5 \times 25 = 70.5 \end{cases}$$

We have $-29.5 < 0$, as no count of text messages could be negative, there could not be any outliers on the lower end. On the other hand, we have $70.5 < 81$, there is at least one outlier in the study. ∎

As a measure of spread, it is important to determine the numerical difference between a data value and the mean of the data set, known as its deviation from the mean, and then find the arithmetic mean of all deviations associated with all data in the set. This can obviously provide a single number by which actual data values are varying from the mean. However, this number, representing the mean deviation, is always equal to zero, thus failing to provide insight into the spread of the distribution. In order to avoid this problem, it is customary to determine the arithmetic mean of all squared deviations, known as the *sample variance*. The sample variance is always nonnegative and reflects the spread of the distribution, i.e. its randomness. The sample variance is defined as follows:

$$\sigma_n^2 = \left(\frac{1}{n}\right) \sum_{i=1}^{n} (X_i - \overline{X}_n)^2 \tag{8.2a}$$

Another definition of sample variance, which is in fact more common, is as follows:

$$\sigma_n^2 = \left(\frac{1}{n-1}\right) \sum_{i=1}^{n} (X_i - \overline{X}_n)^2 \tag{8.2b}$$

Note that the squares of the deviations of the observations from their mean is averaged over n in (8.2a) and over $n-1$ in (8.2b), the latter is known as an unbiased estimate of variance, and the former as a biased estimate of variance. In any event, when n is a very large integer, the difference between these two variances becomes very insignificant.

If we take the positive square root of the sample variance σ_n^2, we then have the *sample standard deviation*, also known as the *sample root mean square deviation*. The standard deviation has the same unit as the data values have. Standard deviation describes the degree of variation from the mean. A low standard deviation reflects that data values are all similar and close to the mean, while a high standard deviation reflects that the data are more spread out. The standard deviation is used to provide bounds around the sample mean that describe data positions within multiples of standard deviations (i.e. $\pm m\sigma_n$, $m = 1, 2, \ldots$). It is also common that any data value that lies below $\overline{X}_n - 3\sigma_n$ or above $\overline{X}_n + 3\sigma_n$ may be considered as an outlier.

Example 8.4
In a graduate course, there were 10 students, whose final marks, out of 100, were as follows: 98, 86, 64, 88, 53, 71, 64, 69, 79, and 58. Determine the biased estimate of variance and standard deviation.

Solution
We first need to determine the mean, so-called class average. Noting that $n = 10$, we then have the following:

$$\overline{X}_{10} = \frac{98 + 86 + 64 + 88 + 53 + 71 + 64 + 69 + 79 + 58}{10} = 73$$

We also have the following variance:

$$\sigma_{10}^2 = \left(\frac{1}{10}\right)\left((98 - 73)^2 + (86 - 73)^2 + (64 - 73)^2 + (88 - 73)^2\right.$$
$$+ (53 - 73)^2 + (71 - 73)^2 + (64 - 73)^2$$
$$\left. + (69 - 73)^2 + (79 - 73)^2 + (58 - 73)^2\right) = 186$$

The standard deviation is thus as follows:

$$\sigma_{10} = \sqrt{186} \cong 13.6$$

■

In a data set, standard deviation is a commonly used measure to assess the variation in the data set. To make comparison between different data sets in terms of their variability expressed relative to their sample mean, standard deviation by itself may not be the right measure, as their means may be different. It is generally useful to compare each data set's relative variation, rather than the individual standard deviations. To this end, the **coefficient of variation**, defined as the ratio of the standard deviation to the mean, for each data set needs to be calculated and then compared. The coefficient of variation is thus defined as follows:

$$\text{Coefficient of variation} = \frac{\sigma_n}{\overline{X}_n} \tag{8.3}$$

A larger coefficient of variation highlights a greater dispersion relative to the sample mean. The coefficient of variation, which has no unit, reflects the variations of samples, even if the two data sets are not measured in the same units.

Example 8.5
The height of a bridge on a highway is measured using two different equipment. Measurements made with equipment A have a mean of 3.92 m and a standard deviation of 1.53 cm, whereas measurements made with equipment B have a mean of 12.93 ft and a standard deviation of 0.59 in. Which of these two equipment is relatively more precise?

Solution
Noting 1 cm = 0.01 m, for equipment A, the coefficient of variation is as follows:

$$\frac{1.53 \times 10^{-2}}{3.92} \times 100 \cong 0.39\%$$

Noting 1 ft $= 12$ in, for equipment B, the coefficient of variation is as follows:

$$\frac{0.59}{12.93 \times 12} \times 100 \cong 0.38\%$$

We can conclude that the measurements made with equipment B are relatively more precise. ∎

It is important to highlight that if all measurements made are shifted by a positive or negative constant, then the measure of central tendency gets shifted by the same constant, but the measure of variation is unaffected by any shift in measurements. This is in contrast to the fact that if all measurements are scaled by a constant other than 1, then the measures of central tendency and variation are both affected.

8.5 Measures of Shape

A probability distribution can provide a general idea of the shape, but there are two *measures of shape* that give a more precise evaluation, namely, skewness that indicates the amount and direction of skew of a distribution (departure from horizontal symmetry), and kurtosis that characterizes the length and thinness of tails of a distribution, relative to those of a Gaussian distribution. One major application is to test if the distribution of interest is Gaussian, as many statistical inferences require that a distribution be Gaussian or nearly Gaussian. A Gaussian distribution has zero skewness as well as excess kurtosis of zero, so if the distribution is close to those values, then it is probably close to Gaussian.

Skewness is the degree of asymmetry, or departure from symmetry, of a probability distribution. Sometimes a distribution is not symmetric about its mean, but instead has one of its tails longer than the other, as shown in Figure 8.5. The third central moment divided by the third power of the standard deviation (to make this measure independent of scale or origin) is used to describe the *coefficient of skewness*, which is a dimensionless quantity. The coefficient of skewness is positive or negative if the distribution is skewed to the right or left, respectively. For a symmetric distribution, such as the Gaussian distribution, the coefficient of skewness is zero. Note that it is possible to define coefficient of skewness using other measures of central tendency.

If the longer and flatter tail of a distribution occurs to the right, it is said to be *skewed to the right* or *positively skewed*, as its mean is greater than its mode, such a probability distribution appears as a left-leaning curve. For example, the distribution of incomes for employees in a large company is generally skewed to the right, as few people in a given company earn high salaries. If the longer and flatter tail of a distribution occurs to the left, it is said to be *skewed to the left* or *negatively skewed*, as its mean is less than its mode, such a probability distribution appears as right-leaning curve. For example, the distribution of

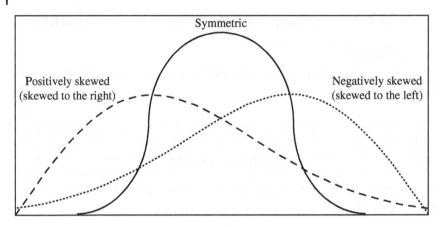

Figure 8.5 Skewness of distribution.

grades on an easy test is typically skewed to the left, as more students get high marks. Sometimes, a well-known distribution is always skewed to the left or to the right, such as the Rayleigh distribution which is always skewed to the right, and sometimes, a widely used distribution can be skewed to the left or to the right, depending on some of its parameters, such as the beta distribution.

Kurtosis is a measure of the tailedness of a probability distribution, as shown in Figure 8.6. There are various measures to define the coefficient of kurtosis. The most commonly ***coefficient of kurtosis*** is defined as the fourth central moment divided by the fourth power of the standard deviation. This measure is invariant under a change of scale or origin. The coefficient of kurtosis of any Gaussian distribution is 3, with ***excess kurtosis*** defined as the coefficient of kurtosis minus three, a Gaussian distribution has then an excess kurtosis equal to 0. It is common to compare the excess kurtosis of a probability distribution to 0.

For many probability distributions with practical applications, a positive excess kurtosis corresponds to a taller and sharper peak with longer and fatter tails than the Gaussian distribution and a negative excess kurtosis corresponds to a shorter and broader peak with shorter and thinner tails than the Gaussian distribution. To this effect, kurtosis has been mistakenly regarded as a measure of peakedness. Kurtosis is related to the tails of the distribution, not its degree of peakedness. In fact, a distribution with a flat top may have a large positive excess kurtosis, while one with extreme peakedness may have a large negative excess kurtosis. The coefficient of kurtosis and in turn the excess kurtosis provides a measure of outliers in a distribution. A larger kurtosis generally indicates a more serious outlier problem. In other words, a larger kurtosis is the result of rare extreme deviations as opposed to frequent modest deviations.

Distributions with zero or around zero excess kurtosis are said to be ***mesokurtic***. An example of mesokurtic distribution is a binomial distribution with $p \cong 0.5$. Distributions with negative excess kurtosis are said to be

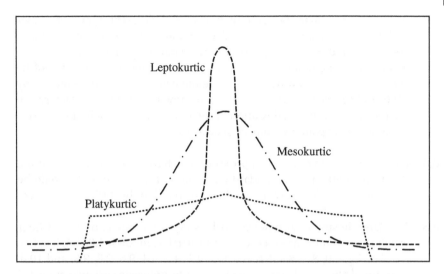

Figure 8.6 Tailedness of distribution.

platykurtic. This means that the distribution produces fewer and less extreme outliers than does the Gaussian distribution. An example of a platykurtic distribution is the uniform distribution, which does not produce outliers. Distributions with positive excess kurtosis are said to be *leptokurtic*. An example of a leptokurtic distribution is the Laplace distribution, which has tails that asymptotically approach zero more slowly than a Gaussian, and therefore produces more outliers than the Gaussian distribution.

8.6 Summary

Since in most applications, the distribution of a random variable is not known, the collection of a data set relating to the random variable and presenting and analyzing the data is of great importance. The descriptive statistics, through measures of central tendency, measures of dispersion, and measures of shape, were discussed in this chapter.

Problems

8.1 Give an example of a data set of nine observations that have the same mean, median, and mode.

8.2 Find the unbiased estimate of variance of the data 3, 4, 5, 6, 6, and 7, representing the number of cars sold by a random sample of six salespersons in a city during a sale period.

8.3 A snowmobile manufacturer makes several different types of snow-mobiles. The mean and standard deviation of prices of snowmobiles are $5000 and $1250 respectively. In response to significant increase in snowfalls, the manufacturer has decided to increase the prices of its snowmobiles in two ways: (i) increase the prices of each snowmobile by $500, and (ii) increase the price of each snow mobile by 10%. Determine the mean and standard deviation of the new prices for each of these two schemes and explain the impact of each scheme.

8.4 If the scores for an exam are approximated to have a Gaussian distribution, the mean is 50 and the standard deviation is 15, a score of 35 would be what percentile? What score would correspond to the 99.9% percentile?

8.5 In a large hospital, 20 newborn babies are randomly chosen and their weights, in pounds, are recorded and sorted as follows: 1.8, 2.6, 3.6, 5.0, 5.2, 5.6, 7.2, 8.0, 8.2, 8.4, 8.6, 8.6, 9.2, 9.2, 9.2, 9.4, 9.6, 9.8, 9.8, and 10.0. Determine, the sample mean, median, mode, unbiased variance, standard deviation, quartiles, and coefficient of variation.

9

Estimation

Estimation has a host of applications in engineering, business, medicine, and economics. As shown in Figure 9.1, we briefly discuss parameter estimation as well as estimation of a random variable. Parameter estimation is concerned with the estimation of a parameter of a random variable, such as its mean and variance. Estimation of a random variable deals with the value of an inaccessible random variable in terms of the observation of an accessible random variable, such as the estimation of the communication channel input based on the observed channel output.

9.1 Parameter Estimation

Parameter estimation is a process through which either a single value or a range of values is obtained to approximate the actual value for a parameter of a random variable based on the available data. Note that the unknown parameter of the random variable that has to be estimated can be either a deterministic (non-random) quantity or a random quantity with a known prior distribution. An estimate that specifies a single value is called a *point estimate* and an estimate that specifies a range of values is called an *interval estimate*, where an interval, known as the confidence interval, covers the true value of the parameter with a specified probability. For instance, in estimating the average height of adults in a country, 173 cm represents a point estimate, but the range [168 cm, 178 cm] with 95% confidence, on the other hand, represents an interval estimate.

To estimate an unknown parameter, we need to collect some data. Let X be a random variable, and the set of independent, identically-distributed (iid) random variables $X_1, X_2, ..., X_n$ is called a *random sample* of size n of X. Any real-valued function of a random sample $X_1, X_2, ..., X_n$ is called a *statistic*. A statistic that is a function of the random vector $X = (X_1, X_2, ..., X_n)$ and is used to estimate an unknown parameter θ is called an *estimator* $\hat{\Theta}$. Therefore,

Probability, Random Variables, Statistics, and Random Processes: Fundamentals & Applications,
First Edition. Ali Grami.
© 2020 John Wiley & Sons, Inc. Published 2020 by John Wiley & Sons, Inc.
Companion Website: www.wiley.com/go/grami/PRVSRP

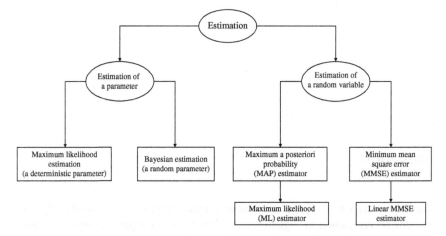

Figure 9.1 A simple breakdown of estimation problems.

the estimator $\widehat{\Theta}$ itself is a random variable because it depends on the random sample. After making n observations, the set $x_1, x_2, \ldots, x_n$ represents the values of the observed data. The realized value of an estimator for a particular sample of data, $X_1 = x_1, X_2 = x_2, \ldots, X_n = x_n$, is called an ***estimate*** of θ, denoted by $\widehat{\theta}$. Note that the hat notation distinguishes the sample-based quantity from the parameter of interest. Simply put, a caret placed over a parameter signifies a statistic used as an estimate of the parameter. An estimator is a measured value based on a random sample to estimate a parameter, as its value highly depends on which particular sample is obtained. Estimates can only be as good as the data set from which they are calculated. Note that it is customary to apply the term statistics to both estimates and estimators.

9.2 Properties of Point Estimators

A point estimate of an unknown parameter θ is a statistic $\widehat{\theta}$ that represents the best guess at the value of θ. There may be more than one sensible point estimate of a parameter. Figure 9.2 shows the relationship between a point estimate $\widehat{\theta}$ and an unknown parameter θ. A good estimator should have properties such as being unbiased, efficient, and consistent.

An estimator for a population parameter is unbiased if the difference between the expected value of the estimator and the true value of the parameter being estimated is zero; otherwise the estimator is said to be biased. Thus, the estimator $\widehat{\Theta}$ is an ***unbiased estimator*** of the parameter θ, if we have

$$E[\widehat{\Theta}] = \theta \qquad\qquad (9.1a)$$

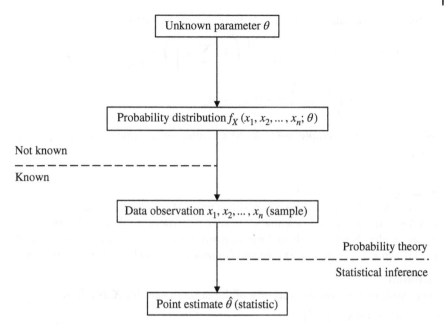

Figure 9.2 Relationship between a point estimate $\hat{\theta}$ and an unknown parameter θ.

In other words, an unbiased estimator yields the actual value of the unknown parameter, on the average. For an unbiased estimator $\hat{\Theta}$, the **mean square estimation error** equals its variance, that is we have

$$e_{\text{MSE}} = E[(\hat{\Theta} - \theta)^2] = \sigma_{\hat{\Theta}}^2 \tag{9.1b}$$

Unbiasedness is a desirable property for a point estimate to possess. However, biased estimates are not unusual, as they are often relatively easy to obtain.

Example 9.1
Let $X = (X_1, X_2, \ldots, X_n)$ be a random sample of the random variable X with unknown mean μ_X, where $X_1, X_2, \ldots, X_n$ are independent, identically-distributed random variables. Show that the estimator of μ_X, known as the sample mean and denoted by $\overline{X}_n$, is an unbiased estimator of μ_X.

Solution
The **sample mean** is given by:

$$\overline{X}_n = \left(\frac{1}{n}\right) \sum_{i=1}^{n} X_i$$

The expected value of the sample mean is thus as follows:

$$E[\overline{X}_n] = E\left[\left(\frac{1}{n}\right)\sum_{i=1}^{n} X_i\right] = \frac{1}{n}\left(E\left[\sum_{i=1}^{n} X_i\right]\right) = \frac{1}{n}\left(\sum_{i=1}^{n} E[X_i]\right)$$

$$= \frac{1}{n}\left(\sum_{i=1}^{n} \mu_X\right) = \frac{1}{n}(n\mu_X) = \mu_X$$

We thus have the sample mean $\overline{X}_n$ as an unbiased estimator of the expected value μ_X. ∎

Example 9.2

Let $X = (X_1, X_2, ..., X_n)$ be a random sample of the random variable X with unknown mean μ_X and unknown variance σ_X^2, where $X_1, X_2, ..., X_n$ are independent identically-distributed random variables. Show that the estimator of σ_X^2, known as the sample variance and denoted by σ_n^2, is a biased estimator of σ_X^2.

Solution

By definition, we have the variance of the random variable X as follows:

$$\sigma_X^2 = E[(X - \mu_X)^2]$$

The estimator of σ_X^2, also known as the **sample variance** and defined as the arithmetic average of the square variation about the sample mean, is as follows:

$$\sigma_n^2 = \left(\frac{1}{n}\right)\sum_{i=1}^{n}(X_i - \overline{X}_n)^2$$

Noting that the **variance of the sample mean** is as follows:

$$\mathrm{Var}[\overline{X}_n] = \left(\frac{1}{n^2}\right)\left(\sum_{i=1}^{n}\mathrm{Var}[X_i]\right) = \left(\frac{1}{n^2}\right)(n)\,\mathrm{Var}[X_i] = \left(\frac{1}{n}\right)\sigma_X^2 = \frac{\sigma_X^2}{n}$$

the expected value of the sample variance can then be simplified as follows:

$$E[\sigma_n^2] = E\left[\frac{1}{n}\sum_{i=1}^{n}(X_i - \overline{X}_n)^2\right] = E\left[\frac{1}{n}\sum_{i=1}^{n}((X_i - \mu_X) - (\overline{X}_n - \mu_X))^2\right]$$

$$= \left(\frac{1}{n}\right)E\left[\sum_{i=1}^{n}\left((X_i - \mu_X)^2 - 2(X_i - \mu_X)(\overline{X}_n - \mu_X)\right.\right.$$

$$\left.\left. + (\overline{X}_n - \mu_X)^2\right)\right]$$

$$= \left(\frac{1}{n}\right)E\left[\sum_{i=1}^{n}(X_i - \mu_X)^2\right] - \left(\frac{2}{n}\right)E\left[\sum_{i=1}^{n}(X_i - \mu_X)(\overline{X}_n - \mu_X)\right]$$

$$+ \left(\frac{1}{n}\right)E\left[\sum_{i=1}^{n}(\overline{X}_n - \mu_X)^2\right]$$

$$= \left(\frac{1}{n}\right) \sum_{i=1}^{n} E[(X_i - \mu_X)^2] - \left(\frac{2}{n}\right) E\left[(\overline{X}_n - \mu_X) \sum_{i=1}^{n}(X_i - \mu_X)\right]$$

$$+ \left(\frac{1}{n}\right) \sum_{i=1}^{n} E[(\overline{X}_n - \mu_X)^2]$$

$$= \left(\frac{1}{n}\right) \sum_{i=1}^{n} E[(X_i - \mu_X)^2] - \left(\frac{2}{n}\right) E[(\overline{X}_n - \mu_X)(n\overline{X}_n - n\mu_X)]$$

$$+ \left(\frac{1}{n}\right)(n)E[(\overline{X}_n - \mu_X)^2]$$

$$= \left(\frac{1}{n}\right) \sum_{i=1}^{n} E[(X_i - \mu_X)^2] - E[(\overline{X}_n - \mu_X)^2]$$

$$= \left(\frac{1}{n}\right)(n)\sigma_X^2 - \frac{\sigma_X^2}{n} = \left(\frac{n-1}{n}\right)\sigma_X^2$$

Therefore, σ_n^2 is a biased estimator of σ_X^2. Note that it is possible to obtain an unbiased estimator of σ_X^2 for a finite sample size n, if the estimator of σ_X^2 is initially defined as follows:

$$\sigma_n^2 = \left(\frac{1}{n-1}\right) \sum_{i=1}^{n} (X_i - \overline{X}_n)^2 \rightarrow E[\sigma_n^2] = \sigma_X^2 \qquad \blacksquare$$

In point estimation, the property of being unbiased is necessary, but not essential, for it may be outweighed by other factors. A shortcoming for the criterion of unbiasedness is that it may not provide a unique statistic for a given problem of estimation. For instance, for a population with the Gaussian distribution, the sample mean and the sample median are both unbiased, but for the same sample size, the variance of the sample median is greater than the variance of the sample mean. This in turn suggests that another criterion may be needed to decide which unbiased estimator is best for estimating a given parameter. An estimator is a more ***efficient estimator*** of the parameter if its variance is smaller than that of another estimator, provided that both estimators are unbiased. Note that an estimator is the most efficient unbiased estimator of the parameter if it has the minimum variance. In practice, inefficient estimators are often used, as they are rather easy to obtain.

Example 9.3

Suppose there is a population with a Poisson distribution whose parameter λ, representing its mean, is unknown. Show that the sample mean is an unbiased estimator of λ and by increasing the sample size, the estimator becomes more efficient.

Solution

The expected value of the sample mean is as follows:

$$E[\overline{X}_n] = E\left[\left(\frac{1}{n}\right)\sum_{i=1}^{n}X_i\right] = \frac{1}{n}\left(E\left[\sum_{i=1}^{n}X_i\right]\right) = \frac{1}{n}\left(\sum_{i=1}^{n}E[X_i]\right)$$

$$= \frac{1}{n}\left(\sum_{i=1}^{n}\lambda\right) = \frac{1}{n}(n\lambda) = \lambda$$

Thus, the sample mean $\overline{X}_n$ is an unbiased estimator of λ. The variance of $\overline{X}_n$ is as follows:

$$\mathrm{Var}[\overline{X}_n] = \left(\frac{1}{n^2}\right)\left(\sum_{i=1}^{n}\mathrm{Var}[X_i]\right) = \left(\frac{1}{n^2}\right)(n)\,\mathrm{Var}[X_i] = \left(\frac{1}{n^2}\right)n\lambda = \frac{\lambda}{n}$$

Clearly by increasing the sample size n, the estimator of λ becomes more efficient. This confirms our intuition that using more data to estimate the mean corresponds in mathematical terms to obtaining a point estimate with a smaller variance. ∎

Another measure of quality for an estimator is its consistency. A consistent estimator becomes more efficient as the sample size increases. The estimator $\hat{\Theta}_n$, which is based on a random sample of size n, is said to be a **consistent estimator**, if, for any small $\varepsilon > 0$, we have the following:

$$\lim_{n\to\infty} P(|\hat{\Theta}_n - \theta| \geq \varepsilon) = 0 \tag{9.2}$$

For a biased estimator or an asymptomatically biased estimator to be consistent, it is sufficient to show that the variance asymptotically goes to zero as the sample size increases.

Example 9.4

Suppose there is a population with a uniform distribution over the interval [0, a], where a is a positive real number. Show that the sample mean is a consistent estimator for the mean $\frac{a}{2}$.

Solution

The mean and variance of the sample mean are respectively as follows:

$$E[\overline{X}_n] = E\left[\left(\frac{1}{n}\right)\sum_{i=1}^{n}X_i\right] = \left(\frac{1}{n}\right)\left(E\left[\sum_{i=1}^{n}X_i\right]\right)$$

$$= \left(\frac{1}{n}\right)\left(\sum_{i=1}^{n}E[X_i]\right) = \left(\frac{1}{n}\right)\left(\sum_{i=1}^{n}\frac{a}{2}\right) = \left(\frac{1}{n}\right)\frac{na}{2} = \frac{a}{2}$$

and

$$\mathrm{Var}[\overline{X}_n] = \left(\frac{1}{n^2}\right)\left(\sum_{i=1}^{n}\mathrm{Var}[X_i]\right) = \left(\frac{1}{n^2}\right)(n)\,\mathrm{Var}[X_i] = \left(\frac{1}{n^2}\right)(n)\frac{a^2}{12} = \frac{a^2}{12n}$$

Using Chebyshev's inequality, we have the following:

$$
P\left(\left|\overline{X}_n - \frac{a}{2}\right| \ge \varepsilon\right) \le \frac{E\left[\left(\overline{X}_n - \frac{a}{2}\right)^2\right]}{\varepsilon^2}
$$

$$
= \frac{1}{\varepsilon^2} E\left[\left(\overline{X}_n - E[\overline{X}_n] + E[\overline{X}_n] - \frac{a}{2}\right)^2\right]
$$

$$
= \frac{1}{\varepsilon^2} E\left[(\overline{X}_n - E[\overline{X}_n])^2 + \left(E[\overline{X}_n] - \frac{a}{2}\right)^2\right.
$$

$$
\left. + 2(\overline{X}_n - E[\overline{X}_n])\left(E[\overline{X}_n] - \frac{a}{2}\right)\right]
$$

$$
= \frac{1}{\varepsilon^2}\left(\operatorname{Var}[\overline{X}_n] + E\left[\left(E[\overline{X}_n] - \frac{a}{2}\right)^2\right]\right.
$$

$$
\left. + 2E\left[(\overline{X}_n - E[\overline{X}_n])\left(E[\overline{X}_n] - \frac{a}{2}\right)\right]\right)
$$

$$
= \frac{1}{\varepsilon^2}\left(\frac{a^2}{12n} + E\left[\left(\frac{a}{2} - \frac{a}{2}\right)^2\right]\right.
$$

$$
\left. + 2E\left[\left(\overline{X}_n - \frac{a}{2}\right)\left(\frac{a}{2} - \frac{a}{2}\right)\right] = \frac{1}{\varepsilon^2}\left(\frac{a^2}{12n}\right)\right)
$$

We thus have

$$
\lim_{n\to\infty} P\left(\left|\overline{X}_n - \frac{a}{2}\right| \ge \varepsilon\right) = \lim_{n\to\infty} \frac{1}{\varepsilon^2}\left(\frac{a^2}{12n}\right) = 0
$$

Hence the sample mean $\overline{X}_n$ is a consistent estimator for the mean. ∎

9.3 Maximum Likelihood Estimators

In maximum likelihood estimation, considered as a classical approach to parameter estimation, the parameter of interest is assumed to be a non-random, but unknown, constant. Maximum likelihood method finds a point estimator for an unknown parameter. Maximum likelihood estimation selects as its estimate the unknown, but deterministic, value of the parameter that maximizes the probability of the observed data. Maximum likelihood estimators are not optimal in general. It is, however, the most widely-used approach to obtain practical estimators, as it can be implemented for complex estimation problems and it can also yield optimal performance when the data set is large enough, i.e. it is asymptotically efficient. If an efficient estimator exists, the maximum likelihood estimation will yield it.

Suppose we have a sequence of n iid observed values of a particular random sample, $x_1, x_2, \ldots, x_n$, for the random variable X, and assume θ is the parameter of interest to be estimated. We now define the ***likelihood function*** $L(\theta)$

that is a function of both the n observed sample values and the unknown, but deterministic, parameter θ, as follows:

$$L(\theta) = \begin{cases} f_X(x_1, x_2, \ldots, x_n; \theta) = f_X(x_1; \theta)f_X(x_2; \theta)\cdots f_X(x_n; \theta) = \prod_{i=1}^{n} f_X(x_i; \theta) \\ \qquad \text{where } X \text{ is a continuous random variable} \\ p_X(x_1, x_2, \ldots, x_n; \theta) = p_X(x_1; \theta)p_X(x_2; \theta)\cdots p_X(x_n; \theta) = \prod_{i=1}^{n} p_X(x_i; \theta) \\ \qquad \text{where } X \text{ is a discrete random variable} \end{cases}$$

(9.3a)

Since $L(\theta)$ is a product of either probability density functions (pdf's) or probability mass functions (pmf's), it is always positive. It is thus possible to work with its logarithm rather than the function itself, as we can then work with the sum of terms instead of the product of terms, while noting that both the logarithm of a function and the function itself have their maximum values at the same value. We now define the ***log-likelihood function*** as follows:

$$l(\theta) = \ln L(\theta) = \begin{cases} \sum_{i=1}^{n} \ln f_X(x_i; \theta) \text{ where } X \text{ is a continuous random variable} \\ \sum_{i=1}^{n} \ln p_X(x_i; \theta) \text{ where } X \text{ is a discrete random variable} \end{cases}$$

(9.3b)

The maximum likelihood method selects the estimator value $\widehat{\Theta}_{ML} = \widehat{\theta}$, where $\widehat{\theta}$ is the parameter value that maximizes the value of $l(\theta)$ or $L(\theta)$, whichever is easier to maximize. In other words, the ***maximum likelihood estimation*** helps select the parameter value $\widehat{\theta}$ that maximizes the probability of the particular observation sequence $x_1, x_2, \ldots, x_n$. It is important to note that should the likelihood or log-likelihood function consist of more than one unknown parameter, we then need to take the partial derivatives of the likelihood or log-likelihood function with respect to each parameter, set each derivative to zero, and solve the resulting equations simultaneously to determine the parameters of interest. The maximum likelihood estimation in most cases converges to the true parameter value when the number of samples approaches infinity (i.e. as $n \to \infty$).

Example 9.5
Suppose we have a sequence of n independent, identically-distributed observed values of a random sample, $x_1, x_2, \ldots, x_n$, for an exponential random variable X with unknown parameter λ. Determine the maximum-likelihood estimate for λ.

Solution
The likelihood function is as follows:

$$L(\lambda) = f_X(x_1, x_2, \ldots, x_n; \lambda) = \prod_{i=1}^{n} f_X(x_i; \lambda) = \prod_{i=1}^{n} \lambda e^{-\lambda x_i} = \lambda^n e^{-\lambda(x_1 + x_2 + \ldots + x_n)}$$

The log-likelihood function is then as follows:

$$l(\lambda) = \ln L(\lambda) = n \ln \lambda - \lambda \sum_{i=1}^{n} x_i$$

The derivative of the log-likelihood function with respect to λ is set equal to zero to determine the maximum likelihood estimate of λ, we thus have the following:

$$\frac{d}{d\lambda} l(\lambda) = \frac{n}{\lambda} - \sum_{i=1}^{n} x_i = 0 \rightarrow \hat{\lambda} = \frac{n}{\displaystyle\sum_{i=1}^{n} x_i} = \frac{1}{\dfrac{\displaystyle\sum_{i=1}^{n} x_i}{n}}$$

Hence the maximum likelihood estimator of λ is as follows:

$$\hat{\Lambda}_{\text{ML}} = \frac{1}{\dfrac{\displaystyle\sum_{i=1}^{n} X_i}{n}} = \frac{1}{\overline{X}_n}$$

where $\overline{X}_n$ is the sample mean. Obtaining $\overline{X}_n = (\hat{\Lambda}_{\text{ML}})^{-1}$ makes sense, as the expected value of the exponential random variable X is λ^{-1}. ∎

Example 9.6
Suppose we have a sequence of n independent, identically-distributed observed values of a random sample, $x_1, x_2, \ldots, x_n$, for a Poisson random variable X with unknown parameter λ. Determine the maximum-likelihood estimate for λ.

Solution
The likelihood function is as follows:

$$L(\lambda) = f_X(x_1, x_2, \ldots, x_n; \lambda) = \prod_{i=1}^{n} f_X(x_i; \lambda) = \prod_{i=1}^{n} \frac{e^{-\lambda} \lambda^{x_i}}{x_i!} = \frac{e^{-n\lambda} \lambda^{(x_1 + x_2 + \ldots + x_n)}}{\displaystyle\prod_{i=1}^{n} x_i!}$$

The log-likelihood function is then as follows:

$$l(\lambda) = \ln L(\lambda) = -n\lambda + \left(\sum_{i=1}^{n} x_i \right) \ln \lambda - \sum_{i=1}^{n} \ln(x_i!)$$

The derivative of the log-likelihood function with respect to λ is set equal to zero to determine the maximum likelihood estimate of λ, we thus have the following:

$$\frac{d}{d\lambda} l(\lambda) = -n + \left(\sum_{i=1}^{n} x_i \right) \frac{1}{\lambda} = 0 \rightarrow \hat{\lambda} = \frac{1}{n} \sum_{i=1}^{n} x_i$$

Hence the maximum likelihood estimator of λ is as follows:

$$\hat{\Lambda}_{\mathrm{ML}} = \frac{1}{n} \sum_{i=1}^{n} X_i = \overline{X}_n$$

where $\overline{X}_n$ is the sample mean. Obtaining $\overline{X}_n = \hat{\Lambda}_{\mathrm{ML}}$ makes sense, as the expected value of the Poisson random variable X is λ. ∎

Example 9.7

Suppose we have a sequence of n independent, identically-distributed observed values of a random sample, $x_1, x_2, \ldots, x_n$, for a Bernoulli random variable X with unknown probability of success p. Determine the maximum-likelihood estimate for p.

Solution

The likelihood function is as follows:

$$L(p) = f_X(x_1, x_2, \ldots, x_n; p) = \prod_{i=1}^{n} f_X(x_i; p) = \prod_{i=1}^{n} p^{x_i} (1-p)^{1-x_i}$$

$$= p^{(x_1 + x_2 + \cdots + x_n)} (1-p)^{n - (x_1 + x_2 + \cdots + x_n)}$$

The log-likelihood function is then given by:

$$l(p) = \ln L(p) = \left(\sum_{i=1}^{n} x_i \right) \ln p + \left(n - \sum_{i=1}^{n} x_i \right) \ln(1-p)$$

The derivative of the log-likelihood function with respect to p is set equal to zero to determine the maximum likelihood estimate of p, we thus have the following:

$$\frac{d}{dp} l(p) = \left(\sum_{i=1}^{n} x_i \right) \left(\frac{1}{p} \right) - \left(n - \left(\sum_{i=1}^{n} x_i \right) \right) \left(\frac{1}{1-p} \right) = 0 \rightarrow \hat{p} = \frac{1}{n} \sum_{i=1}^{n} x_i$$

Hence the maximum likelihood estimator of p is as follows:

$$\hat{P}_{\mathrm{ML}} = \frac{1}{n} \sum_{i=1}^{n} X_i = \overline{X}_n$$

where $\overline{X}_n$ is the sample mean. For a Bernoulli random variable X, the expected value is p, so $\hat{P}_{\mathrm{ML}} = \overline{X}_n$ makes sense. ∎

Example 9.8

Suppose we have a sequence of n independent, identically-distributed observed values of a random sample, $x_1, x_2, \ldots, x_n$, for a Gaussian random variable X with unknown mean μ_X and unknown variance σ_X^2. Determine the maximum-likelihood estimate for μ_X and σ_X^2.

Solution

The likelihood function is as follows:

$$L(\mu_X, \sigma_X^2) = f_X(x_1, x_2, \ldots, x_n; \mu_X, \sigma_X^2) = \prod_{i=1}^{n} f_X(x_i; \mu_X, \sigma_X^2)$$

$$= \prod_{i=1}^{n} (2\pi\sigma_X^2)^{-\frac{1}{2}} \exp\left(-\frac{(x_i - \mu_X)^2}{2\sigma_X^2}\right)$$

The log-likelihood function is then as follows:

$$l(\mu_X, \sigma_X^2) = \ln L(\mu_X, \sigma_X^2) = \ln\left((2\pi\sigma_X^2)^{-\frac{n}{2}}\right) - \sum_{i=1}^{n}\left(\frac{(x_i - \mu_X)^2}{2\sigma_X^2}\right)$$

$$= -\frac{n}{2}\ln(2\pi\sigma_X^2) - \frac{1}{2\sigma_X^2}\sum_{i=1}^{n}(x_i - \mu_X)^2$$

The partial derivatives of the log-likelihood function with respect to μ_X and σ_X^2 are set equal to zero to determine the maximum likelihood estimates of μ_X and σ_X^2, we thus have the following:

$$\begin{cases} \frac{\partial}{\partial \mu_X} l(\mu_X, \sigma_X^2) = \frac{2}{2\sigma_X^2}\sum_{i=1}^{n}(x_i - \mu_X) = 0 \\ \frac{\partial}{\partial \sigma_X^2} l(\mu_X, \sigma_X^2) = -\frac{n}{2}\frac{2\pi}{2\pi\sigma_X^2} + \frac{1}{2\sigma_X^4}\sum_{i=1}^{n}(x_i - \mu_X)^2 = 0 \end{cases} \rightarrow \begin{cases} \hat{\mu}_X = \frac{1}{n}\sum_{i=1}^{n} x_i \\ \widehat{\sigma_X^2} = \frac{1}{n}\sum_{i=1}^{n}(x_i - \hat{\mu}_X)^2 \end{cases}$$

Hence the maximum likelihood estimators of μ_X and σ_X^2 are as follows:

$$\begin{cases} \widehat{M}_{ML} = \frac{1}{n}\sum_{i=1}^{n} X_i = \overline{X}_n \\ \widehat{\Sigma^2}_{ML} = \frac{1}{n}\sum_{i=1}^{n}(X_i - \overline{X}_n)^2 = \sigma_n^2 \end{cases}$$

where $\overline{X}_n$ is the sample mean and σ_n^2 is the sample variance. For a Gaussian random variable X, the sample mean estimator is unbiased, but the sample variance estimator is biased. ∎

It is important to point that sometimes setting the derivative to zero may not lead to the maximum value. For instance, when the parameter of interest is an integer-value or when the maximum is at the end points of the acceptable range, the maximum value must therefore be found in another way.

9.4 Bayesian Estimators

In a Bayesian estimation, the parameter to be estimated is a random variable Θ with a known prior distribution. A random experiment is performed to determine the value of $\Theta = \theta$ that is present, but cannot be observed directly. We thus have a sequence of n independent, identically-distributed observed values of a random sample, $x_1, x_2, \ldots, x_n$, for the random variable X. The prior pdf $f_\Theta(\theta)$ represents information about θ before having the observed values $x_1, x_2, \ldots, x_n$, and the posterior pdf $f_\Theta(\theta \mid x_1, x_2, \ldots, x_n)$ represents the information about θ after having observed the random sample. Assuming the cost function of interest is the squared error, it can be shown that the best estimate is the conditional mean of the random variable Θ, given $x_1, x_2, \ldots, x_n$, i.e. we have $\hat{\theta} = E[\Theta \mid x_1, x_2, \ldots, x_n]$. The *optimum Bayesian estimator* is thus as follows:

$$\hat{\Theta}_B = E[\Theta \mid X_1, X_2, \ldots, X_n] \tag{9.4}$$

Example 9.9
Suppose we have a sequence of n independent, identically-distributed observed values of a random sample, $x_1, x_2, \ldots, x_n$, for a Bernoulli random variable X where the probability of success p is a uniform random variable over the interval $[0, 1]$. Assuming the squared error cost function, determine the Bayesian estimator of p.

Solution
The prior pdf of the parameter p is as follows:

$$f_P(p) = \begin{cases} 1 & 0 \le p \le 1 \\ 0 & \text{elsewhere} \end{cases}$$

The posterior pdf of the parameter p is as follows:

$$\begin{aligned} f_{P|X}(p \mid x_1, x_2, \ldots, x_n) &= \frac{f_{X,P}(x_1, x_2, \ldots, x_n; p)}{f_X(x_1, x_2, \ldots, x_n)} \\ &= \frac{f_{X|P}(x_1, x_2, \ldots, x_n \mid p) f_P(p)}{\int_{-\infty}^{\infty} f_{X|P}(x_1, x_2, \ldots, x_n \mid p) f_P(p) dp} \end{aligned}$$

where we have

$$\begin{aligned} f_{X|P}(x_1, x_2, \ldots, x_n \mid p) &= (p^{x_1}(1-p)^{1-x_1}) \ldots (p^{x_n}(1-p)^{1-x_n}) \\ &= p^{x_1 + \cdots + x_n}(1-p)^{n-(x_1 + \cdots + x_n)} \\ &= p^k(1-p)^{n-k} \end{aligned}$$

and $k = x_1 + \ldots + x_n$ is the number of successes in n independent trials. We also have

$$\int_{-\infty}^{\infty} f_{X|P}(x_1, x_2, \ldots, x_n \mid p) f_P(p) dp = \int_0^1 p^k(1-p)^{n-k} \, dp$$

Using calculus, this joint pdf can be mathematically manipulated to be in the following compact form:

$$\int_{-\infty}^{\infty} f_{X|P}(x_1, x_2, \ldots, x_n \mid p) f_P(p) dp = \frac{k!(n-k)!}{(n+1)!}$$

The posterior pdf of the parameter p thus becomes as follows:

$$f_{P|X}(p \mid x_1, x_2, \ldots, x_n) = \frac{p^k(1-p)^{n-k} \times 1}{\dfrac{k!(n-k)!}{(n+1)!}} = \frac{(n+1)! p^k(1-p)^{n-k}}{k!(n-k)!}$$

Note that the posterior pdf of the parameter p depends only on the observed number of successes k. The Bayesian estimate of p in the mean-square sense is given by the conditional mean of p given $x_1, x_2, \ldots, x_n$. Using calculus, it can be shown that we have

$$\hat{p}_B = E[p \mid X] = \int_0^1 p f_{P|X}(p \mid x_1, x_2, \ldots, x_n) dp$$

$$= \int_0^1 p \frac{(n+1)! p^k(1-p)^{n-k}}{k!(n-k)!} dp$$

$$= \frac{(n+1)!}{k!(n-k)!} \int_0^1 p^{k+1}(1-p)^{n-k} dp$$

$$= \frac{(n+1)!}{k!(n-k)!} \frac{(k+1)!(n-k)!}{(n+2)!} = \frac{k+1}{n+2}$$

Noting that we have

$$k = \sum_{i=1}^n x_i$$

the Bayesian estimator of p is thus as follows:

$$\hat{P}_B = \frac{n}{n+2}\left(\overline{X}_n + \frac{1}{n}\right)$$

Note that as $n \to \infty$, we then have $\hat{P}_B \to \overline{X}_n$. ∎

Example 9.10

Suppose we have a sequence of n independent, identically-distributed observed values of a random sample, $x_1, x_2, \ldots, x_n$, for a Gaussian random variable X with unknown mean μ_X and $\sigma_X^2 = 1$. Suppose that μ_X is itself a Gaussian random variable with mean 0 and variance 1. Assuming the squared error cost function, determine the Bayesian estimator of μ_X.

Solution

The prior pdf of the parameter μ_X is as follows:

$$f_M(\mu_X) = \frac{1}{\sqrt{2\pi}} \exp\left(-\frac{(\mu_X)^2}{2}\right)$$

The posterior pdf of the parameter μ_X is as follows:

$$f_{M|X}(\mu_X \mid x_1, x_2, \ldots, x_n) = \frac{f_{X,M}(x_1, x_2, \ldots, x_n; \mu_X)}{f_X(x_1, x_2, \ldots, x_n)}$$

$$= \frac{f_{X|M}(x_1, x_2, \ldots, x_n \mid \mu_X) f_M(\mu_X)}{\int_{-\infty}^{\infty} f_{X|M}(x_1, x_2, \ldots, x_n \mid \mu_X) f_M(\mu_X) d\mu_X}$$

where we have

$$f_{X|M}(x_1, x_2, \ldots, x_n \mid \mu_X) = \frac{1}{(2\pi)^{\frac{n}{2}}} \exp\left(-\sum_{i=1}^{n} \frac{(x_i - \mu_X)^2}{2}\right)$$

The posterior pdf of the parameter μ_X thus becomes as follows:

$$f_{M|X}(\mu_X \mid x_1, x_2, \ldots, x_n)$$

$$= \frac{\left(\frac{1}{(2\pi)^{\frac{n}{2}}} \exp\left(-\sum_{i=1}^{n} \frac{(x_i - \mu_X)^2}{2}\right)\right)\left(\frac{1}{\sqrt{2\pi}} \exp\left(-\frac{\mu_X^2}{2}\right)\right)}{\int_{-\infty}^{\infty} \left(\frac{1}{(2\pi)^{\frac{n}{2}}} \exp\left(-\sum_{i=1}^{n} \frac{(x_i - \mu_X)^2}{2}\right)\right)\left(\frac{1}{\sqrt{2\pi}} \exp\left(-\frac{\mu_X^2}{2}\right)\right) d\mu_X}$$

After some mathematical manipulation, it can be shown that the pdf of μ_X given $x_1, x_2, \ldots, x_n$ is a Gaussian random variable whose mean is given by

$$\hat{\mu}_X = \left(\frac{1}{n+1}\right)\left(\sum_{i=1}^{n} x_i\right)$$

The Bayesian estimator of μ_X is thus given by:

$$\hat{M}_B = \left(\frac{n}{n+1}\right)\overline{X}_n$$

Note that as $n \to \infty$, we then have $\hat{M}_B \to \overline{X}_n$. ∎

9.5 Confidence Intervals

As discussed earlier, an estimate that specifies a single value of the parameter is called a point estimate. The single value gives no indication as to the accuracy of the estimate and fails short for providing the confidence placed on it.

An estimate that specifies a range of values is called an *interval estimate*. In interval estimation, the set of values is expected to contain the true value of the parameter with a high degree of confidence. Assuming with a high probability of $1 - \alpha$, the true value lies in the estimated range of the possible values for the parameter, $1 - \alpha$ is then called the *confidence level* or *confidence coefficient*. When we use the observed data to determine an interval that contains the true value of the parameter of interest with probability $1 - \alpha$, then such an interval is called a $(1 - \alpha) \times 100\%$ *confidence interval*. The end-points of a confidence interval are called the *confidence limits*.

It is important to note that the confidence intervals and confidence levels are numerical measures of the accuracy and reliability of a statistical estimate respectively, where a narrower confidence interval reflects a more accurate estimate and a higher confidence level indicates a more reliable estimate.

As the sample mean is an important statistical measure, we now focus on confidence intervals for the sample mean $\overline{X}_n$ and the parameters that shape these intervals. When a sample mean is estimated using a method of estimation, the chances are quite slim that the estimate will actually equal the population mean μ_X. It is thus important to accompany such a point estimate of μ_X with a statement reflecting how close we expect the estimate to be. The error $\overline{X}_n - \mu_X$ is the difference between the estimator and the true value of the parameter. We assume that the sampling is either from an infinite population or with replacement from a finite population so as to make the arithmetic simple. We also assume either the observed samples $X_1, X_2, \ldots, X_n$ are iid Gaussian random variables or n is large enough to allow us to approximate the sample mean as a Gaussian random variable using the central limit theorem.

Noting that the sample mean $\overline{X}_n$ is a Gaussian random variable with unknown mean μ_X and known variance $\frac{\sigma_X^2}{n}$, we then have $\frac{\overline{X}_n - \mu_X}{\sigma_X/\sqrt{n}}$ as the standard (zero-mean, unit-variance) Gaussian random variable. We need to determine $z_{\alpha/2}$ in such a way that we have the following:

$$P\left(-z_{\alpha/2} < \frac{\overline{X}_n - \mu_X}{\sigma_X/\sqrt{n}} < z_{\alpha/2}\right) = 1 - \alpha \tag{9.5}$$

Noting that $z_{\alpha/2}$ is selected in such a way that the area to its right under the standard Gaussian curve, i.e. $Q(z_{\alpha/2})$, equals $\frac{\alpha}{2}$, we then have the following:

$$P\left(\overline{X}_n - \frac{\sigma_X z_{\alpha/2}}{\sqrt{n}} < \mu_X < \overline{X}_n + \frac{\sigma_X z_{\alpha/2}}{\sqrt{n}}\right) = 1 - 2Q(z_{\alpha/2}) \tag{9.6}$$

The $(1 - \alpha)$ confidence interval for the mean μ_X is thus given by

$$\left[\overline{X}_n - \frac{\sigma_X z_{\alpha/2}}{\sqrt{n}}, \overline{X}_n + \frac{\sigma_X z_{\alpha/2}}{\sqrt{n}}\right] \tag{9.7}$$

It is thus evident that the sample size n, the confidence level $1 - \alpha$, and the variance σ_X^2 each can have a direct and significant impact on the confidence interval $\overline{X}_n \pm z_{\alpha/2} \times \frac{\sigma_X}{\sqrt{n}}$, where $\overline{X}_n$ is the point estimate, $z_{\alpha/2}$ is known as the *reliability coefficient*, and $\frac{\sigma_X}{\sqrt{n}}$ is the error of the estimator.

Example 9.11

Consider a Gaussian random variable with known variance σ_X^2 and unknown mean μ_X. Find the $(1 - \alpha)$ confidence interval for the sample mean based on a random sample size of n. Note that the possible values of the variance σ_X^2 are 0.01 and 1, those of confidence level $(1 - \alpha)$ are 95% and 99%, and those of sample size n are 10 and 100. Comment on the results.

Solution

There are 8 ($= 2 \times 2 \times 2$) different cases, and the results are provided in the following table:

n	σ_X^2	$1-\alpha$	$1-Q(z_{\alpha/2})$	$z_{\alpha/2}$	$\dfrac{\sigma_X z_{\alpha/2}}{\sqrt{n}}$	$\left(\overline{X}_n - \dfrac{\sigma_X z_{\alpha/2}}{\sqrt{n}},\ \overline{X}_n + \dfrac{\sigma_X z_{\alpha/2}}{\sqrt{n}}\right)$
10	0.01	95%	97.5%	1.96	0.062	$(\overline{X}_n - 0.062, \overline{X}_n + 0.062)$
10	0.01	99%	99.5%	2.57	0.081	$(\overline{X}_n - 0.081, \overline{X}_n + 0.081)$
10	1	95%	97.5%	1.96	0.620	$(\overline{X}_n - 0.620, \overline{X}_n + 0.620)$
10	1	99%	99.5%	2.57	0.813	$(\overline{X}_n - 0.813, \overline{X}_n + 0.813)$
100	0.01	95%	97.5%	1.96	0.020	$(\overline{X}_n - 0.020, \overline{X}_n + 0.020)$
100	0.01	99%	99.5%	2.57	0.026	$(\overline{X}_n - 0.026, \overline{X}_n + 0.026)$
100	1	95%	97.5%	1.96	0.196	$(\overline{X}_n - 0.196, \overline{X}_n + 0.196)$
100	1	99%	99.5%	2.57	0.257	$(\overline{X}_n - 0.257, \overline{X}_n + 0.257)$

The table highlights the fact that an increase in the number of samples n or a decrease in the variance σ_X^2 or a decrease in the confidence level $1 - \alpha$ can all make the confidence interval narrower. ∎

9.6 Estimation of a Random Variable

The objective is to estimate the value of an inaccessible random variable in terms of the observation of an accessible random variable. The inaccessibility of the random variable of interest may be because it is impossible to measure, such as the surface temperature of the sun, or because it is obscure to measure, such as a received signal corrupted by the channel degradations in a communication system, or because it is not available soon enough, such as the temperature in a certain location at a certain time in future.

The inaccessible random variable is represented by Y and the accessible random variable by X. The estimator $\hat{Y}$ is given by a function of the observation X that is by $g(X)$. The random variable $Y - \hat{Y} = Y - g(X)$ is called the **estimation error**. As the estimation error is generally nonzero, there is a cost associated with the error, and $c(Y - g(X))$ represents the **cost function**, which, in turn, is a random variable. The goal is to determine the function $g(X)$ that minimizes the expected value of the cost function, $E[c(Y - g(X))]$, known as the **Bayes risk**. This average cost measures the performance of a given estimator.

Different cost functions result in different estimators. If the cost function is the all-or-none (the hit-or-miss), i.e. a cost that is unity when the error is outside of a very small tolerance band, and is zero within the band, then the estimator is known as the **maximum a posteriori probability (MAP) estimator**. The MAP estimator is the mode of the posterior probability density function. If the cost function is the square of the error, then the estimator is known as the **minimum mean square error (MMSE) estimator**. The MMSE estimator is the mean of the posterior probability density function. For pdf's whose mean and mode are the same, such as the Gaussian pdf, the MMSE and MAP estimators will be identical. If the cost function is the absolute value of the error, then the estimator is known as the **minimum mean absolute error (MMAE) estimator**. The MMAE estimator penalizes errors proportionally, and is the median of the posterior probability density function. In short, the mean, mode and median of the posterior probability density function, that is the conditional mean, mode, and median of Y given $X = x$, are well-known estimators minimizing the Bayes risk. Our brief focus in this chapter is on the MAP and MMSE estimation techniques.

9.7 Maximum a Posteriori Probability Estimation

If X and Y are continuous random variables, the MAP estimate maximizes the probability of choosing the correct estimate. The **MAP estimator** for Y given the observation X is represented by the mode of the following pdf:

$$\text{MAP Estimator } \hat{Y}_{MAP}\text{: Mode of } f_{Y|X}(y \mid x) = \text{Mode of } \frac{f_{X|Y}(x \mid y)f_Y(y)}{f_X(x)}$$

$$= \text{Mode of } f_{X|Y}(x \mid y)f_Y(y) \tag{9.8}$$

where $f_X(x)$ is disregarded as it does not depend on the random variable Y. The MAP estimator is the globally optimal solution to the estimation problem, i.e. there is no other estimator that can do better.

Example 9.12

Determine the MAP estimator of Y in terms of X when the random variables X and Y have the following joint pdf:

$$f_{X,Y}(x, y) = y e^{-y(1+x)} \quad x > 0 \text{ and } y > 0$$

Solution

The conditional pdf of Y given X is given by

$$f_{Y|X}(y \mid x) = \frac{f_{X,Y}(x, y)}{f_X(x)} = \frac{ye^{-y(1+x)}}{\int_{-\infty}^{\infty} f_{X,Y}(x, y) dy} = \frac{ye^{-y(1+x)}}{\int_0^{\infty} ye^{-y(1+x)} dy} = \frac{ye^{-y(1+x)}}{\frac{1}{(1+x)^2}}$$

$$= (1 + x)^2 ye^{-y(1+x)}$$

where integration by parts was used. In order to determine the MAP estimator, the conditional pdf of Y given X needs to be maximized:

$$\frac{d}{dy}(1 + x)^2 ye^{-y(1+x)} = 0 \rightarrow \hat{y} = \frac{1}{1 + x} \rightarrow \hat{Y}_{MAP} = \frac{1}{1 + X} \qquad \blacksquare$$

The ***maximum likelihood estimator*** for Y given the observation X is represented by the mode of the following pdf:

$$\text{ML Estimator } \hat{Y}_{ML}: \text{ Mode of } f_{X|Y}(x \mid y) \tag{9.9}$$

The maximum likelihood (ML) estimate therefore does not use information about the a priori pdf of the random variable Y.

Example 9.13

Determine the ML estimator of Y in terms of X when the random variables X and Y are related as follows:

$$X = Y + N$$

where Y and N are both independent of one another and each has a Gaussian distribution with zero-mean and unit-variance.

Solution

Noting that the random variable N has a Gaussian distribution with zero-mean and unit-variance, then for a given $Y = y$, the conditional pdf $f_{X|Y}(x|y)$ is also a Gaussian distribution, but with mean y and unit-variance. We thus have the following conditional probability density function:

$$f_{X|Y}(x \mid y) = \frac{1}{\sqrt{2\pi}} \exp\left(-\frac{(x-y)^2}{2}\right)$$

The ML estimate of Y, given $X = x$, is the value of y that maximizes the above conditional pdf. To maximize this function, the exponent must be zero, we thus have

$$\hat{y} = x \rightarrow \hat{Y}_{ML} = X \qquad \blacksquare$$

Note that the ML estimate is the same as the MAP estimate when all possible values of Y are equally likely, that is when the random variable Y has a uniform distribution.

9.8 Minimum Mean Square Error Estimation

In the *MMSE estimation*, also known as the *least mean square (LMS) estimation* or simply *Bayes' estimation*, when X and Y are continuous random variables, the mean square error is often used as the cost function:

$$e_{\text{MMSE}} = E[(Y - g(X))^2] = \int_{-\infty}^{\infty} \int_{-\infty}^{\infty} (y - g(x))^2 f_{X,Y}(x, y) \, dx \, dy \qquad (9.10)$$

It can be shown that when the optimality criterion is the minimization of the mean square value of the estimation error, the estimator $\hat{Y}$ is as follows:

$$\hat{Y}_{\text{MMSE}} = g(X) = \text{Mean of} f_{Y|X}(y \mid x) = \int_{-\infty}^{\infty} y f_{Y|X}(y \mid x) dy = E[Y \mid X]$$

$$(9.11)$$

We thus conclude that the MMSE estimate of Y given X is the conditional expectation of Y based on the available value of X. The MMSE estimator is an unbiased estimator of Y, and also has the advantage that it is mathematically tractable. The MMSE is the variance of the conditional probability density function $f_{Y|X}(y \mid x)$. This estimator is called the *regression curve*, which traces the conditional expected value of Y given the observation X.

If the random variables X and Y are statistically independent, then we have $E[Y \mid X] = E[Y]$, which is a constant. In other words, knowledge of X has no effect on the estimate of Y, as the observation X provides no information about Y. Also, if $Y = f(X)$, then $E[Y \mid X] = f(X)$, and the resulting mean square error is zero. This is an obvious result, because if X is observed and $Y = f(X)$, then Y is determined uniquely, resulting in no estimation error.

Example 9.14
Suppose the random variables X and Y have the following joint probability density function:

$$f_{X,Y}(x, y) = \begin{cases} \frac{3}{4} + xy & 0 < x < 1 \text{ and } 0 < y < 1 \\ 0 & \text{elsewhere} \end{cases}$$

Find the minimum mean square error estimate of Y given X.

Solution
To determine the mean square error estimator of Y given X, we first find the conditional probability density function $f_{Y|X}(y\,|\,x)$, which is as follows:

$$f_{Y|X}(y\,|\,x) = \frac{f_{X,Y}(x,y)}{f_X(x)} = \frac{\dfrac{3}{4} + xy}{\int_{-\infty}^{\infty} f_{X,Y}(x,y)dy} = \frac{\dfrac{3}{4} + xy}{\int_0^1 \left(\dfrac{3}{4} + xy\right)dy}$$

$$= \frac{3 + 4xy}{3 + 2x} \quad 0 < y < 1$$

We thus have the following:

$$\hat{y} = \int_0^1 y\frac{3 + 4xy}{3 + 2x}dy = \frac{\dfrac{3}{2} + \dfrac{4x}{3}}{3 + 2x} = \frac{9 + 8x}{6(3 + 2x)} \rightarrow \hat{Y}_{\text{MMSE}} = \frac{9 + 8X}{6(3 + 2X)} \quad \blacksquare$$

When an inference about the random variable Y needs to be made in the absence of observations, the mean square error estimation is called the **blind estimation**, where the estimation of the random variable Y is made by a constant k. It can be shown that the resulting mean square error estimate is as follows:

$$k = E[Y] \tag{9.12}$$

Note that the minimum mean square error is the variance of the random variable Y. In short, before the experiment is performed, the best estimate of the random variable Y, in the mean square sense, is its mean value. For example, before a fair die is rolled, the minimum mean square error estimate of the outcome is 3.5, even though it is not in the range of Y.

Example 9.15
Suppose the random variables Y is an exponential random variable with the following probability density function:

$$f_Y(y) = \begin{cases} 2\exp(-2y) & y > 0 \\ 0 & \text{elsewhere} \end{cases}$$

Find the blind estimate of Y.

Solution
The blind estimator of Y is as follows:

$$k = E[Y] = \int_{-\infty}^{\infty} y f_Y(y)dy = \int_0^{\infty} y\,(2e^{-2y})dy = \frac{1}{2} \qquad \blacksquare$$

9.9 Linear Minimum Mean Square Error Estimation

When an inference about the random variable Y in the context of mean square error estimation needs to be made in terms of a linear function of the random variable X, it is called the ***linear minimum mean square error (LMMSE) estimation***, that is we have

$$\widehat{Y}_{LMMSE} = g(X) = a\,X + b \tag{9.13}$$

The parameters a and b are determined in such a way that the mean square error is minimum. By differentiating the cost function with respect to a and b, and set them equal to zero, we then have the following:

$$\begin{cases} a = \rho_{X,Y}\dfrac{\sigma_Y}{\sigma_X} \\ b = \mu_Y - a\,\mu_X \end{cases} \tag{9.14}$$

where μ_X and σ_X are the mean and standard deviation of X, respectively, μ_Y and σ_Y are those of Y, $\rho_{X,Y}$ is the correlation coefficient between X and Y, and the MMSE is $\sigma_Y^2(1 - \rho_{X,Y}^2)$. Note that in the linear mean square estimation, the estimation error $Y - \widehat{Y} = Y - aX - b$ is orthogonal to the observation X that is we must have

$$E[(Y - aX - b)X] = 0 \tag{9.15}$$

This is known as the ***orthogonality principle*** and the line $Y = a\,X + b$ is called a ***regression line***.

Note that if X and Y are uncorrelated (i.e. $\rho_{X,Y} = 0$), then the best estimate for Y is its mean μ_Y and the MMSE is σ_Y^2, and if $|\rho_{X,Y}| = 1$, the mean square error is zero, that is Y is essentially a linear function of X. The magnitude of the correlation coefficient reflects the degree upon which observing X can improve our knowledge about Y. Linear estimates are easy to compute, and it is not necessary to know the complete probability model of X and Y, instead only the expected values and variances of X and Y and the covariances of X and Y are required to evaluate an optimum linear estimator. In general, a linear estimator has a larger mean square error than the nonlinear estimator.

Example 9.16
Suppose that the random variables X and Y have the joint probability density function given by

$$f_{X,Y}(x,y) = \begin{cases} \dfrac{x+y}{8} & 0 \le x \le 2 \text{ and } 0 \le y \le 2 \\ 0 & \text{elsewhere} \end{cases}$$

Find the linear minimum mean square error estimator for Y given X.

Solution
We first need to find the marginal pdf's of the random variables X and Y and their means and variances:

$$f_X(x) = \int_0^2 f_{X,Y}(x,y)dy = \frac{1}{8}\int_0^2 (x+y)dy = \begin{cases} \dfrac{(x+1)}{4} & 0 \le x \le 2 \\ 0 & \text{elsewhere} \end{cases}$$

$$\rightarrow \begin{cases} E[X] = \dfrac{1}{4}\displaystyle\int_0^2 x(x+1)dx = \dfrac{7}{6} \\ E[X^2] = \dfrac{1}{4}\displaystyle\int_0^2 x^2(x+1)dx = \dfrac{5}{3} \end{cases}$$

$$\rightarrow \sigma_X^2 = E[X^2] - (E[X])^2 = \frac{11}{36} \rightarrow \sigma_X = \frac{\sqrt{11}}{6}$$

$$f_Y(y) = \int_0^2 f_{X,Y}(x,y)dx = \frac{1}{8}\int_0^2 (x+y)dx = \begin{cases} \dfrac{(y+1)}{4} & 0 \le y \le 2 \\ 0 & \text{elsewhere} \end{cases}$$

$$\rightarrow \begin{cases} E[Y] = \dfrac{1}{4}\displaystyle\int_0^2 y(y+1)dy = \dfrac{7}{6} \\ E[Y^2] = \dfrac{1}{4}\displaystyle\int_0^2 y^2(y+1)dy = \dfrac{5}{3} \end{cases}$$

$$\rightarrow \sigma_Y^2 = E[Y^2] - (E[Y])^2 = \frac{11}{36} \rightarrow \sigma_Y = \frac{\sqrt{11}}{6}$$

We now find $E[XY]$ in order to determine the correlation coefficient $\rho_{X,Y}$:

$$E[XY] = \frac{1}{8}\int_0^2\int_0^2 (xy)(x+y)dy\,dx = \frac{4}{3} \rightarrow$$
$$COV(X,Y) = E[XY] - E[X]E[Y]$$
$$= \frac{4}{3} - \frac{7}{6}\times\frac{7}{6} = -\frac{1}{36} \rightarrow$$

$$\rho_{X,Y} = \frac{COV(X,Y)}{\sigma_X\sigma_Y} = \frac{-\frac{1}{36}}{\frac{\sqrt{11}}{6}\times\frac{\sqrt{11}}{6}} = -\frac{1}{11}$$

We thus get the following:

$$\begin{cases} a = \rho_{X,Y}\dfrac{\sigma_Y}{\sigma_X} = \left(-\dfrac{1}{11}\right)\dfrac{\frac{\sqrt{11}}{6}}{\frac{\sqrt{11}}{6}} = -\dfrac{1}{11} \\ b = \mu_Y - a\,\mu_X = \dfrac{7}{6} + \dfrac{1}{11}\times\dfrac{7}{6} = \dfrac{14}{11} \end{cases} \rightarrow \hat{Y}_{LMMSE} = aX + b = -\frac{1}{11}X + \frac{14}{11}$$

∎

Example 9.17

Let X and Y be jointly Gaussian (normal) random variables, where μ_X and μ_Y are the mean values, and σ_X and σ_Y are the standard deviations of the random variables X and Y, respectively, and $\rho_{X,Y}$ is the correlation coefficient between X and Y. Determine the maximum a posteriori estimator, the maximum likelihood estimator, the minimum mean square error estimator, and the linear minimum mean square error estimator for Y given the observation X.

Solution

Note that since the joint pdf of X and Y has the Gaussian distribution, their conditional pdfs are also Gaussian. The MAP estimator is the maximum of the following conditional pdf of Y given X:

$$f_{Y|X}(y \mid x) = \frac{1}{\sqrt{2\pi\sigma_Y^2(1 - \rho_{X,Y}^2)}}$$

$$\times \exp\left(-\frac{1}{2\sigma_Y^2(1 - \rho_{X,Y}^2)}\left(y - \left(\rho_{X,Y}\frac{\sigma_Y}{\sigma_X}(x - \mu_X) + \mu_Y\right)\right)^2\right)$$

The value of y for which the exponent is zero gives rise to the MAP estimator:

$$y - \left(\rho_{X,Y}\frac{\sigma_Y}{\sigma_X}(x - \mu_X) + \mu_Y\right) = 0 \rightarrow \hat{y} = \rho_{X,Y}\frac{\sigma_Y}{\sigma_X}(x - \mu_X) + \mu_Y$$

$$\rightarrow \hat{Y}_{\text{MAP}} = \rho_{X,Y}\frac{\sigma_Y}{\sigma_X}(X - \mu_X) + \mu_Y$$

The ML estimator is the maximum of the following conditional pdf of X given Y:

$$f_{X|Y}(x \mid y) = \frac{1}{\sqrt{2\pi\sigma_X^2(1 - \rho_{X,Y}^2)}}$$

$$\times \exp\left(-\frac{1}{2\sigma_X^2(1 - \rho_{X,Y}^2)}\left(x - \left(\rho_{X,Y}\frac{\sigma_X}{\sigma_Y}(y - \mu_y) + \mu_X\right)\right)^2\right)$$

The value of y for which the exponent is zero gives rise to the ML estimator

$$x - \left(\rho_{X,Y}\frac{\sigma_X}{\sigma_Y}(\hat{y} - \mu_y) + \mu_X\right) = 0 \rightarrow \hat{y} = \frac{\sigma_Y}{\rho_{X,Y}\sigma_X}(x - \mu_X) + \mu_Y$$

$$\rightarrow \hat{Y}_{\text{ML}} = \frac{\sigma_Y}{\rho_{X,Y}\sigma_X}(X - \mu_X) + \mu_Y$$

It is important to note that the MAP estimator and the ML estimator are not equal, as knowledge of the a priori probability of X makes a difference. The minimum mean square error estimator of Y in terms of X using the above conditional pdf $f_{Y|X}(y \mid x)$ is as follows:

$$\hat{Y}_{\text{MMSE}} = E[Y \mid X] = \rho_{X,Y}\frac{\sigma_Y}{\sigma_X}(X - \mu_X) + \mu_Y$$

$$= \left(\rho_{X,Y}\frac{\sigma_Y}{\sigma_X}\right)X + \left(\mu_Y - \rho_{X,Y}\frac{\sigma_Y}{\sigma_X}\mu_X\right)$$

As reflected above, this best estimator is a linear mean square estimator. This shows that the optimum nonlinear mean square error estimator of Y in terms of X when X and Y are jointly Gaussian random variables is identical to the best linear estimator. Moreover, the performance of MAP estimator is identical to that of the MMSE estimator. ∎

9.10 Linear MMSE Estimation Using a Vector of Observations

The MAP and MMSE estimators can be extended from a single observation to where a vector of observations is available, and can also be generalized from estimation of a single random variable to that of a vector of random variables. In fact, we can estimate multiple random variables based on multiple observations. However, our focus in this section is solely on linear mean square estimation of a single random variable given an observation of a random vector, i.e. given multiple measurements.

We want to determine the MMSE estimate of the random variable Y as a linear combination of observations $(X_1, X_2, \ldots, X_n)$. The set of coefficients for the optimum linear estimator of a random variable, i.e. $a_1, a_2, \ldots, a_n$, is the solution to a set of linear equations completely defined by the first and second moments. Suppose Y is a random variable with expected value $E[Y]$, X is an n-dimensional random column vector of observations with expected value $E[X]$ and an $n \times n$ *covariance matrix* C_{XX}, C_{YX} is the $1 \times n$ *cross-covariance vector* of Y and X, and a is the column vector representing the optimum coefficients. It can be shown that the linear MMSE estimator of Y given the random vector X is as follows:

$$\hat{Y}_{\text{LMMSE}} = C_{YX}\,C_{XX}^{-1}(X - E[X]) + E[Y] = a^T(X - E[X]) + E[Y]$$

$$= \sum_{k=1}^{n} a_k X_k + \left(E[Y] - \sum_{k=1}^{n} a_k E[X_k]\right) = \sum_{k=1}^{n} a_k X_k + b \tag{9.16}$$

From the above equation, we first conclude that the estimator adjusts the mean value $E[Y]$ of the random variable being estimated, by a linear combination of the deviations $X_k - E[X_k]$ between the measured random variables and their respective means. Secondly, we note that b is a scalar to make the linear estimator unbiased, that is, the expected value of the estimator is the expected value of the random variable for which an estimate is sought. Therefore, if the means of the random variable Y and the random vector X are all zero, we then have $b = 0$.

In order to produce the linear estimator with the MMSE, the coefficients a_k, where $k = 1, \ldots, n$, must be such that the following **orthogonality conditions** are satisfied:

$$E\left[\left((Y - E[Y]) - \sum_{k=1}^{n} a_k(X_k - E[X_k])\right)(X_j - E[X_j])\right] = 0 \quad j = 1, 2, \ldots, n$$

(9.17a)

Using matrix algebra, the above set of equations can be expressed in the following compact from:

$$\begin{pmatrix} COV(X_1,X_1) & \cdots & COV(X_1,X_n) \\ \vdots & \ddots & \vdots \\ COV(X_n,X_1) & \cdots & COV(X_n,X_n) \end{pmatrix}\begin{pmatrix} a_1 \\ \vdots \\ a_n \end{pmatrix} = \begin{pmatrix} COV(X_1,Y) \\ \vdots \\ COV(X_n,Y) \end{pmatrix} \to a = C_{XX}^{-1}C_{XY}$$

(9.17b)

where $COV(W, Z) = E[WZ] - E[W]E[Z]$ is the covariance of the random variables W and Z, and the vector C_{XY} is the transpose of the vector C_{YX}. Note that the estimation error $Y - \hat{Y}_{LMMSE}$ is uncorrelated with the elements of the vector X and the MMSE is as follows:

$$e_{LMMSE} = \sigma_Y^2 - C_{YX}\,a = \sigma_Y^2 - C_{YX}\,C_{XX}^{-1}C_{XY}$$

(9.18)

In many signal processing applications, the vector X represents a collection of observed samples and the vector $a^T = (a_1, a_2, \ldots, a_n)$ is a representation of a linear filter. The linear MMSE estimator is optimal when the MMSE happens to be linear, which is the case for the Gaussian random variables. Otherwise better estimators exist, though they will be nonlinear. Since the linear MMSE estimator relies on the correlation between random variables, a random variable uncorrelated with the available random vector cannot be linearly estimated.

Example 9.18
In a cellular radio communication system, the radio receiver in a base station has two identical, but spatially distanced, antennas to receive two noisy versions of the transmitted signal S that has zero mean and variance σ_S^2. The received signals are $R_1 = S + N_1$ and $R_2 = S + N_2$, where S, N_1, and N_2 are all independent random variables. Assume that the random variables N_1 and N_2, representing the additive noise, both have zero mean and unit variance. Determine the linear mean square error estimation of S based on two antennas. Comment on the results.

Solution
The random variables S, N_1, and N_2, which are independent, all have zero means. Based on the two random variables R_1 and R_2, the covariance terms of

interest are as follows:

$$COV(R_1, R_1) = E[R_1 R_1] = \sigma_S^2 + \sigma_{N_1}^2 = \sigma_S^2 + 1$$
$$COV(R_2, R_2) = E[R_2 R_2] = \sigma_S^2 + \sigma_{N_2}^2 = \sigma_S^2 + 1$$
$$COV(R_1, R_2) = E[R_1 R_2] = \sigma_S^2$$
$$COV(R_2, R_1) = E[R_2 R_1] = \sigma_S^2$$
$$COV(R_1, S) = E[R_1 S] = \sigma_S^2$$
$$COV(R_2, S) = E[R_2 S] = \sigma_S^2$$

After solving the following matrix equation, we obtain the coefficients of interest as follows:

$$\begin{pmatrix} a_1 \\ a_2 \end{pmatrix} = \begin{pmatrix} \sigma_S^2 + 1 & \sigma_S^2 \\ \sigma_S^2 & \sigma_S^2 + 1 \end{pmatrix}^{-1} \begin{pmatrix} \sigma_S^2 \\ \sigma_S^2 \end{pmatrix} \rightarrow \begin{pmatrix} a_1 \\ a_2 \end{pmatrix} = \frac{\sigma_S^2}{2\sigma_S^2 + 1} \begin{pmatrix} 1 \\ 1 \end{pmatrix}$$

The estimate of the signal and the MMSE for two antennas are respectively as follows:

$$\widehat{S}_{\text{LMMSE(2-antennas)}} = \frac{\sigma_S^2}{2\sigma_S^2 + 1}(R_1 + R_2) = \frac{2\sigma_S^2}{2\sigma_S^2 + 1}\left(S + \frac{N_1 + N_2}{2}\right)$$

and

$$e_{\text{LMMSE(2-antennas)}} = \frac{\sigma_S^2}{2\sigma_S^2 + 1}$$

For a single antenna, the coefficient of interest is as follows:

$$a_1 = \frac{\sigma_S^2}{\sigma_S^2 + 1}$$

The estimate of the signal and the MMSE are respectively as follows:

$$\widehat{S}_{\text{LMMSE(1-antenna)}} = \frac{\sigma_S^2}{\sigma_S^2 + 1}(R_1) = \frac{\sigma_S^2}{\sigma_S^2 + 1}(S + N_1)$$

and

$$e_{\text{LMMSE(1-antenna)}} = \sigma_S^2\left(1 - \frac{\sigma_S^2}{\sigma_S^2 + 1}\right) = \frac{\sigma_S^2}{\sigma_S^2 + 1}$$

The ratio representing the MMSE for the two-antenna system over the MMSE for one-antenna system is always less than one, this points to the fact that the MMSE for two antenna system is always smaller than that for a one-antenna system. An increase in the variance of the transmitted signal reduces this ratio. Moreover, in a two-antenna system, the receiver adds the two signals and then scales down the sum and the two noise components are averaged out. ∎

9.11   Summary

In this chapter, a brief overview of classical estimation theory was provided. First, parameter estimation was discussed, where the unknown parameter could be a deterministic or a random quantity. The focus then turned toward estimation of an inaccessible random variable, in terms of the observation of an accessible random variable, where various estimation techniques were briefly discussed.

Problems

9.1   Let $\widehat{\Theta}_1$ and $\widehat{\Theta}_2$ be unbiased estimators for the parameter θ. Show that the estimator $\widehat{\Theta}_3 = p\widehat{\Theta}_1 + (1 - p)\widehat{\Theta}_2$ is also an unbiased estimator for θ, where $0 \leq p \leq 1$.

9.2   In a communication system, the phase of the received sinusoidal signal is $\Omega = \Theta + \Gamma$, where Θ is the phase of the transmitted sinusoidal signal and Γ is the phase noise that is uniformly distributed in the interval [0, 2π]. Suppose the signal is transmitted n times and that the noise terms are iid random variables. Show that the sample mean of the output is a biased estimator of Θ.

9.3   Suppose we have a sequence of n independent, identically-distributed observed values of a random sample, $x_1, x_2, \ldots, x_n$, for the binomial random variable X, with known parameter m, but unknown probability p. Determine the maximum-likelihood estimate for p.

9.4   Suppose we have a sequence of n independent, identically-distributed observed values of a random sample, $x_1, x_2, \ldots, x_n$, for the Pareto random variable X, with unknown parameter $\alpha > 1$ and known parameter $k > 0$. Determine the maximum-likelihood estimate for α.

9.5   Suppose we have a sequence of n independent, identically-distributed observed values of a random sample, $x_1, x_2, \ldots, x_n$, for the Rayleigh random variable X, with unknown parameter α^2. Determine the maximum-likelihood estimate for α^2.

9.6   Suppose we have a sequence of n independent, identically-distributed observed values of a random sample, $x_1, x_2, \ldots, x_n$, for the uniform random variable X over the interval [0, θ]. Determine the maximum-likelihood estimate for θ.

9.7 Suppose we have a sequence of n independent, identically-distributed observed values of a random sample, $x_1, x_2, ..., x_n$, for the exponential random variable X with unknown parameter λ. Suppose that λ is itself an exponential random variable with mean α. Find the Bayesian estimator of λ, assuming the squared error cost function.

9.8 Consider a Gaussian random variable X with variance 100 and unknown mean μ_X. What sample size n is required such that the width of 95% confidence interval for the sample mean to be 5?

9.9 Find the mean square estimator of a random variable Y by a function $g(X)$ of the random variable X.

9.10 Suppose the random variables X and Y have the following joint probability density function:

$$f_{X,Y}(x, y) = \begin{cases} 2 & 0 < y \leq x < 1 \\ 0 & \text{elsewhere} \end{cases}$$

Find the best mean square error estimator of Y given X.

9.11 Determine the mean square estimate of a random variable Y by a constant k.

9.12 Determine the linear mean square error estimator of Y given X.

9.13 Suppose X is a uniform random variable over the interval $[-1, 1]$ and we have $Y = X^2$. Determine the best mean square estimator and linear mean square estimator of Y in terms of X.

9.14 Determine the MAP estimator of Y in terms of X when the random variables X and Y are related as follows:

$$X = Y + N$$

where Y and N are both independent of one another and each has a Gaussian distribution with zero-mean and unit-variance.

10

Hypothesis Testing

When the decision is based on probabilities, the reasoning is referred to as *statistical inference*. The fundamentals of probability theory are employed to make accurate decisions in the presence of uncertainty. The steps in a statistical inference are thus as follows: perform an experiment, collect and analyze data, and state a decision. This chapter briefly highlights the significance testing, where a given hypothesis is accepted or rejected, and testing simple hypotheses, where testing of one hypothesis against another is considered using various decision rules and metrics. For example, if we have 100 independent, identically-distributed (iid) samples of a Gaussian random variable with unknown mean, the question that can be addressed by significance testing is whether or not we should accept the hypothesis that the sample mean is, say, 2, whereas the question addressed by hypothesis testing is that if the sample mean is equal to, say, 1 or 2. The focus of this chapter is on binary hypothesis testing.

10.1   Significance Testing

In order to make decisions, assumptions are made about the populations, which may or may not be true. An assumption made about a population parameter is called a *statistical hypothesis* or simply a *hypothesis*. A statistical hypothesis is formulated for the sole purpose of rejecting (nullifying) it. *Significance testing* is a process through which a formally stated hypothesis about the population parameter can be accepted or rejected. Through significance testing, the credibility (plausibility) of a specific statement can thus be assessed. Unfortunately, in general, it is not possible to make correct decisions at all times, as the available data may be limited and/or distorted, i.e. the observed data may not be sufficient and/or accurate.

In a statistical hypothesis, the set of possible outcomes is divided into a set of values for which we accept the hypothesis and another set of values for which

Probability, Random Variables, Statistics, and Random Processes: Fundamentals & Applications,
First Edition. Ali Grami.
© 2020 John Wiley & Sons, Inc. Published 2020 by John Wiley & Sons, Inc.
Companion Website: www.wiley.com/go/grami/PRVSRP

we reject it. There are many ways to partition the observation space into two regions; hence, there is a need for a criterion. If possible, hypotheses are formulated to be tested as a single value for a parameter. For instance, to determine whether one pain medication is stronger than another or one educational approach is more effective than another, we hypothesize that there is no difference in the strength of the two pain medications or no difference in the effectiveness of the two educational approaches.

The two hypotheses required for hypothesis testing are the *null hypothesis*, denoted by H_0, which is a statement that generally specifies a particular value for the parameter being studied, and an *alternative hypothesis*, denoted by H_1, which is a statement that specifies either an alternative value or a range of alternative values. The two hypotheses H_0 and H_1 are logical opposites, that is if one is rejected, the other has been effectively accepted. When a hypothesis is specified exactly, it is a *simple hypothesis*, and when it is not completely specified, it is a *composite hypothesis*. The null hypothesis is in most cases simple.

The asymmetry between the null and alternative hypotheses is due to the fact that we assume the null hypothesis, and then see if there is sufficient evidence in the observed data to reject this assumption in favor of a conjectured alternative hypothesis. When the goal of a study is to establish an assertion, the negation of the assertion should be taken as the null hypothesis, and the assertion thus becomes the alternative hypothesis. In a way, the term null hypothesis is used for any hypothesis set up primarily to see whether it can be rejected. The alternative hypothesis usually specifies that the population parameter of interest is not greater than or equal to or less than the value assumed under the null hypothesis.

Sample observations are collected to determine if the data support H_0 or H_1, not to determine whether H_0 or H_1 is true. The decision between the two hypotheses is made on the basis of the value of a statistic called the *test statistic*, which is a function of the observations in a random sample. If the results observed in a random sample are considerably different from the results expected under the hypothesis, then the observed differences are significant.

Suppose we have a random sample $X = (X_1, X_2, ..., X_n)$ of a random variable X whose pdf $f_X(x; \theta)$ depends on a population parameter θ, which is unknown but not random, and the observed data represented by the vector $x = (x_1, x_2, ..., x_n)$. We want to test a hypothesis about a parameter θ of the random variable X. The objective of a significance test is to accept or reject the null hypothesis H_0. In order to specify the decision rule, the observation space, which is an n-dimensional Euclidean space, is partitioned into two disjoint regions R_0 and R_1. If we have $x \in R_0$, we will then decide on H_0. The region R_0 is thus known as the *acceptance region* or *region of nonsignificance*. If we have $x \in R_1$, we will then decide on H_1. The region R_1 is thus known as the *rejection region* or *critical region* or *region of significance*. The boundaries that mark the beginning of the critical region are called **critical values**. The decision rule is thus as follows:

Accept H_0 if $x \in R_0$

Reject H_0 if $x \in R_1$ $\hspace{4cm}$ (10.1)

Note that not rejecting the hypothesis is not the same as accepting the hypothesis. When we do not reject the hypothesis, we are saying the hypothesis is plausible, but it is essential to note that there can be other equally plausible hypotheses.

There are two types of errors, known as **Type I error** and **Type II error** that can occur when the above decision rule is followed. They are as follows:

Type I error: Reject H_0 when H_0 is true

Type II error: Accept H_0 when H_0 is false $\hspace{3cm}$ (10.2)

The probabilities of Type I and Type II errors are defined, respectively, as follows:

$P(x \in R_1 \mid H_0) = \alpha$

$P(x \in R_0 \mid H_1) = \beta$ $\hspace{5cm}$ (10.3)

where α is called the **level of significance** or **size of the rejection region**, and $(1 - \beta)$ is known as the **power of the test**. Table 10.1 highlights the four distinct decisions and their corresponding probabilities.

The level of significance α, which is the probability of rejecting the null hypotheses, given that it was true, is generally specified before any samples are taken so that the results obtained will not compromise the objectivity of the testing method. The level of significance of a test is considered to be an important design criterion for testing. Note there is an interesting relationship between estimation theory and detection theory, in that the sum of the level of significance and the confidence level is 100%, as the confidence interval is the acceptance region. If the null hypothesis is true, the probability of a Type I

Table 10.1 Decisions and errors in significance testing.

		Decision based on the data	
		Reject H_0	Do not reject H_0
Truth (unknown)	H_0 is true	Type I error (false rejection)	Correct decision
		Probability = α	Probability = $1 - \alpha$
	H_0 is false	Correct decision	Type II error (false acceptance)
		Probability = $1 - \beta$	Probability = β

error (false rejection) can be evaluated as follows:

$$\alpha = \int_{R_1} f_X(x \mid H_0)dx \qquad (10.4)$$

However, in the absence of a probability model for the condition H_0 being false, the probability of a Type II error (false acceptance) cannot be evaluated. In order to have good decision rules (tests of hypotheses), it is desirable to have both α and β as small as possible. However, they are not independent of each other, and in general, for a given sample size n, a decrease in one type of error yields an increase in the other type. In some applications, such as digital transmission, both types of errors are of equal importance. However, in most applications, such as the medical diagnostic testing and radar systems, as one error may be more serious than the other, a compromise should be made in favor of a limitation of the more serious error. As neither types of errors are independent of the sample size n, the only viable way to simultaneously bring a reduction to both types of errors is to increase the sample size n, which may not always be possible.

Example 10.1
Discuss the significance testing in the context of the criminal court of law, where on the one hand, the accused may be guilty beyond a reasonable doubt or not guilty, and on the other hand, the jury may find the defendant guilty or not guilty.

Solution
The null hypothesis H_0 states that the accused is not guilty. The accused is found not guilty unless the null hypothesis of not guilty is disproved. This does not imply that the defendant has been proved innocent if found not guilty, it just implies that the accused has not been proved guilty. Note that if proof of guilt is not established, the accused is freed, and we act as if the null hypothesis of innocence was accepted. The following table summarizes the possible scenarios for the outcome of a hypotheses test:

	Jury's verdict: accept H_0	Jury's verdict: reject H_0
H_0 is true: Defendant is not guilty	Decision is correct Probability = $1 - \alpha$	Type I error Probability = α
H_0 is not true: Defendant is guilty	Type II error Probability = β	Decision is correct Probability = $1 - \beta$

∎

In summary, in order to decide what statement goes in the null hypothesis and what statement goes in the alternative hypothesis, the following rules of thumb should be followed:

(i) What is expected as a result of the test usually should be placed in the alternative hypothesis.

(ii) The null hypothesis should contain a statement of equality.

(iii) The null hypothesis is the hypothesis that is tested.

(iv) The null and alternative hypotheses are complementary.

10.2 Hypothesis Testing for Mean

As mentioned earlier, in hypothesis testing, a decision is made between two hypotheses on the basis of the value of a test statistic, which is a function of the observations in a random sample. The following is a general formula for a test statistic that will be applicable in many of the hypothesis tests discussed in this chapter:

$$\text{Test statistics} = \frac{\text{Relevant statistics} - \text{Hypothesized parameter}}{\text{Standard error of the relevant statistics}} \quad (10.5a)$$

As we want to consider the testing of a hypothesis about a population mean for the random sample of size n, the relevant statistics is the sample mean, the hypothesized parameter is the population mean, and the standard error of the relevant statistics is represented by the standard deviation of the sample mean using the population variance, if available, otherwise, the sample variance is used instead. The test statistics for the mean is thus as follows:

$$\text{Test statistics} = \begin{cases} \dfrac{\overline{X}_n - \mu_X}{\sigma_X/\sqrt{n}} & \text{if } \sigma_X \text{ is known} \\ \dfrac{\overline{X}_n - \mu_X}{\sigma_n/\sqrt{n}} & \text{if } \sigma_X \text{ is not known} \end{cases} \quad (10.5b)$$

Our focus is on the following three distinct cases as how to determine the standardized test statistics:

Case-I: If the population has a Gaussian distribution with known variance σ_X^2, then the standardized test statistic is as follows:

$$Z = \frac{\overline{X}_n - \mu_X}{\sigma_X/\sqrt{n}} \quad (10.6a)$$

where Z is the standard Gaussian random variable. Note that the population variance is not generally known in practice.

Case-II: If the sample size n is large, the population variance is not known, but the sample variance σ_n^2 is known, then for any non-Gaussian population distribution, the standardized test statistic is as follows:

$$Z = \frac{\overline{X}_n - \mu_X}{\sigma_n/\sqrt{n}} \quad (10.6b)$$

where Z, due to the central limit theorem, is the standard Gaussian random variable.

Case-III: If the population has a Gaussian distribution with unknown variance, but the sample variance σ_n^2 is known, then the standardized test statistic is as follows:

$$T = \frac{\overline{X}_n - \mu_X}{\sigma_n / \sqrt{n}} \tag{10.6c}$$

where T is a random variable having the Student's t-distribution with $n-1$ degrees of freedom.

All possible values that the test statistic can have are points on the horizontal axis of the distribution of the test statistic and are partitioned into two regions. One region constitutes what is known as the rejection region and the other region makes up the nonrejection region. The values of the test statistic forming the rejection region are those values that are less likely to occur if the null hypothesis is true, while the values making up the acceptance region are more likely to occur if the null hypothesis is true. Based on the decision rule, we reject the null hypothesis if the value of the test statistic computed from the samples is one of the values in the rejection region and do not reject the null hypothesis if the value of the test statistic is one of the values in the nonrejection region.

A test with the rejection region in both tails of the distribution of a statistic is known as a ***two-tailed test*** or a ***two-sided test*** and the area under each is equal to half of the level of significance, i.e. $\frac{\alpha}{2}$. A test with the rejection region in one tail of the distribution of a statistic is known as a ***one-tailed test*** or a ***one-sided test***, whose area is equal to the level of significance α. Moreover, if the rejection region is in the right tail of the distribution, it is then known as a ***right-tailed test*** or a ***right-sided test***, and if it is in the left tail of the distribution, it is known as a ***left-tailed test*** or a ***left-sided test***.

Assuming the null hypothesis H_0 is $\mu_X = \mu_0$, Table 10.2 highlights the critical regions for testing $\mu_X = \mu_0$.

Table 10.3 shows some critical values for both the one-tailed test (z_α) and the two-tailed test $(z_{\alpha/2})$ in tests involving the Gaussian distribution, and Figure 10.1 shows the rejection and nonrejection regions for the right-tailed, left-tailed, and two-tailed tests.

Table 10.2 Statistic for mean.

Alternative Hypothesis H_1	Test Type	Conditions to reject H_0 Case-I	Case-II	Case-III
$\mu_X < \mu_0$	Left-tailed	$Z \leq -z_\alpha$	$Z \leq -z_\alpha$	$T \leq -t_\alpha$
$\mu_X > \mu_0$	Right-tailed	$Z \geq z_\alpha$	$Z \geq z_\alpha$	$T \geq t_\alpha$
$\mu_X \neq \mu_0$	Two-tailed	$|Z| \geq z_{\alpha/2}$	$|Z| \geq z_{\alpha/2}$	$|T| \geq t_{\alpha/2}$

Table 10.3 Critical values of Z.

Level of significance α	Critical values of z_α for one-tailed tests	Critical values of $z_{\alpha/2}$ for two-tailed tests
0.10 (=10%)	$z_\alpha \leq -1.28$ or $z_\alpha \geq 1.28$	$|z_{\alpha/2}| \geq 1.65$
0.05 (=5%)	$z_\alpha \leq -1.65$ or $z_\alpha \geq 1.65$	$|z_{\alpha/2}| \geq 1.96$
0.02 (=2%)	$z_\alpha \leq -2.06$ or $z_\alpha \geq 2.06$	$|z_{\alpha/2}| \geq 2.33$
0.01 (=1%)	$z_\alpha \leq -2.33$ or $z_\alpha \geq 2.33$	$|z_{\alpha/2}| \geq 2.58$
0.005 (=0.5%)	$z_\alpha \leq -2.58$ or $z_\alpha \geq 2.58$	$|z_{\alpha/2}| \geq 2.81$
0.002 (=0.2%)	$z_\alpha \leq -2.88$ or $z_\alpha \geq 2.88$	$|z_{\alpha/2}| \geq 3.09$
0.001 (=0.1%)	$z_\alpha \leq -3.09$ or $z_\alpha \geq 3.09$	$|z_{\alpha/2}| \geq 3.29$
0.0005 (=0.05%)	$z_\alpha \leq -3.29$ or $z_\alpha \geq 3.29$	$|z_{\alpha/2}| \geq 3.48$
0.0002 (=0.02%)	$z_\alpha \leq -3.54$ or $z_\alpha \geq 3.54$	$|z_{\alpha/2}| \geq 3.72$
0.0001 (=0.01%)	$z_\alpha \leq -3.72$ or $z_\alpha \geq 3.72$	$|z_{\alpha/2}| \geq 3.89$

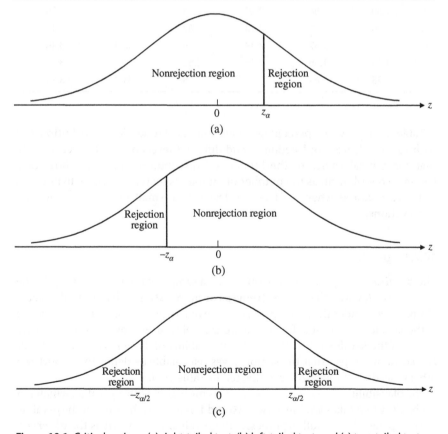

Figure 10.1 Critical regions: (a) right-tailed test; (b) left-tailed test; and (c) two-tailed test.

Table 10.4 Critical values of T.

n	Percentage points of Student's t distribution (t_α)					
	$\alpha = 0.10$	$\alpha = 0.05$	$\alpha = 0.01$	$\alpha = 0.005$	$\alpha = 0.001$	$\alpha = 0.0005$
1	3.08	6.31	31.8	63.7	318	637
5	1.48	2.02	3.36	4.03	5.89	6.87
10	1.37	1.81	2.76	3.17	4.14	4.59
15	1.34	1.75	2.60	2.95	3.73	4.07
20	1.32	1.72	2.53	2.85	3.55	3.85
25	1.32	1.71	2.48	2.79	3.44	3.72
30	1.31	1.70	2.46	2.75	3.39	3.65
35	1.31	1.69	2.44	2.72	3.34	3.59
40	1.30	1.68	2.42	2.70	3.31	3.55
45	1.30	1.68	2.41	2.69	3.28	3.52
50	1.30	1.68	2.40	2.68	3.26	3.50
55	1.30	1.67	2.40	2.67	3.25	3.48
60	1.30	1.67	2.39	2.66	3.23	3.46
120	1.29	1.66	2.36	2.62	3.15	3.37
∞	1.28	1.65	2.33	2.58	3.09	3.29

Table 10.4 provides percentage points of the Student's t-distribution (t_α) with various degrees of freedom n and different levels of significance α. Note that the critical values of the Student's t distribution approach those of a Gaussian distribution as the number of samples is increased. These hypothesis tests are required when we have to deal with a small number of Gaussian observations.

10.2.1 p-Value

The **p-value** is the probability of observing a sample statistic (obtaining a value for the test statistic) that is as extreme as or more extreme than the value actually observed under the assumption that the null hypothesis is true. The p-value is the smallest value of α for which we can reject a null hypothesis. In other words, if the p-value is less than α (the probability of Type I error), then reject H_0, i.e. do not reject that the sample gives reasonable evidence to support H_1; otherwise, do not reject H_0, as reflected in Table 10.5.

The plausibility of a null hypothesis H_0 is measured with a p-value, which is a probability that takes a value between 0 and 1. Note that the smaller the p-value, the less plausible the null hypothesis. The p-value, also known as the **observed**

Table 10.5 Decision rules.

p-value $< \alpha$	$\rightarrow$	H_0 is rejected
p-value $> \alpha$	$\rightarrow$	H_0 is accepted

Table 10.6 Significance of p-value.

$0 < p \leq 0.001$	$\rightarrow$	Very highly significant
$0.001 \leq p \leq 0.01$	$\rightarrow$	Highly significant
$0.01 \leq p \leq 0.05$	$\rightarrow$	Statistically significant
$0.05 \leq p < 1$	$\rightarrow$	Not statistically significant

level of significance or the *attained significant level*, can be constructed from a data set under the assumption that H_0 is true. If the p-value is small, there is strong evidence against H_0, and it should be rejected. If the p-value is large, there is strong evidence for accepting H_0 (but not proving H_0 to be true). A result has statistical significance when it is very unlikely to have occurred given the null hypothesis. Table 10.6 provides a general outline in assessing the significance of a p-value for most, but not all, applications.

A step-by-step outline as how to proceed with the problem of significance testing is as follows:

(1) Formulate a null hypothesis (which is the negation of the claim) and an appropriate alternative hypothesis (which is accepted when the null hypothesis is rejected) using parameters of interest.
(2) Specify the probability of Type I error, also called the level of significance α.
(3) Set up a criterion for testing the null hypothesis against the alternative hypothesis using an appropriate statistic.
(4) Calculate from the data the value of the statistic on which the decision is to be used, and compute the p-value.
(5) Decide whether to reject the null hypothesis or whether not to reject it by comparing the value of α and the p-value.

Example 10.2

The mean price of cars produced by a car manufacturer is \$36 000 and the standard deviation is \$2000. It is claimed that the prices of new cars have increased. To test this claim, a sample of 50 new cars is tested, and it is found that the mean price is \$37 000. Can we support the claim at a 0.01 level of significance?

Solution

We have to decide between the two hypotheses:

H_0: $\mu_X = 36\,000$, and there is no change in price
H_1: $\mu_X > 36\,000$, and there is an increase in price

As we are interested in determining values higher than $36\,000$, it is thus a right-tailed test, and the rejection region is above the acceptance region. Noting that the population size is large and the population variance is known, the standardized test statistic is as follows:

$$Z = \frac{\overline{X}_n - \mu_X}{\sigma_X/\sqrt{n}} = \frac{37\,000 - 36\,000}{2000/\sqrt{50}} \cong 3.54$$

There are two ways to draw the same conclusion:

$$\alpha = 0.01 \rightarrow z_\alpha = Q^{-1}(0.01) = \Phi^{-1}(0.99) \cong 2.33 < Z \cong 3.54$$
$$\rightarrow H_0 \text{ is rejected}$$

or

$$Z \cong 3.54 \rightarrow p\text{-value} \cong Q(3.54) = 1 - \Phi(3.54) \cong 0.0002 < \alpha = 0.01$$
$$\rightarrow H_0 \text{ is rejected}$$

Since the z-score 3.54 lies inside the range $(2.33, \infty)$ or equivalently the p-value 0.0002 is less than the level of significance 0.01, the hypothesis H_1 that is the claim ($\mu_X > 36\,000$) should be accepted. ∎

Example 10.3

The mean lifetime of a sample of 100 batteries produced by a company is computed to be 3140 hours with a standard deviation of 240 hours. If μ_X is the mean lifetime of all the batteries produced by the company, test the null hypothesis $\mu_X = 3200$ hours against the alternative hypothesis $\mu_X \neq 3200$ hours, using levels of significance of 0.05 and 0.01.

Solution

We have to decide between the two hypotheses:

H_0: $\mu_X = 3200$, and there is no change in lifetime
H_1: $\mu_X \neq 3200$, and there is a change in lifetime

We choose a two-tailed test, as we are interested in determining values both larger and smaller than 3200. Noting that the population size is large, and the population variance is not known, the standardized test statistic is as follows:

$$Z = \frac{\overline{X}_n - \mu_X}{\sigma_n/\sqrt{n}} = \frac{3140 - 3200}{240/\sqrt{100}} \cong -2.5$$

We consider two different levels of significance, though the approaches are the same. Assuming the level of significance is 0.05, there are two ways to draw the same conclusion:

$$\alpha = 0.05 \rightarrow -z_{\alpha/2} = -Q^{-1}(0.025) = -\Phi^{-1}(0.975) \cong -1.96 > Z \cong -2.5$$
$$\rightarrow H_0 \text{ is rejected}$$

or

$$Z \cong -2.5 \rightarrow p\text{-value} \cong 2 \times Q(2.5) = 2(1 - \Phi(2.5)) \cong 0.012 < \alpha = 0.05$$
$$\rightarrow H_0 \text{ is rejected}$$

Since the z-score -2.5 lies outside the range $(-1.96, 1.96)$ or equivalently the p-value 0.012 is less than the level of significance 0.05, the hypothesis H_1 that is the claim ($\mu_X \neq 3200$) should be accepted. Assuming the level of significance is 0.01, there are two ways to draw the same conclusion:

$$\alpha = 0.01 \rightarrow -z_{\alpha/2} = -Q^{-1}(0.005) = -\Phi^{-1}(0.995) \cong -2.58 < Z \cong -2.5$$
$$\rightarrow H_0 \text{ is accepted}$$

or

$$Z \cong -2.5 \rightarrow p\text{-value} \cong 2 \times Q(2.5) = 2(1 - \Phi(2.5)) \cong 0.012 > \alpha = 0.01$$
$$\rightarrow H_0 \text{ is accepted}$$

Since the z-score -2.5 lies inside the range $(-2.58, 2.58)$ or equivalently the p-value 0.012 is greater than the level of significance 0.01, the hypothesis H_1 that is the claim ($\mu_X \neq 3200$) should be rejected. ∎

Example 10.4
A pharmaceutical company claimed that a certain medicine was 90% effective. In a sample of 200 people who took this medicine, the medicine was effective for 160 people. Determine whether the claim is legitimate, assuming a level of significance of 0.01.

Solution
We have to decide between the two hypotheses:

$H_0: p = 0.9$, and the claim is correct
$H_1: p < 0.9$, and the claim is false

We choose a left-tailed test, as we are interested in determining values lower than 0.9. Noting that the population size is large and the population variance is not known, and using the Gaussian approximation to the binomial distribution, the standardized test statistic is as follows

$$Z = \frac{\overline{X}_n - np}{\sqrt{np(1-p)}} = \frac{160 - 200 \times 0.9}{\sqrt{200 \times 0.9 \times 0.1}} \cong -4.714$$

There are two ways to draw the same conclusion:

$$\alpha = 0.01 \rightarrow -z_\alpha = -Q^{-1}(0.01) = -\Phi^{-1}(0.99) \cong -2.33 > Z \cong -4.714$$
$$\rightarrow H_0 \text{ is rejected}$$

or

$$Z \cong -4.714 \rightarrow p\text{-value} \cong 1 - Q(-4.714) = Q(4.714) \cong 1\text{E} - 06 < \alpha = 0.01$$
$$\rightarrow H_0 \text{ is rejected}$$

Since the z-score -4.714 lies inside the range $(-\infty, -2.33)$ or equivalently the p-value 0.000001 is less than the level of significance 0.01, the hypothesis H_1 that is the claim ($p < 0.9$) should be accepted. ∎

Example 10.5

Suppose we have a random sample $X = (X_1, X_2, ..., X_n)$ of a Gaussian random variable X with mean μ_X and variance σ_X^2, where n is the sample size. We have to decide between the two hypotheses:

$$H_0: \mu_X = \mu_0$$
$$H_1: \mu_X = \mu_1(> \mu_0)$$

As a decision procedure, we reject H_0 if the sample mean $\overline{X}_n \geq m$, where we have $\mu_1 > m > \mu_0$. Determine the probability of a Type I error, i.e. α. Find the probability of a Type II error, i.e. β. What is the impact of n on α and β?

Solution

The sample mean $\overline{X}_n$ is a Gaussian random variable whose mean and variance are μ_X and $\frac{\sigma_X^2}{n}$, respectively. Since the test calls for the rejection of H_0 when $\overline{X}_n \geq m$, the probability of rejecting H_0 given H_0 is true, i.e. α, is given by

$$\alpha = P(\overline{X}_n \geq m \mid H_0) = \int_{\frac{m-\mu_0}{\sigma_X/\sqrt{n}}}^{\infty} \left(\frac{1}{\sqrt{2\pi}}\right) \exp\left(-\frac{x^2}{2}\right) dx = Q\left(\frac{m - \mu_0}{\sigma_X/\sqrt{n}}\right)$$

Since the test calls for the acceptance of H_0 when $\overline{X}_n \leq m$, the probability of accepting H_0 given H_1 is true, i.e. β, is given by

$$\beta = P(\overline{X}_n \leq m \mid H_1) = \int_{-\infty}^{\frac{m-\mu_1}{\sigma_X/\sqrt{n}}} \left(\frac{1}{\sqrt{2\pi}}\right) \exp\left(-\frac{x^2}{2}\right) dx$$
$$= 1 - Q\left(\frac{m - \mu_1}{\sigma_X/\sqrt{n}}\right) = Q\left(\frac{\mu_1 - m}{\sigma_X/\sqrt{n}}\right)$$

It is imperative to note that with an increase in the sample size n, both α and β decrease, as reflected in Figure 10.2. ∎

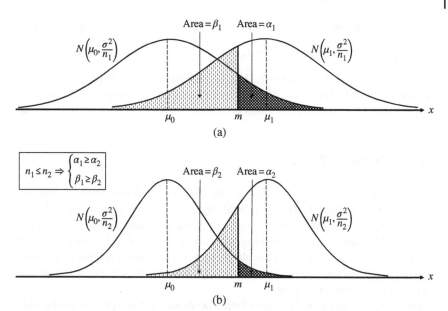

Figure 10.2 (a, b) Probabilities of Types I and II errors for sample means with various sample sizes.

Example 10.6

In the past, a machine has produced washers having a mean thickness of 0.1 cm. To determine whether the machine is in proper working order, a sample of nine washers is chosen for which the mean thickness is $\frac{7}{60}$ cm and the variance is 0.01 cm. Assume the samples are Gaussian, but the variance is unknown. Test the hypothesis that the machine is in proper working order using a level of significance of 0.05.

Solution

We have to decide between the two hypotheses:

$H_0: \mu_X = 0.1$, and the machine is in proper working order
$H_1: \mu_X \neq 0.1$, and the machine is not in proper working order

We choose a two-tailed test, as we are interested in determining values both larger and smaller than 0.1. Noting that the size of Gaussian samples is small and the population variance is not known, the standardized test statistic is thus as follows:

$$T = \frac{\overline{X}_n - \mu_X}{\sigma_n/\sqrt{n}} = \frac{\frac{7}{60} - 0.1}{\sqrt{0.01/9}} = 0.5$$

Assuming the level of significance is 0.05, and there are nine samples, we have

$$\frac{\alpha}{2} = 0.025 \text{ and } n - 1 = 8 \rightarrow t_{\alpha/2, n-1} = 2.31 > T \cong 0.5 \rightarrow H_0 \text{ is accepted}$$

As 0.5 lies inside the range $(-2.31, 2.31)$, the hypothesis H_1 that is the claim $(\mu_X \neq 0.1)$ should be rejected. The null hypothesis is thus accepted, and the data support the assertion that the mean is 0.1 cm. ∎

10.3 Decision Tests

The solution to a hypothesis testing problem is specified in terms of a decision rule; different decision tests thus provide different decision rules and yield different performances. We want to briefly introduce various decision tests in the context of binary hypothesis testing with continuous-valued observations, as shown in Figure 10.3. Note that multiple hypothesis testing is a generalization of binary hypothesis testing, and the decision tests for the continuous case can carry over to the discrete case.

Assuming the null and alternative hypotheses are both simple, the probability of a Type I error and the probability of a Type II error, which in turn provide us with two criteria in the design of a hypothesis test, can be evaluated as follows:

$$\alpha = P(x \in R_1 \mid H_0) = \int_{R_1} f_X(x \mid H_0) dx$$

$$\beta = P(x \in R_0 \mid H_1) = \int_{R_0} f_X(x \mid H_1) dx \qquad (10.7)$$

Note that the vector $X = (X_1, X_2, ..., X_n)$ represents a random sample of the random variable X and $x = (x_1, x_2, ..., x_n)$ represents the observation vector. It is customary in hypothesis testing to focus on the false positive probability α, known as the **size of a test**, and the true positive probability $1 - \beta$, referred to as the **power of a test**. It is the primary goal to have a large $1 - \beta$ and a small α, however, these are competing objectives, that is one is at the expense of the other.

Example 10.7
In a radar signal detection system, the objective is to distinguish between the presence and the absence of a target. Suppose we have n observations x_i, $i = 1$, 2, ..., n, of a radar signal, and x_is are Gaussian iid random variables under each hypothesis. Under the hypothesis H_0, that is when there exists no target, observations x_i, $i = 1, 2, ..., n$, have mean μ_0 and variance σ_0^2, while under the hypothesis H_1, that is when target is present, observations x_i, $i = 1, 2, ..., n$, have mean μ_1 and variance σ_1^2, and $\mu_1 > \mu_0$ and $\sigma_1^2 > \sigma_0^2$. We take the n observation samples to form the sample mean $\overline{X}_n$, upon which a decision is made.

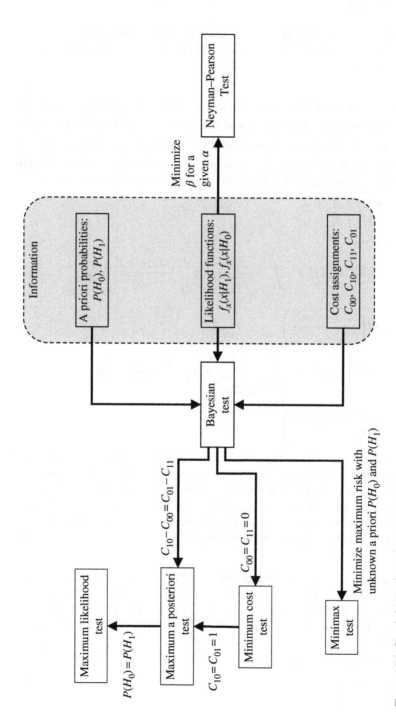

Figure 10.3 Simple binary hypothesis tests.

Determine the probability of a Type I error α, also known as the false alarm probability, and the probability of a Type II error β, often referred to as the miss probability.

Solution

If H_0 is true, then the sample mean of n independent observations $\overline{X}_n$ is a Gaussian random variable with mean μ_0 and variance $\frac{\sigma_0^2}{n}$, and if H_1 is true, then the sample mean of n independent observations $\overline{X}_n$ is a Gaussian random variable with mean μ_1 and variance $\frac{\sigma_1^2}{n}$. For some suitable constant $\mu_0 < \rho < \mu_1$, through which the rejection region is defined, the decision rule is then as follows:

Accept H_0 if $\overline{X}_n \le \rho$.
Accept H_1 if $\overline{X}_n > \rho$.

We thus have

$$\alpha = P(\overline{X}_n > \rho \mid H_0) = \int_\rho^\infty \frac{\sqrt{n}}{\sqrt{2\pi\sigma_0^2}} \exp\left(-\frac{n(x-\mu_0)^2}{2\sigma_0^2}\right) dx$$

$$= Q\left(\frac{\sqrt{n}(\rho - \mu_0)}{\sigma_0}\right)$$

$$\beta = P(\overline{X}_n \le \rho \mid H_1) = \int_{-\infty}^\rho \frac{\sqrt{n}}{\sqrt{2\pi\sigma_1^2}} \exp\left(-\frac{n(x-\mu_1)^2}{2\sigma_1^2}\right) dx$$

$$= Q\left(\frac{\sqrt{n}(\mu_1 - \rho)}{\sigma_1}\right)$$

Note that there is an interesting trade-off between the two types of errors. As ρ increases, α decreases from 1 to 0, while β increases from 0 to 1. It is imperative to note that an increase in the number of observation samples n can bring about a simultaneous reduction in both α and β. The value of the threshold ρ and the number of observation samples n, individually or both together, can help provide a right balance between the two types of errors. ∎

There are various sets of information that can be used in solving a binary hypothesis testing problem. One is the set of a priori probabilities, should that be known. It can summarize the state of knowledge about the applicable hypothesis (probability model) before any outcome (observed data) is available, as given by

$$P_0 = P[H_0]$$

$$P_1 = P[H_1] \tag{10.8}$$

where $P_0 + P_1 = 1$.

Another is the set of conditional probability density functions, referred to as *likelihood functions*, corresponding to the probability density functions for the n dimensional observation vector x conditioned on each of the two hypotheses, as given by

$$H_0 : f_X(x \mid H_0)$$
$$H_1 : f_X(x \mid H_1) \tag{10.9}$$

Finally, there is the set of costs associated with the outcomes, as the costs and benefits of uncertain outcomes can be quantified, and thus measured. In a binary hypothesis testing problem, there are four possible outcomes of the hypothesis test, where a cost to each outcome, as a measure of its relative importance, is assigned. The *costs* are defined as follows:

$$C_{ij} = \text{Cost of accepting } H_i \text{ when } H_j \text{ is true} \quad i = 0, 1 \quad \text{and} \quad j = 0, 1 \tag{10.10}$$

It is reasonable to assume that making a correct decision is always much less costly than making a mistake (erroneous decision), i.e. $C_{00} < C_{10}$ and $C_{11} < C_{01}$. Note that C_{10} and C_{01} are the costs associated with Type I error α and Type II error β, respectively.

10.4 Bayesian Test

Assuming D_i denotes the event that the decision is made to accept H_i, $P(D_i, H_j)$ represents the probability that we accept H_i when H_j is true, and C_{ij} reflects the cost associated with $P(D_i, H_j)$, we define the *average cost* $\overline{C}$, also known as the *Bayes' risk*, as follows:

$$\overline{C} = C_{00}P(D_0, H_0) + C_{10}P(D_1, H_0) + C_{01}P(D_0, H_1) + C_{11}P(D_1, H_1) \tag{10.11}$$

The test that minimizes the average cost $\overline{C}$ is called the *Bayesian test*. It can be shown that the optimum decision rule in the Bayesian test, based on the observation vector x, in terms of the *likelihood ratio test* (LRT), is as follows:

$$\Lambda(x) = \frac{f_X(x \mid H_1)}{f_X(x \mid H_0)} \overset{H_1}{\underset{H_0}{\gtrless}} \frac{(C_{10} - C_{00})P(H_0)}{(C_{01} - C_{11})P(H_1)} = \eta \rightarrow \ln \Lambda(x) \overset{H_1}{\underset{H_0}{\gtrless}} \ln \eta \tag{10.12}$$

A careful examination of the above test gives rise to the following interesting observations:

- The left-hand side of the test is a function of the observed data constructed from the measureable model, and its right side is a precomputable threshold determined from the a priori probabilities and costs.

- The likelihood ratio $\Lambda(x)$ is a scaler-valued function regardless of the dimension n of the observed data x.
- A function of the observed data $f(x)$ that is as good as knowing x is known as a **sufficient statistic**.
- The likelihood ratio is a one-dimensional random variable, thus taking on a different value in each experiment, and its probability density function oftentimes is of interest.
- The likelihood ratio is a nonnegative quantity. It is generally assumed that it is a finite positive quantity, however, if it is zero or infinite, we know with certainty what the correct hypothesis is.

A curve representing the true positive probability $1 - \beta$ versus the false positive probability α is referred to as the **receiver operating characteristic** (ROC). An example of ROC curve is shown in Figure 10.4. For an ideal test, we have $\alpha = 0$ and $1 - \beta = 1$, which is the upper left-hand corner in an ROC plot. The closer the ROC curve is to the upper left-hand corner, the better the test is. Each choice of the threshold η completely specifies a decision rule, with which a particular operating point of the ROC curve is associated. Note that as $\eta \to 0$, we have $\alpha \to 1$ and $\beta \to 0$, and as $\eta \to \infty$, we have $\alpha \to 0$ and $\beta \to 1$. In the Bayesian test, the threshold η is selected, and the system then operates at that particular point on the ROC curve. In general, the context of the hypothesis testing problem suggests how to select the costs C_{ij}. We now introduce some widely known binary hypothesis testing methods in relation to the Bayesian test.

10.4.1 Minimum Cost Test

When we have $C_{00} = C_{11} = 0$, and C_{10}, the cost of a false alarm (i.e. deciding H_1 when H_0 is correct), and C_{01}, the cost of a miss (i.e. deciding H_0 when H_1 is correct), are not the same, we have the **minimum cost test**. The **LRT** is as follows:

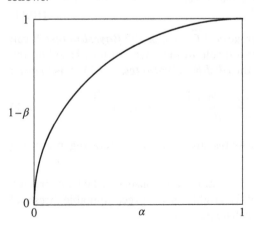

Figure 10.4 An ROC curve.

$$\Lambda(x) = \frac{f_X(x \mid H_1)}{f_X(x \mid H_0)} \underset{H_0}{\overset{H_1}{\gtrless}} \eta = \frac{C_{10}P(H_0)}{C_{01}P(H_1)} \tag{10.13}$$

Note that only the relative cost $\frac{C_{10}}{C_{01}}$ influences the test, not the individual costs.

10.4.2   Maximum a Posteriori Probability (MAP) Test

When $C_{10} - C_{00} = C_{01} - C_{11}$, the Bayesian test becomes the widely known *maximum a posteriori probability (MAP) test*. For instance, in a binary digital transmission system, a decision rule that minimizes the probability of an erroneous decision warrants symmetric cost functions, where we have $C_{00} = C_{11} = 0$ and $C_{01} = C_{10} = 1$ (i.e. the two types of errors, α and β, have the same cost). In other words, the minimum average cost $\overline{C}$ is the probability of making an incorrect decision. In a MAP test or equivalently a minimum probability of error test, we thus have

$$\Lambda(x) \underset{H_0}{\overset{H_1}{\gtrless}} \eta = \frac{P(H_0)}{P(H_1)} \tag{10.14}$$

10.4.3   Maximum-Likelihood (ML) Test

When we have $P(H_0) = P(H_1)$, the MAP test becomes *maximum-likelihood (ML) test*. In the absence of a priori information, the ML test, which is based on the largest a posteriori probability, in place of the MAP test, must be used. The LRT for ML test is as follows:

$$\Lambda(x) \underset{H_0}{\overset{H_1}{\gtrless}} 1 \tag{10.15}$$

Example 10.8

Suppose the observations consist of a set of n random sample values $x = (x_1, x_2, \ldots, x_n)$. Under both hypotheses, all sample values are iid Gaussian random variables. Under H_1, each sample has a mean μ_1 and a variance σ_1^2. Under H_0, each sample has a mean μ_0 and a variance σ_0^2. Determine the likelihood ratio test and the sufficient statistic.

Solution

The probability density functions for the n dimensional observation vector X conditioned on each of the two hypotheses is as follows:

$$f_X(x \mid H_1) = \prod_{i=1}^{n} \left(\frac{1}{\sqrt{2\pi\sigma_1^2}} \right) \exp\left(-\frac{(x_i - \mu_1)^2}{2\sigma_1^2} \right)$$

and

$$f_X(\pmb{x} \mid H_0) = \prod_{i=1}^{n} \left(\frac{1}{\sqrt{2\pi\sigma_0^2}} \right) \exp\left(-\frac{(x_i - \mu_0)^2}{2\sigma_0^2} \right)$$

The likelihood ratio is thus as follows:

$$\Lambda(\pmb{x}) = \frac{\prod_{i=1}^{n} \left(\frac{1}{\sqrt{2\pi\sigma_1^2}} \right) \exp\left(-\frac{(x_i - \mu_1)^2}{2\sigma_1^2} \right)}{\prod_{i=1}^{n} \left(\frac{1}{\sqrt{2\pi\sigma_0^2}} \right) \exp\left(-\frac{(x_i - \mu_0)^2}{2\sigma_0^2} \right)} \rightarrow$$

$$\ln \Lambda(\pmb{x}) = \ln\left[\frac{\prod_{i=1}^{n} \left(\frac{1}{\sqrt{2\pi\sigma_1^2}} \right) \exp\left(-\frac{(x_i - \mu_1)^2}{2\sigma_1^2} \right)}{\prod_{i=1}^{n} \left(\frac{1}{\sqrt{2\pi\sigma_0^2}} \right) \exp\left(-\frac{(x_i - \mu_0)^2}{2\sigma_0^2} \right)} \right] \rightarrow$$

$$\ln \Lambda(\pmb{x}) = n \ln\left(\frac{1}{\sqrt{2\pi\sigma_1^2}} \right) - \sum_{i=1}^{n} \frac{(x_i - \mu_1)^2}{2\sigma_1^2} - n$$

$$\times \ln\left(\frac{1}{\sqrt{2\pi\sigma_0^2}} \right) + \sum_{i=1}^{n} \frac{(x_i - \mu_0)^2}{2\sigma_0^2} \rightarrow$$

$$\ln \Lambda(\pmb{x}) = \sum_{i=1}^{n} \left(\frac{(x_i - \mu_0)^2}{2\sigma_0^2} - \frac{(x_i - \mu_1)^2}{2\sigma_1^2} \right) + n\ln\left(\frac{\sigma_0}{\sigma_1} \right) \rightarrow$$

$$\ln \Lambda(\pmb{x}) = \sum_{i=1}^{n} \left(\frac{x_i^2 - 2x_i\mu_0 + \mu_0^2}{2\sigma_0^2} - \frac{x_i^2 - 2x_i\mu_1 + \mu_1^2}{2\sigma_1^2} \right) + n\ln\left(\frac{\sigma_0}{\sigma_1} \right) \rightarrow$$

$$\ln \Lambda(\pmb{x}) = \sum_{i=1}^{n} \left(\frac{x_i^2 - 2x_i\mu_0}{2\sigma_0^2} - \frac{x_i^2 - 2x_i\mu_1}{2\sigma_1^2} \right)$$

$$+ \sum_{i=1}^{n} \left(\frac{\mu_0^2}{2\sigma_0^2} - \frac{\mu_1^2}{2\sigma_1^2} \right) + n\ln\left(\frac{\sigma_0}{\sigma_1} \right) \rightarrow$$

$$\ln \Lambda(x) = \sum_{i=1}^{n} \left(\frac{x_i^2(\sigma_1^2 - \sigma_0^2) - 2x_i(\sigma_1^2 \mu_0 - \sigma_0^2 \mu_1)}{2\sigma_0^2 \sigma_1^2} \right)$$

$$+ \sum_{i=1}^{n} \left(\frac{\sigma_1^2 \mu_0^2 - \sigma_0^2 \mu_1^2}{2\sigma_0^2 \sigma_1^2} \right) + n \ln \left(\frac{\sigma_0}{\sigma_1} \right)$$

The likelihood ratio test is thus as follows:

$$\ln \Lambda(x) \underset{H_0}{\overset{H_1}{\gtrless}} \ln \eta \rightarrow f(x_1, x_2, \ldots, x_n) \underset{H_0}{\overset{H_1}{\gtrless}} \gamma$$

where the sufficient statistic is as follows:

$$f(x_1, x_2, \ldots, x_n) = \sum_{i=1}^{n} (x_i^2(\sigma_1^2 - \sigma_0^2) - 2x_i(\sigma_1^2 \mu_0 - \sigma_0^2 \mu_1))$$

and the new threshold is as follows:

$$\gamma = -2\sigma_0^2 \sigma_1^2 \left(\sum_{i=1}^{n} \left(\frac{\sigma_1^2 \mu_0^2 - \sigma_0^2 \mu_1^2}{2\sigma_0^2 \sigma_1^2} \right) + n \ln \left(\frac{\sigma_0}{\sigma_1} \right) - \ln \eta \right) \qquad \blacksquare$$

10.4.4 Minimax Test

When the a priori probabilities cannot be known or are difficult to determine, and we do not want to choose them arbitrarily, as their incorrect values can then negatively impact the performance, we need to employ the minimax test. The *minimax test* makes the Bayesian hypothesis testing robust with respect to uncertainty in the a priori probabilities. In the minimax test, we assume the least favorable choice of a priori probability, for which the Bayes risk is minimized. This approach designs conservatively a decision rule that yields the best possible worst-case performance. This minimization of the maximum possible risk guarantees a minimum level of performance independent of the a priori probabilities.

10.5 Neyman–Pearson Test

There are many situations in which a priori probabilities and the relative costs associated with the two types of errors $\alpha(=P(D_1 | H_0))$ and $\beta(=P(D_0 | H_1))$ are not known. In such cases, the Neyman–Pearson test provides an alternative workable solution that is based on minimization of β for a given α. In other words, the *Neyman–Pearson test* maximizes the power of the test $1 - \beta$ for a given level of significance α. In the Neyman–Pearson test, the critical (or rejection) region R_1 is selected such that $1 - \beta$ is maximum subject to the constraint $\alpha = \alpha_0$. This is a classical constrained optimization problem, which can be solved using the Lagrange multiplier method. It can be shown that the Neyman–Pearson test leads to the following decision rule:

$$\Lambda(x) \overset{H_1}{\underset{H_0}{\gtrless}} \lambda \tag{10.16}$$

where λ is chosen such that

$$\int_{\Lambda(x) \geq \lambda} f_X(x \mid H_0)dx = \alpha \tag{10.17}$$

In the Neyman–Pearson test, an increase in λ decreases β, but increases α. Note that the maximum likelihood test for a simple binary hypothesis can be obtained as the special case where $\lambda = 1$.

Example 10.9

In a radar signal detection system, the received signal x is assumed to be a unit-variance Gaussian random variable, whose mean is either zero or one, depending if there exists no target (hypothesis H_0) or if target is present (hypothesis H_1). Assuming $\alpha = 0.01$, use the Neyman–Pearson test to determine the decision rule.

Solution

The test is then as follows:

$$\Lambda(x) = \frac{f_X(x \mid H_1)}{f_X(x \mid H_0)} = \frac{\dfrac{1}{\sqrt{2\pi}} \exp\left(-\dfrac{(x-1)^2}{2}\right)}{\dfrac{1}{\sqrt{2\pi}} \exp\left(-\dfrac{x^2}{2}\right)}$$

$$= \exp(x - 0.5) \overset{H_1}{\underset{H_0}{\gtrless}} \lambda \rightarrow x \overset{H_1}{\underset{H_0}{\gtrless}} 0.5 + \ln \lambda$$

The critical region R_1 is thus as follows:

$$R_1 = \{x \mid x > 0.5 + \ln \lambda\}$$

We now determine λ such that $\alpha = 0.01$, that is we must have

$$0.01 = \int_{R_1} f_X(x \mid H_0)dx = \int_{0.5 + \ln \lambda}^{\infty} \frac{1}{\sqrt{2\pi}} \exp\left(-\frac{x^2}{2}\right)dx$$

$$= Q(0.5 + \ln \lambda) \rightarrow 2.33 \cong 0.5 + \ln \lambda \rightarrow \lambda \cong 6.23 \qquad \blacksquare$$

10.6 Summary

An introductory view of the subject of hypothesis testing containing the areas of significance testing and decision tests in the context of simple binary decision theory was briefly discussed. First, the concept of significance testing in the context of the sample mean was introduced and then Bayesian and Neyman–Pearson tests were briefly presented.

Problems

10.1 Consider the significance test for H_0: coin is fair that is the probability of a head is the same as the probability of a tail. Assuming there are 100 tosses, determine a test at a significant level of 1%.

10.2 A diverse group of 50 university students were asked how many tweets they send per month. The mean and standard deviation of the number of tweets for these students are 304.6 and 101.5, respectively, whereas the expected number of tweets for all who tweet is 325. Test $H_0 : \mu_X = 325$ versus $H_1 : \mu_X \neq 325$. Determine the p-value for the test of hypothesis, and draw a conclusion for $\alpha = 5\%$.

10.3 A diverse group of 50 high school students was asked how many hours a week they listen to music. The mean and standard deviation of the number of hours for these students are 5.97 and 1.016, respectively, whereas the expected number of hours for all who listen to music is 5.5. Test $H_0 : \mu_X = 5.5$ versus $H_1 : \mu_X > 5.5$. Determine the p-value for the test of hypothesis, and draw a conclusion for $\alpha = 1\%$.

10.4 A university professor claims that the average number of hours her students study for her course is no more than 15 h/wk. To investigate her claim, 36 students of hers are randomly selected and interviewed. The mean and variance of the sampled data are 17 and 9, respectively. Does the sample evidence contradict the professor's claim at 5% significant level?

10.5 Consider a binary decision problem with the following conditional pdfs:

$$H_0 : f_X(x \mid H_0) = \left(\frac{1}{\sqrt{2\pi}}\right) \exp\left(-\frac{x^2}{2}\right)$$

$$H_1 : f_X(x \mid H_1) = \left(\frac{1}{\sqrt{2\pi}}\right) \exp\left(-\frac{(x-1)^2}{2}\right)$$

We take n independent samples to form the sample mean $\overline{X}_n$. Determine the number of samples n required in order to have $\alpha = \beta = 1\%$.

10.6 Consider a binary decision problem with the following conditional pdfs:

$$H_0: f_N(n \mid H_0) = \frac{\lambda_0^n e^{-\lambda_0}}{n!}$$

$$H_1: f_N(n \mid H_1) = \frac{\lambda_1^n e^{-\lambda_1}}{n!}$$

Determine the maximum likelihood ratio test and solve for n.

10.7 In a binary communication system, a 0 is transmitted with probability p and a 1 with probability $1 - p$. If a 0 is transmitted, the receive signal is $R = -1 + N$ and if a 1 is transmitted, the received signal is $R = 1 + N$, where N is a zero-mean σ_X^2-variance Gaussian random variable. Given the received signal R, determine the decision rule for the minimum probability of error and the corresponding bit error rate.

10.8 In a binary communication system, during every symbol period T, one of the two equally likely signals $s_0(t)$ and $s_1(t)$ is transmitted and that n iid observation samples of the received signal $r(t)$ are made during a symbol period, where

$$r(t) = s_i(t) + n(t) \quad i = 0, 1$$

and $n(t)$ is the channel noise, which is a zero-mean unit-variance Gaussian random process. Let n samples of $s_0(t)$ and $s_1(t)$ be represented, respectively, by the two n-dimensional vectors $(s_{01}, s_{02}, \ldots, s_{0n})$ and $(s_{11}, s_{12}, \ldots, s_{1n})$. Determine the MAP test.

10.9 Consider a binary decision problem with the following conditional pdfs:

$$H_0: f_X(x \mid H_0) = 0.5 e^{-|x|}$$

$$H_1: f_X(x \mid H_1) = e^{-2|x|}$$

Assuming we have $\eta = 1$, determine the Bayes risk.

10.10 Prove that the minimum average cost, known as the Bayes' risk, can be expressed as follows:

$$\frac{f_X(x \mid H_1)}{f_X(x \mid H_0)} \underset{H_0}{\overset{H_1}{\gtrless}} \frac{(C_{10} - C_{00})P(H_0)}{(C_{01} - C_{11})P(H_1)}$$

Part IV
Random Processes

11

Introduction to Random Processes

In many practical applications, from the transmission of time-varying signals over telecommunications networks to the observation of fluctuating stock prices over a long period of time, it is necessary to deal with functions of time. These functions are unpredictable; otherwise, their study would serve no purpose, and thus become unnecessary. As the study of random processes is important, this chapter introduces the fundamentals of random processes in a probabilistic sense. The Gaussian process and the Poisson process are also briefly discussed.

11.1 Classification of Random Processes

Suppose to every outcome (sample point) ω in the sample space Ω of a random experiment, according to some rule, a function of time t is assigned. The ensemble or collection of all such functions that result from a random experiment, denoted by $X(t, \omega)$, is a *random process* or a *stochastic process*.

The function $X(t, \omega)$ versus t, for ω fixed, is a deterministic function and is called a *realization*, *sample path*, *ensemble member*, or *sample function* of the random process. For a given $\omega = \omega_i$, a specific function of time t, i.e. $X(t, \omega_i)$, is thus produced, and denoted by $x(t)$. For a specific time $t = t_k$, $X(t_k, \omega)$ is a random variable, and is called a *time sample*. For a specific ω_i and a specific t_k, $X(t_k, \omega_i)$ is simply a nonrandom constant. It is common to suppress ω, and simply let $X(t)$ denote a random process.

The notion of a random process is an extension of the random variable concept. The difference is that in random processes, the mapping is done onto functions of time rather than onto constants (real numbers). The basic connection between the concept of a random process and the concept of a random variable is that at any time instant (for each choice of observation instant) the random process defines a random variable. However, the distributions at various times could be potentially different.

Probability, Random Variables, Statistics, and Random Processes: Fundamentals & Applications,
First Edition. Ali Grami.
© 2020 John Wiley & Sons, Inc. Published 2020 by John Wiley & Sons, Inc.
Companion Website: www.wiley.com/go/grami/PRVSRP

In a random process, before conducting an experiment, it is not possible to define the sample functions that will be observed in the future exactly. A random process can be described by the form of its sample functions. In *nondeterministic random processes*, the future values of any sample function cannot be exactly predicted from the observed past values, such as a seismic signal. In *deterministic random processes*, the future values of any sample function can be predicted completely from its past values, such as a sinusoidal signal with a random initial phase.

It is important to note that two random processes are considered to be equal if their respective sample functions are identical for each outcome in the same random experiment. Although a random process can be complex, all random processes discussed in this book are assumed to be real random processes.

The classification of a random process depends inherently on three measures: the state space, the index (time) parameter, and the statistical dependencies among the random variables $X(t)$ for different values of the index parameter t. Our focus in this section is to classify the random processes according to the characteristics of time t and the random variable $X(t)$ at time t.

11.1.1 State Space

The set of possible values that the random process $X(t)$ may take on is called its *state space*. If the possible values are finite or countably infinite, the random process $X(t)$ is then called a *discrete-state random process* or a *discrete-valued random process*. For such a random process, $X(t)$ at any given time t is a discrete random variable. Examples of such a random process may include the gear positions in a manual transmission car while it is being driven in a city during a day and the number of passengers on a tour bus during a weekend.

If the possible values of the random process $X(t)$ are part of a finite or infinite continuous interval (or set of such intervals), the random process $X(t)$ is then called a *continuous-state random process* or a *continuous-valued random process*. For such a random process, $X(t)$ at any given time t is a continuous random variable. Examples of such a random process may include wind velocity on a mountain top during the climbing season and the amount of rainfall in a city.

It is also possible to have a *mixed-valued random process*, where possible values of $X(t)$ at any time can be continuous or discrete. Examples of such a random process may include the signal transmitted over a walkie-talkie radio system and the current flowing in an ideal rectifier in an electronic device.

11.1.2 Index (Time) Parameter

In a *continuous-time random process*, the time parameter t is continuous that is it can take real values in an interval or set of intervals on the real line and thus comes from an uncountably infinite set. Such a random process thus consists of an uncountable number of random variables. One example of such a process is the channel noise in a communication system, as noise is present at all times.

In a ***discrete-time random process***, the time parameter t is a countable set. A discrete-time random process is often called a ***random sequence*** and generally denoted by $X(n)$, where n is an integer. One example of such a sequence can be formed by periodically sampling a continuous-time random process.

Our focus in this book is mostly on continuous-time, continuous-valued random processes, though other types of random processes may be briefly discussed as well.

Example 11.1

Depending on the two possible values of the state space and the two possible types of the index (time) parameter, provide real-life examples of sample functions for the four possible cases.

Solution

Figure 11.1 shows some functions of time for the four cases, where randomness, induced by ω, is evident in all sample functions. Noting T and S represent the time parameter and the state space, respectively, we can have the following four cases:

(i) If both T and S are continuous, the random process $X(t)$ is then called a ***continuous random process***, and an example representing $X(t)$ may be the sunlight intensity in a certain location on the Earth's land surface at time t, where $T = [0, \infty)$, and, say, $S = [0, 120\ 000]$ lux.

(ii) If T is discrete and S is continuous, the random process $X(n)$ is then called a ***continuous random sequence***, and an example representing $X(n)$ may be the daily maximum temperature in a certain place on the Earth's land surface on the nth day, where $T = \{1, 2, 3, ...\}$ and, say, $S = [-90, 60]$ degrees Celsius.

(iii) If T is continuous and S is discrete, the random process $X(t)$ is then called a ***discrete random process***, and an example representing $X(t)$ may be the number of people in an emergency department of a hospital at time t, where $T = [0, \infty)$ and $S = \{0, 1, 2, ...\}$.

(iv) If both T and S are discrete, the random process $X(n)$ is then called a ***discrete random sequence***, and an example representing $X(n)$ may be the outcome of the nth toss of a typical die, where $T = \{1, 2, 3, ...\}$ and $S = \{1, 2, 3, 4, 5, 6\}$. ∎

Example 11.2

Suppose we have $X(t) = A \cos(2\pi f_c t + \Theta)$, where the amplitude A and the frequency f_c are nonrandom nonzero constants, and Θ is a discrete random variable with $P(\Theta = 0) = \frac{1}{2}$ and $P(\Theta = \pi) = \frac{1}{2}$. Is $X(t)$ a random process?

Solution

$X(t)$ is a continuous random process because it comprises random time-varying functions. The value of the signal is a function of time and the random variable Θ represents the initial phase of the sinusoidal signal. The two sample functions of this random process are as follows:

$$X(t) = A \cos(2\pi f_c t)$$

and

$$X(t) = A \cos(2\pi f_c t + \pi) = -A \cos(2\pi f_c t) \qquad \blacksquare$$

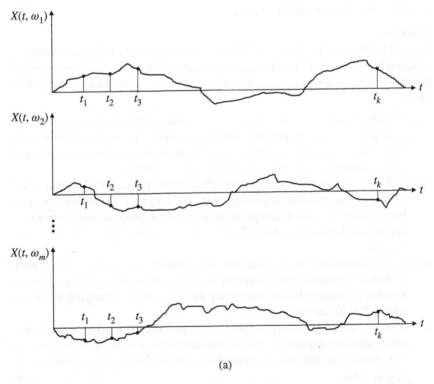

(a)

Figure 11.1 Classification of random processes: (a) continuous-time continuous-valued random process (continuous random process); (b) discrete-time continuous-valued random process (continuous random sequence); (c) continuous-time discrete-valued random process (discrete random process); and (d) discrete-time discrete-valued random process (discrete random sequence).

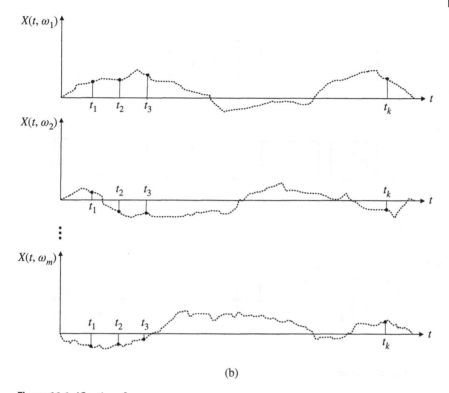

(b)

Figure 11.1 (*Continued*)

Example 11.3

Suppose we have $X(t) = A \cos\left(2\pi f_c t + \frac{\pi}{2}\right)$, where A is a uniform random variable in the interval $[-1, 1]$ and the frequency f_c is a nonrandom nonzero constant. Is $X(t)$ a random process?

Solution

$X(t)$ is a continuous random process because it comprises random time-varying functions. The value of the signal is a function of time and the random variable A represents the random amplitude of the sinusoidal signal. This random process has infinitely many sample functions, as there are infinitely many possible values for A in the range $[-1, 1]$. Some sample functions are as follows:

$$A = -1 \rightarrow X(t) = -1 \cos\left(2\pi f_c t + \frac{\pi}{2}\right) = \sin(2\pi f_c t)$$

$$A = 0 \rightarrow X(t) = 0$$

$$A = 1 \rightarrow X(t) = \cos\left(2\pi f_c t + \frac{\pi}{2}\right) = -\sin(2\pi f_c t)$$

■

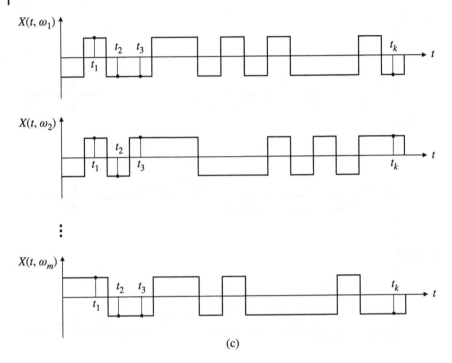

(c)

Figure 11.1 (*Continued*)

11.2 Characterization of Random Processes

The distinguishing feature of a random process is the relationship of the random variables at various times. As a random process at a given time is a random variable, it can thus be described with a probability distribution. In general, the form of the distribution of a random process is different for different instants of time. In most cases, it is not possible to determine the distribution of a random process at a certain time or the bivariate distribution at any two different times or joint distribution at many different times, because all sample functions are hardly ever known. As a matter of fact, from practical consideration standpoint, more often than not, only one sample function of finite duration is all that is ever available.

11.2.1 Joint Distributions of Time Samples

Let $X(t_1)$, ..., $X(t_k)$ be the k random variables obtained by sampling the random process $X(t)$ at time instants t_1, ..., t_k, where k is a positive integer. The ***kth-order joint cdf of a random process*** is then defined as follows:

$$F_{X(t_1),...,X(t_k)}(x_1, \ldots, x_k) = P[X(t_1) \leq x_1, \ldots, X(t_k) \leq x_k] \qquad (11.1)$$

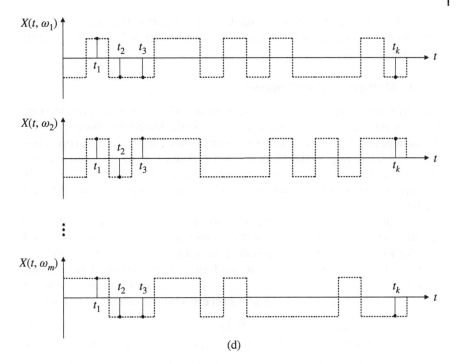

(d)

Figure 11.1 (Continued)

The joint probability density functions (pdf's) or probability mass functions (pmf's) depending on whether the random process $X(t)$ is continuous-valued or discrete-valued can be obtained using the joint cumulative distribution function (cdf). The complete characterization of the random process $X(t)$ requires the massive knowledge of all the distributions as $k \to \infty$. Fortunately, often much less is sufficient, as most useful models of random processes can be obtained through a small set of effective methods, already developed for random variables.

11.2.2 Independent Identically Distributed Random Process

A discrete-time random process consisting of a sequence of independent identically-distributed random variables with common cdf $F_X(x)$ is called the *independent identically distributed random process*. The joint cdf of such a random process is then equal to the product of the individual cdf's, i.e. we have

$$F_{X(t_1),\ldots,X(t_k)}(x_1, \ldots, x_k) = F_X(x_1) \times F_X(x_2) \times \cdots \times F_X(x_k) \qquad (11.2)$$

Thus, a first-order distribution is sufficient to characterize an independent identically distributed random process $X(t)$. The above equation clearly implies that if the sequence is discrete-valued, the joint pmf is then the product of individual pmf's, and if it is continuous-valued, the joint pdf is then the product of individual pdf's.

11.2.3 Multiple Random Processes

In many practical applications, we often need to deal with more than one random process at a time as we may need to assess the inter-relatedness (i.e. positive or negative correlation) between the processes, say, between two processes $X(t)$ and $Y(t)$. Examples may include when $X(t)$ and $Y(t)$ are the input and output of an engineering system, or represent the temperature and humidity in a certain geographical location, or reflect the market values of two different stocks.

Two random processes $X(t)$ and $Y(t)$ are specified by their joint cdf for all possible choices of time samples of the processes. Let $X(t_1), \ldots, X(t_k)$ be the k random variables obtained by sampling the random process $X(t)$ at times $t_1, \ldots, t_k$, and $Y(t_1'), \ldots, Y(t_m')$ be the m random variables obtained by sampling the random process $Y(t)$ at times $t_1', \ldots, t_m'$, where both k and m are positive integers. The **joint cdf of two random processes** is then defined as follows:

$$F_{X(t_1),\ldots,X(t_k),Y(t_1'),\ldots,Y(t_m')}(x_1, \ldots, x_k, y_1, \ldots, y_m)$$
$$= P[X(t_1) \le x_1, \ldots, X(t_k) \le x_k, Y(t_1') \le y_1, \ldots, Y(t_m') \le y_m] \qquad (11.3)$$

Note that the above definition can be easily extended to multiple random processes, and from the joint cdf of multiple random processes, the cdf for each random process can be obtained.

11.2.4 Independent Random Processes

The random processes $X(t)$ and $Y(t)$ are **independent random processes** if the random vectors $\{X(t_1), \ldots, X(t_k)\}$ and $\{Y(t_1'), \ldots, Y(t_m')\}$ are independent for positive integers k and m, and for all $t_1, \ldots, t_k$ and $t_1', \ldots, t_m'$, i.e. the joint cdf's is equal to the product of the individual cdf's. In other words, we have

$$F_{X(t_1),\ldots,X(t_k),Y(t_1'),\ldots,Y(t_m')}(x_1, \ldots, x_k, y_1, \ldots, y_m)$$
$$= F_{X(t_1),\ldots,X(t_k)}(x_1, \ldots, x_k) \times F_{Y(t_1'),\ldots,Y(t_m')}(y_1, \ldots, y_m) \qquad (11.4)$$

In many applications, it is rather easy to see whether the two random processes are independent. Note that the concept of independent random processes can be extended to more than two random processes.

11.3 Moments of Random Processes

The moments of random processes provide valuable insights into their joint behavior. Our main focus is on the first-order and second-order moments of a continuous random process or two continuous random processes, as they are important and useful, yet rather easy to determine.

11.3.1 Mean and Variance Functions of a Random Process

The ***mean and variance functions of the random process*** $X(t)$ are, in general, both deterministic functions of time, and defined, respectively, as follows:

$$\mu_X(t) = E[X(t)] = \int_{-\infty}^{\infty} x f_{X(t)}(x) \, dx$$

$$\sigma_X^2(t) = E[(X(t) - E[X(t)])^2] = \int_{-\infty}^{\infty} (x - \mu_X(t))^2 f_{X(t)}(x) dx$$

$$= \int_{-\infty}^{\infty} x^2 f_{X(t)}(x) dx - (\mu_X(t))^2 \tag{11.5}$$

where $f_{X(t)}(x)$ is the first-order pdf of the random process $X(t)$ at time t. The mean function $\mu_X(t)$ specifies the mean value of $X(t)$ at different instants of time and the variance $\sigma_X^2(t)$ reflects the variation of $X(t)$ at different time instants.

11.3.2 Autocorrelation and Autocovariance Functions of a Random Process

The ***autocorrelation and autocovariance functions*** of the random process $X(t)$ are, in general, functions of t_1 and t_2, and are defined, respectively, as follows:

$$R_X(t_1, t_2) = E[X(t_1)X(t_2)] = \int_{-\infty}^{\infty} \int_{-\infty}^{\infty} x_1 x_2 f_{X(t_1),X(t_2)}(x_1, x_2) \, dx_1 dx_2$$

$$C_X(t_1, t_2) = E[(X(t_1) - \mu_X(t_1))(X(t_2) - \mu_X(t_2))]$$

$$= \int_{-\infty}^{\infty} \int_{-\infty}^{\infty} (x_1 - \mu_X(t_1))(x_2 - \mu_X(t_2)) f_{X(t_1),X(t_2)}(x_1, x_2) \, dx_1 dx_2$$

$$= R_X(t_1, t_2) - \mu_X(t_1)\mu_X(t_2) \tag{11.6}$$

where $f_{X(t_1),X(t_2)}(x_1, x_2)$ is the second-order joint pdf of the random variables $X(t_1)$ and $X(t_2)$. Note that we have $R_X(t, t) = E[X^2(t)]$ and $C_X(t, t) = \sigma_X^2(t)$. For real random processes, the autocorrelation function $R_X(t_1, t_2)$ and the autocovariance function $C_X(t_1, t_2)$ are both symmetric, i.e. we have $R_X(t_1, t_2) = R_X(t_2, t_1)$ and $C_X(t_1, t_2) = C_X(t_2, t_1)$.

The ***correlation coefficient*** – defined as a measure of the extent to which a random variable $X(t_1)$ can be predicted as a linear function of the random variable $X(t_2)$ – is given by

$$\rho_X(t_1, t_2) = \frac{C_X(t_1, t_2)}{\sqrt{C_X(t_1, t_1)}\sqrt{C_X(t_2, t_2)}} = \frac{C_X(t_1, t_2)}{\sigma_X(t_1)\sigma_X(t_2)} \tag{11.7}$$

The correlation coefficient is symmetric, i.e. $\rho_X(t_1, t_2) = \rho_X(t_2, t_1)$, and we have $|\rho_X(t_1, t_2)| \leq 1$. If $C_X(t_1, t_2) > 0$, then $X(t_1)$ and $X(t_2)$ are positively correlated, if $C_X(t_1, t_2) < 0$, then $X(t_1)$ and $X(t_2)$ are negatively correlated, and if $C_X(t_1, t_2) = 0$, then $X(t_1)$ and $X(t_2)$ are uncorrelated, i.e. there is no linear correlation. The mean, variance, autocorrelation, and autocovariance functions of two different random processes can be the same, as these functions can provide only partial descriptions.

Example 11.4

Suppose we have the random process $X(t) = R\cos(2\pi t)$, where R is a random variable whose mean and variance are $\mu_R \neq 0$ and $\sigma_R^2 \neq 0$, respectively. Determine the mean, variance, autocorrelation, and autocovariance functions, as well as the correlation coefficient of $X(t)$.

Solution

At time t, $\cos(2\pi t)$ is a nonrandom constant, and $X(t)$ is thus a random variable. The mean and variance functions of $X(t)$ are, respectively, as follows:

$$\mu_X(t) = E[X(t)] = E[R\cos(2\pi t)] = E[R]\cos(2\pi t) = \mu_R\cos(2\pi t)$$

and

$$\sigma_X^2(t) = E[X^2(t)] - (E[X(t)])^2 = E[R^2\cos^2(2\pi t)] - (E[R])^2\cos^2(2\pi t)$$
$$= E[R^2]\cos^2(2\pi t) - (E[R])^2\cos^2(2\pi t) = (E[R^2] - (E[R])^2)\cos^2(2\pi t)$$
$$= \sigma_R^2\cos^2(2\pi t)$$

Note that the mean and variance of the random process $X(t)$ are both functions of time. The autocorrelation and autocovariance functions of $X(t)$ are, respectively, as follows:

$$R_X(t_1, t_2) = E[X(t_1)X(t_2)] = E[R\cos(2\pi t_1)R\cos(2\pi t_2)]$$
$$= E[R^2]\cos(2\pi t_1)\cos(2\pi t_2)$$
$$= (\sigma_R^2 + \mu_R^2)\cos(2\pi t_1)\cos(2\pi t_2)$$

and

$$C_X(t_1, t_2) = E[(X(t_1) - \mu_X(t_1))(X(t_2) - \mu_X(t_2))]$$
$$= R_X(t_1, t_2) - \mu_X(t_1)\mu_X(t_2)$$
$$= (\sigma_R^2 + \mu_R^2)\cos(2\pi t_1)\cos(2\pi t_2) - (\mu_R\cos(2\pi t_1))(\mu_R\cos(2\pi t_2))$$
$$= \sigma_R^2\cos(2\pi t_1)\cos(2\pi t_2)$$

Note that the autocorrelation and autocovariance functions each is a function of both t_1 and t_2. The correlation coefficient is then as follows:

$$\rho_X(t_1, t_2) = \frac{C_X(t_1, t_2)}{\sqrt{C_X(t_1, t_1)}\sqrt{C_X(t_2, t_2)}}$$

$$= \frac{\sigma_R^2 \cos(2\pi t_1)\cos(2\pi t_2)}{\sqrt{\sigma_R^2 \cos(2\pi t_1)\cos(2\pi t_1)}\sqrt{\sigma_R^2 \cos(2\pi t_2)\cos(2\pi t_2)}}$$

$$= \frac{\cos(2\pi t_1)\cos(2\pi t_2)}{|\cos(2\pi t_1)||\cos(2\pi t_2)|}$$

If t_1 and t_2 are chosen in such a way that $\cos(2\pi t_1)$ and $\cos(2\pi t_2)$ have different signs, then $\rho_X(t_1, t_2) = -1$ and if they have the same signs, then $\rho_X(t_1, t_2) = 1$. ∎

Example 11.5
Suppose we have the random process $X(t) = Y + Zt$, where Y and Z are independent zero-mean unit-variance Gaussian random variables, and t is a non-negative real number representing time t. Determine all sample functions and the pdf of $X(t)$.

Solution
$X(t)$ is a function of time t and its randomness is due to the two random variables Y and Z. The random variables Y and Z each can take any real value, i.e. $y \in (-\infty, \infty)$ and $z \in (-\infty, \infty)$. Therefore, there are infinitely many sample functions, and they are as follows:

$$x(t) = y + zt \quad \forall t \geq 0$$

As a linear combination of independent Gaussian random variables is also a Gaussian random variable, the pdf of $X(t)$ at time t is thus a Gaussian random variable whose mean and variance are, respectively, as follows:

$$\mu_X(t) = E[X(t)] = E[Y + Zt] = E[Y] + E[Zt] = E[Y] + E[Z]t = 0$$

and

$$\sigma_X^2(t) = \sigma_Y^2 + \sigma_Z^2 t^2 = 1 + t^2$$

The pdf of $X(t)$ is thus as follows:

$$f_{X(t)}(x) = \left(\frac{1}{\sqrt{2\pi(t^2+1)}}\right)\exp\left(-\frac{x^2}{2(t^2+1)}\right)$$

Note that $X(t)$ at time t is a zero-mean Gaussian random variable whose variance is a function of time t. ∎

Example 11.6
Consider a random binary waveform $X(t)$ that consists of a sequence of pulses with the following properties: (i) each pulse is of duration T, (ii) pulses are

equally likely to be $\pm A$, (iii) all pulses are statistically independent, and (iv) the starting time of the first pulse, t_d, is equally likely to be between 0 and T. Determine the autocorrelation function of this random process.

Solution

Since the amplitude levels $\pm A$ occur with equal probability, the mean of the random process $X(t)$ is zero, i.e. $E[X(t)] = 0$. Noting that $X(t_1)$ and $X(t_2)$ are random variables obtained by observing the random process $X(t)$ at times t_1 and $t_2 > t_1$, we have the following two cases:

Case-I: We have $|t_2 - t_1| > T$. The random variables $X(t_1)$ and $X(t_2)$ occur in two different pulse intervals and are therefore independent. We thus have

$$E[X(t_1)X(t_2)] = E[X(t_1)]E[X(t_2)] = 0$$

Case-II: We have $|t_2 - t_1| < T$, with $t_2 = 0$. The random variables $X(t_1)$ and $X(t_2)$ occur in the same pulse interval if and only if the starting time of the first pulse, t_d, satisfies the condition $t_d < T - |t_2 - t_1|$. We thus obtain the following conditional expectation:

$$E[X(t_1)X(t_2) \mid t_d] = \begin{cases} A^2 & t_d < T - |t_2 - t_1| \\ 0 & \text{elsewhere} \end{cases}$$

Noting T_d has a uniform distribution over the interval $[0, T]$ and averaging the result over all possible values of t_d, we have

$$E[X(t_1)X(t_2)] = \int_0^{T-|t_2-t_1|} A^2 \left(\frac{1}{T}\right) dt_d = A^2 \left(1 - \frac{|t_2 - t_1|}{T}\right)$$

The autocorrelation function of a random binary wave is thus as follows:

$$R_X(t_1, t_2) = \begin{cases} A^2 \left(1 - \frac{|t_2-t_1|}{T}\right) & |t_2 - t_1| < T \\ 0 & |t_2 - t_1| \geq T \end{cases}$$

Note that the autocorrelation function is a function of only the time difference between t_2 and t_1. ∎

11.3.3 Cross-correlation and Cross-covariance Functions

The ***cross-correlation function*** and ***cross-covariance function*** of two random processes $X(t)$ and $Y(t)$ are defined by

$$R_{XY}(t_1, t_2) = E[X(t_1)Y(t_2)]$$
$$C_{XY}(t_1, t_2) = E[(X(t_1) - \mu_X(t_1))(Y(t_2) - \mu_Y(t_2))]$$
$$= R_{XY}(t_1, t_2) - \mu_X(t_1)\mu_Y(t_2) \tag{11.8}$$

The random processes $X(t)$ and $Y(t)$ are **uncorrelated** if their cross-covariance function is zero, and they are **orthogonal** if their cross-correlation function is zero. In other words, for all t_1 and t_2, we have the following:

$$C_{XY}(t_1, t_2) = 0 \Leftrightarrow \text{Uncorrelated random processes}$$
$$R_{XY}(t_1, t_2) = 0 \Leftrightarrow \text{Orthogonal random processes} \tag{11.9}$$

Example 11.7
Consider a pair of random processes $X(t)$ and $Y(t)$ that are defined as follows:

$$X(t) = \cos(2\pi t + \Theta)$$
$$Y(t) = \sin(2\pi t + \Theta)$$

where the random variable Θ, representing the initial phase, is uniformly distributed over the interval $[0, 2\pi]$. Determine the conditions upon which $X(t)$ and $Y(t)$ can be orthogonal to each other.

Solution
By using a product-to-sum trigonometric identity, we have

$$R_{XY}(t_1, t_2) = E[X(t_1)Y(t_2)] = E[\cos(2\pi t_1 + \Theta)\sin(2\pi t_2 + \Theta)]$$
$$= \frac{1}{2}(E[\sin(2\pi t_1 + 2\pi t_2 + 2\Theta)] - E[\sin(2\pi t_1 - 2\pi t_2)])$$
$$= -\frac{1}{2}\sin(2\pi(t_1 - t_2))$$

In order to have $X(t)$ and $Y(t)$ orthogonal to each other, we must then have

$$R_{XY}(t_1, t_2) = 0 \rightarrow \sin(2\pi(t_1 - t_2)) = 0 \rightarrow 2\pi(t_1 - t_2) = k\pi \rightarrow t_1 - t_2 = \frac{k}{2}$$

where $k = 0, \pm 1, \pm 2, \ldots$, can ensure orthogonality of $X(t_1)$ and $Y(t_2)$. ∎

Example 11.8
Assuming $X(t)$ and $Y(t)$ are two independent random processes, and $W(t)$ is their product, i.e. we have

$$W(t) = X(t)\,Y(t)$$

Determine the mean and autocorrelation functions of the random process $W(t)$.

Solution
Since $X(t)$ and $Y(t)$ are independent, the mean of their product is equal to the product of their means. The mean and autocorrelation functions of the random process $W(t)$ are as follows:

$$E[W(t)] = E[X(t)Y(t)] = E[X(t)]E[Y(t)]$$

and

$$R_W(t_1, t_2) = E[W(t_1)W(t_2)] = E[X(t_1)Y(t_1)X(t_2)Y(t_2)]$$
$$= E[X(t_1)X(t_2)]E[Y(t_1)Y(t_2)]$$
$$= R_X(t_1, t_2)R_Y(t_1, t_2) \qquad\qquad \blacksquare$$

11.4 Stationary Random Processes

Stationary processes are often appropriate for describing random phenomena that occur in communication theory and signal processing. Moreover, the spatial/planar distributions of stars, plants, and animals are often assumed to correspond to a stationary process. In essence, a stationary random process is in probabilistic equilibrium and the particular times at which the process is examined are of no relevance. In other words, its set of characteristics does not change over time.

11.4.1 Strict-Sense Stationary Processes

A random process at a given time is a random variable and, in general, the characteristics of this random variable depend on the time at which the random process is sampled. A continuous-time random process $X(t)$ is said to be a ***stationary process*** or a ***strict-sense stationary process***, if the pdf of any set of samples does not vary with time. In other words, the joint distribution of $X(t_1)$, ..., $X(t_k)$ is the same as the joint distribution of $X(t_1 + \tau)$, ..., $X(t_k + \tau)$, for any positive integer k, any time shift τ, and all choices of t_1, ..., t_k. A ***nonstationary process***, on the other hand, is characterized by a joint distribution that depends on time instants t_1, ..., t_k. In broad terms, a random, but stable, phenomenon that has evolved into a steady-state mode of behavior gives rise to a stationary process, whereas a random, but unstable, phenomenon yields a nonstationary process.

In principle, it is difficult to determine if a process is stationary, as all sample functions are required. In addition, stationary processes cannot occur physically, because in reality, signals begin at some finite time and end later at some finite time. Due to practical considerations, we simply consider the observation interval to be limited, and use the first- and second-order statistics rather than the joint distribution of higher orders. Since it is difficult to determine the distribution of a random process, the focus usually becomes on a partial, yet useful, description of the distribution of the process, such as the mean, autocorrelation, and autocovariance functions of the random process. For a stationary random process, the mean and variance are both constants, i.e. neither of them is a function of time.

11.4.2 Wide-Sense Stationary Processes

As the stationarity requirement is frequently more stringent than necessary for analysis, a more relaxed requirement is introduced. A random process $X(t)$ is a **wide-sense stationary process** or **weakly stationary process**, if it has a constant mean, i.e. it is independent of time, and also its autocorrelation function depends only on the time difference $\tau = t_2 - t_1$ and not on t_1 and t_2 individually. In other words, in a wide-sense stationary process, the mean and autocorrelation functions do not depend on the choice of the time origin. All random processes that are stationary in the strict sense are wide-sense stationary, but the converse, in general, is not true. However, if a Gaussian process is wide-sense stationary, then it is also stationary in the strict sense. If $X(t)$ is wide-sense stationary, we must then have the following:

$$\mu_X(t) = E[X(t)] = \text{Constant}$$
$$R_X(t_1, t_2) = E[X(t_1)X(t_2)] = R_X(t_2 - t_1) = R_X(\tau) \tag{11.10}$$

Note that the time interval τ between samples is called the **lag**. In a wide-sense stationary random process, the autocorrelation function $R_X(\tau)$ has the following properties:

- $R_X(\tau)$ is an even function, i.e. $R_X(\tau) = R_X(-\tau)$.
- $R_X(0) = E[X^2(t)] \geq 0$ gives the mean-square value of the random process.
- The maximum value of $R_X(\tau)$ occurs at $\tau = 0$, i.e. $|R_X(\tau)| \leq R_X(0)$. There may be other lags for which it is just as large, but it cannot be larger.
- The more rapidly $X(t)$ changes with time t, the more rapidly $R_X(\tau)$ decreases from its maximum $R_X(0)$ as τ increases, simply because the autocorrelation function is a measure of the rate of change of a random process and measures the predictability of a random process.
- The autocorrelation function approaches the square of the mean, $(E[X(t)])^2$, as $\tau \to \infty$. For a zero-mean process, this means that samples become uncorrelated for large lags.
- The autocorrelation function does not have to be positive or negative for all values of τ.
- If $R_X(\tau)$ is continuous and has a finite value at $\tau = 0$, then it is continuous at all times.
- The autocorrelation function cannot have an arbitrary shape, as it must have a Fourier transform.
- The autocorrelation function may have the following components:
  o A component that approaches zero as $\tau \to \infty$ (e.g. an exponential function).
  o A periodic component (e.g. a sinusoidal function).
  o A constant component representing the square of the mean (e.g. any nonnegative real number).

It is important to note that a wide-sense stationary random process yields a unique autocorrelation function, but the converse is not true, i.e. two different wide-sense stationary random processes may have the same autocorrelation function.

Example 11.9
Consider a random process $X(t)$ defined as follows:

$$X(t) = A \cos(t) + B \sin(t)$$

where A and B are uncorrelated random variables with zero-mean and unit-variance. Is $X(t)$ wide-sense stationary?

Solution
We have $E[A] = E[B] = 0$. Since A and B are uncorrelated, we have $E[AB] = E[A]E[B] = 0$. Since we have $\sigma_A^2 = E[A^2] - (E[A])^2 = 1$ and $\sigma_B^2 = E[B^2] - (E[B])^2 = 1$, we have $E[A^2] = E[B^2] = 1$. The mean and autocorrelation functions are then as follows:

$$\mu_X(t) = E[X(t)] = E[A \cos(t) + B \sin(t)]$$
$$= E[A \cos(t)] + E[B \sin(t)]$$
$$= E[A] \cos(t) + E[B] \sin(t) = 0$$

and

$$R_X(t, t + \tau) = E[X(t)X(t + \tau)]$$
$$= E[(A \cos(t) + B \sin(t))(A \cos(t + \tau) + B \sin(t + \tau))]$$
$$= E[A^2] \cos(t) \cos(t + \tau) + E[B^2] \sin(t) \sin(t + \tau)$$
$$+ E[AB] \cos(t) \sin(t + \tau) + E[AB] \sin(t) \cos(t + \tau)$$
$$= \cos(t) \cos(t + \tau) + \sin(t) \sin(t + \tau) = \cos(\tau)$$

Since the mean is not a function of time, i.e. it is a constant, and the autocorrelation is a function of the time difference τ, $X(t)$ is a wide-sense stationary random process. ∎

Example 11.10
The autocorrelation function of the wide-sense stationary $X(t)$ is as follows:

$$R_X(\tau) = 100 + \frac{20}{5 + 2\tau^2}$$

Determine the standard deviation of $X(t)$.

Solution

Using the properties of the autocorrelation function, we have the following:

$$\lim_{\tau \to \infty} R_X(\tau) = (E[X(t)])^2 \to (E[X(t)])^2 = \lim_{\tau \to \infty} \left(100 + \frac{20}{5 + 2\tau^2} \right)$$
$$= 100 \to E[X(t)] = 10$$

and

$$R_X(0) = E[X^2(t)] = 104$$

We thus have

$$\sigma_X^2(t) = E[X^2(t)] - (E[X(t)])^2 = 104 - 100 = 4 \to \sigma_X(t) = 2$$

As expected the mean and variance of a wide-sense stationary are both constants, i.e. neither of them is a function of time. ∎

11.4.3 Jointly Wide-Sense Stationary Processes

The random processes $X(t)$ and $Y(t)$ are **jointly wide-sense stationary**, if $X(t)$ and $Y(t)$ are each wide-sense stationary and if their cross-correlation depends only on $\tau = t_2 - t_1$, i.e. we have

$$R_{XY}(t_1, t_2) = R_{XY}(t_2 - t_1) = R_{XY}(\tau) \tag{11.11}$$

We also have $R_{XY}(\tau) = R_{YX}(-\tau)$, which in turn means a shift of $Y(t)$ in one direction (in time) is equivalent to a shift of $X(t)$ in the other direction. The cross-correlation functions have the following properties:

- $R_{XY}(0) = R_{YX}(0)$, but unlike autocorrelation function, they have no particular physical significance, nor do they represent mean-square values.
- The cross-correlation does not necessarily have its maximum at $\tau = 0$, but is upper-bounded by $\sqrt{R_X(0)R_Y(0)}$, and may not achieve this value anywhere.
- If $X(t)$ and $Y(t)$ are statistically independent, then $R_{XY}(\tau) = R_{YX}(\tau) = E[X(t)]E[Y(t)]$.
- The absolute value of the cross-correlation function is upper-bounded, as we have $|R_{XY}(\tau)| \leq 0.5(R_X(0) + R_Y(0))$.

Example 11.11

Show that the sum of two jointly wide-sense stationary random processes $X(t)$ and $Y(t)$, i.e. $Z(t) = X(t) + Y(t)$, is also a wide-sense stationary random process.

Solution

Since $X(t)$ and $Y(t)$ are each wide-sense stationary, we thus have the following:

$$Z(t) = X(t) + Y(t) \to E[Z(t)] = E[X(t) + Y(t)] = E[X(t)] + E[Y(t)] = \mu_X + \mu_Y$$

and

$$R_Z(t_1, t_2) = E[(X(t_1) + Y(t_1))(X(t_2) + Y(t_2))]$$
$$= E[X(t_1)X(t_2)] + E[X(t_1)Y(t_2)]$$
$$+ E[Y(t_1)X(t_2)] + E[Y(t_1)Y(t_2)]$$
$$= R_X(t_2 - t_1) + R_{XY}(t_2 - t_1) + R_{YX}(t_2 - t_1) + R_Y(t_2 - t_1)$$
$$= R_X(\tau) + R_{XY}(\tau) + R_{YX}(\tau) + R_Y(\tau)$$

Since the mean of $Z(t)$ is a constant and its autocorrelation is a function of the time difference τ, $Z(t)$ is then wide-sense stationary. ∎

Example 11.12
In a radar system, the transmitted signal $X(t)$ is returned from a target. The received signal $Y(t)$ is the sum of the attenuated and delayed version of $X(t)$ and the noise $N(t)$, where the signal $X(t)$ and the noise $N(t)$ are statistically independent and both have zero mean. Determine how the round-trip delay can be determined.

Solution
We have

$$Y(t) = a\,X(t - \delta) + N(t)$$

where δ is the propagation time to the target and back, and the scaling factor a is much smaller than 1 as the average power of the returned signal is much smaller than the average power of the noise. The cross-correlation of the transmitted signal and the received signal is as follows:

$$R_{XY}(\tau) = E[X(t)Y(t + \tau)] = E[X(t)(a\,X(t + \tau - \delta) + N(t + \tau))]$$
$$= E[a\,X(t)X(t + \tau - \delta) + X(t)N(t + \tau)] = a\,R_X(\tau - \delta) + R_{XN}(\tau)$$
$$= a\,R_X(\tau - \delta) + E[X(t)]E[N(t)] = a\,R_X(\tau - \delta)$$

The value of τ that maximizes the measured cross-correlation function $R_{XY}(\tau)$ also represents the maximum value of the autocorrelation function $R_X(\tau - \delta)$. Noting that the maximum value of the autocorrelation function $R_X(\tau - \delta)$ occurs at the origin, $\tau = \delta$ represents the round-trip delay and in turn can indicate the distance to the target. ∎

Note that the wide-sense stationary random processes $X(t)$ and $Y(t)$ are **uncorrelated processes** if their cross-correlation function is equal to the product of their means, and they are **orthogonal processes** if their cross-correlation is zero.

11.4.4 Cyclostationary Processes

Some random processes have inherently a periodic nature, in that their statistical properties are repeated every T units of time, where T is the smallest possible positive real number. A random process $X(t)$ is called a **cyclostationary process** if the joint cdf of $X(t_1)$, ..., $X(t_k)$ is the same as the joint cdf of $X(t_1 + mT)$, ..., $X(t_k + mT)$ for all integers m and k, and all choices of sampling instants, t_1, ..., t_k. Note that the sample functions of a cyclostationary random process need not be periodic.

A random process $X(t)$ is called a **wide-sense cyclostationary process** if the mean and autocorrelation functions of $X(t)$ remain the same with respect to time shifts that are multiples of T. In other words, for every integer m, $X(t)$ is a wide-sense cyclostationary process, if we have

$$\mu_X(t + mT) = \mu_X(t)$$
$$R_X(t_1 + mT, t_2 + mT) = R_X(t_1, t_2) \tag{11.12}$$

Note that if a random process is cyclostationary, then it is also wide-sense cyclostationary.

11.4.5 Independent and Stationary Increments

A continuous random process $X(t)$, where $t \geq 0$, is said to have **independent increments** if the set of n random variables $X(t_1) - X(0)$, $X(t_2) - X(t_1)$,...., $X(t_n) - X(t_{n-1})$ are jointly independent for any integer $n \geq 2$ and all time instants $0 \leq t_1 < t_2 < ... < t_n$. In other words, the increments are statistically independent when intervals are nonoverlapping. The independent increment property simplifies analysis, as it makes easy to get the higher-order distributions. For instance, with the independent increments in the context of counting processes, the probability of having m events in the interval $(t_{i-1}, t_i]$ and k events in the interval $(t_{j-1}, t_j]$, when the two intervals are disjoint, is equal to the product of the probability of having m events in the interval $(t_{i-1}, t_i]$ and the probability of having k events in the interval $(t_{j-1}, t_j]$.

A continuous random process $X(t)$, where $t \geq 0$, is said to have **stationary increments** if the two random variables $X(t_2) - X(t_1)$ and $X(t_2 + \tau) - X(t_1 + \tau)$ have the same distributions for all $0 \leq t_1 < t_2$ and all $\tau > 0$. For instance, in the context of counting processes, the distribution of having m events is the same for all intervals of the same duration τ, and does not depend on the exact location of the time interval.

If a continuous random process $X(t)$, where $t \geq 0$ and $X(0) = 0$, is assumed to have the **stationary independent increments** property, we then have the following:

$$\mu_X(t) = E[X(t)] = E[X(1)]\,t = \mu_X(1)\,t$$
$$\sigma_X^2(t) = E[X^2(t)] - (E[X(t)])^2 = \sigma_X^2(1)\,t$$
$$C_X(t_1, t_2) = R_X(t_1, t_2) - E[X(t_1)]E[X(t_2)] = \sigma_X^2(1)\min(t_1, t_2) \qquad (11.13)$$

Obviously, such random processes are nonstationary. The well-known Wiener and Poisson processes both have stationary independent increments.

Example 11.13

Suppose $X(t)$, where $t \geq 0$ and $X(0) = 0$, is a random process with stationary independent increments. Determine the mean and variance of $X(t)$.

Solution

Based on

$$\mu_X(t) = E[X(t)] = E[X(t) - X(0)]$$

and the fact that $X(t)$ has the property of the stationary independent increments, we have

$$\begin{aligned}
\mu_X(t + s) &= E[X(t + s) - X(0)] = E[X(t + s) - X(s) + X(s) - X(0)] \\
&= E[X(t + s) - X(s)] + E[X(s) - X(0)] \\
&= E[X(t) - X(0)] + E[X(s) - X(0)] \\
&= \mu_X(t) + \mu_X(s)
\end{aligned}$$

The only solution to the above functional equation is as follows:

$$\mu_X(t) = \mu_X(1)\,t$$

Based on

$$\sigma_X^2(t) = E[(X(t) - E[X(t)])^2] = E[((X(t) - X(0)) - E[X(t) - X(0)])^2]$$

and the fact that the variance of the sum of independent random variables is the sum of their variances, while using the property of the stationary independent increments, we have

$$\begin{aligned}
\sigma_X^2(t + s) &= E[((X(t + s) - X(0)) - E[X(t + s) - X(0)])^2] \\
&= E[((X(t + s) - X(s) - X(0) + X(s)) \\
&\quad - E[(X(t + s) - X(s) - X(0) + X(s)])^2] \\
&= E[((X(t + s) - X(s)) - E[X(t + s) - X(s)])^2] \\
&\quad + E[((X(s) - X(0)) - E[X(s) - X(0)])^2] \\
&= E[((X(t) - X(0)) - E[X(t) - X(0)])^2] \\
&\quad + E[((X(s) - X(0)) - E[X(s) - X(0)])^2] \\
&= \sigma_X^2(t) + \sigma_X^2(s)
\end{aligned}$$

The only solution to the above functional equation is as follows:

$$\sigma_X^2(t) = \sigma_X^2(1)\, t \qquad\qquad\qquad \blacksquare$$

11.5 Ergodic Random Processes

The subject of ergodicity is very complicated, and it is thus discussed here briefly in the most general form. Ergodicity is a stronger condition than stationarity. In a stationary process, the mean and autocorrelation functions are determined by ensemble averaging, i.e. averaging across all sample functions of the process. This in turn requires complete knowledge of the first-order and second-order joint pdf's of the process. In practice, it is very difficult, if not impossible, to observe all sample functions of a random process at a given time. It is more likely to be able to observe a single sample function for a long period of time. We thus need to define the *time-averaged mean, variance and autocorrelation functions* of the sample function $x(t)$, respectively, as follows:

$$\langle x(t) \rangle = \lim_{T \to \infty} \left(\frac{1}{2T} \right) \int_{-T}^{T} x(t)dt$$

$$\langle (x(t) - \langle x(t) \rangle)^2 \rangle = \lim_{T \to \infty} \left(\frac{1}{2T} \right) \int_{-T}^{T} (x(t) - \langle x(t) \rangle)^2 dt$$

$$\langle x(t)x(t - \tau) \rangle = \lim_{T \to \infty} \left(\frac{1}{2T} \right) \int_{-T}^{T} x(t)x(t - \tau)dt \qquad (11.14)$$

where the symbol $\langle\,\rangle$ denotes the time (temporal) averaging operation. These averages are random variables because their values depend on which sample function of the random process $X(t)$ is used in the time-averaging calculations. Note that the autocorrelation function is an ordinary function of the variable τ, but the mean and the variance both turn out to be real constants. In general, the time averages and ensemble averages are not equal.

11.5.1 Strict-Sense Ergodic Processes

If all the statistical properties of the random process $X(t)$ can be determined from a single sample function, then the random process is said to be a *strict-sense ergodic process*. For ergodic processes, the moments can be determined by time averages as well as ensemble averages, simply put, all time and ensemble averages are interchangeable, not just the mean, variance, and autocorrelation functions.

11.5.2 Wide-Sense Ergodic Processes

In general, we are not interested in estimating all the ensemble averages but rather only in the first and second moments. A random process that

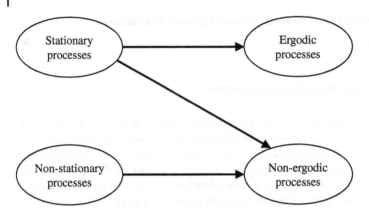

Figure 11.2 Relationship between stationarity and ergodicity.

is ergodic in the mean, variance, and autocorrelation functions is called a *wide-sense ergodic process*. It is generally difficult, if not impossible, to prove that ergodicity exists for a physical process, as only one sample function can be observed. Nevertheless, ergodicity is assumed unless there are compelling physical reasons for not doing so, that is the ergodic property holds whenever needed.

The essence of ergodicity condition is that each sample function of a stationary random process must eventually take on nearly all the modes of behavior of all sample functions if the process is to be ergodic. Figure 11.2 shows the relationship between stationarity and ergodicity. For a random process to be ergodic, it must be stationary, but the converse is not necessarily true. All non-stationary processes are nonergodic, but it is also possible for stationary processes to be nonergodic. For instance, the random process $X(t) = A \cos(t + \Theta)$, where A is a Gaussian random variable, Θ is a uniform random variable, and they are statistically independent, is stationary but nonergodic. The reason lies in the fact that A is a fixed constant in any one sample function, but it is different for other sample functions.

Example 11.14

Suppose we have $X(t) = A \cos(2\pi f_c t + \Theta)$, where the amplitude A and the frequency f_c are nonrandom constants, and Θ is a continuous uniform random variable in the interval $\left[-\frac{\pi}{4}, \frac{\pi}{4}\right]$. Determine if this random process is wide-sense ergodic.

Solution
The expected value of the random process at an arbitrary time t is as follows:

$$E[X(t)] = E[A\cos(2\pi f_c t + \Theta)] = \int_{-\pi/4}^{\pi/4} A\cos(2\pi f_c t + \theta)\left(\frac{2}{\pi}\right)d\theta$$

$$= \frac{2\sqrt{2}A}{\pi}\cos(2\pi f_c t)$$

The mean is a function of time, hence it is not stationary. This process cannot be ergodic, since ergodicity requires stationarity. ∎

Example 11.15
Suppose we have $X(t) = A\cos(2\pi f_c t + \Theta)$, where the amplitude A and the frequency f_c are nonrandom constants, and Θ is a continuous uniform random variable in the interval $[0, 2\pi]$. Determine if this random process is wide-sense ergodic.

Solution
The mean and the autocorrelation functions are as follows:

$$\mu_X(t) = E[X(t)] = 0$$

$$R_X(\tau) = [X(t_1)X(t_2)] = \frac{A^2}{2}\cos(2\pi f_c \tau)$$

It is therefore wide-sense stationary. We now determine the time-averaged mean and autocorrelation functions of $X(t)$, respectively, as follows:

$$\langle x(t) \rangle = \lim_{T\to\infty}\left(\frac{1}{2T}\right)\int_{-T}^{T} x(t)dt = \lim_{T\to\infty}\left(\frac{1}{2T}\right)\int_{-T}^{T} A\cos(2\pi f_c t + \theta)dt$$

$$= \lim_{T\to\infty}\left(\frac{1}{2T}\right)\frac{A}{2\pi f_c}(\sin(2\pi f_c T + \theta) - \sin(2\pi f_c(-T) + \theta))$$

$$= \lim_{T\to\infty}\left(\frac{1}{2T}\right)\frac{A}{2\pi f_c}(2\sin(2\pi f_c T)\cos\theta) = 0$$

$$\langle x(t)x(t-\tau) \rangle = \lim_{T\to\infty}\left(\frac{1}{2T}\right)\int_{-T}^{T} x(t)x(t-\tau)dt$$

$$= \lim_{T\to\infty}\left(\frac{1}{2T}\right)\int_{-T}^{T} A\cos(2\pi f_c t + \theta)A\cos(2\pi f_c(t-\tau) + \theta)dt$$

$$= \lim_{T\to\infty}\left(\frac{A^2}{4T}\right)\int_{-T}^{T} (\cos(4\pi f_c t - 2\pi f_c \tau + 2\theta) + \cos(2\pi f_c \tau))dt$$

$$= \frac{A^2}{2}\cos(2\pi f_c \tau)$$

We showed that the time-averaged mean and autocorrelation functions are identical with the ensemble-averaged mean and autocorrelation functions; hence, the random process is wide-sense ergodic. ∎

11.6 Gaussian Processes

It is of great importance to highlight the fact that, due to the central limit theorem, a Gaussian process is an appropriate model to describe many different physical phenomena, such as noise in a communication system. A random process $X(t)$ is a ***Gaussian process*** if and only if the samples $X(t_1)$, $X(t_2)$, ..., $X(t_n)$ are jointly Gaussian random variables for any integer $n > 0$, and all time instants $t_1, t_2, ..., t_n$. The joint pdf of Gaussian random variables is determined by the ***vector of means*** and the ***covariance matrix*** at the corresponding time instants:

$$\mu_X = \begin{pmatrix} \mu_X(t_1) \\ \vdots \\ \mu_X(t_n) \end{pmatrix}$$

$$K_X = \begin{pmatrix} C_X(t_1, t_1) & \cdots & C_X(t_1, t_n) \\ \vdots & \ddots & \vdots \\ C_X(t_n, t_1) & \cdots & C_X(t_n, t_n) \end{pmatrix} \tag{11.15}$$

For instance, the joint pdf of two random variables resulting from sampling the Gaussian process $X(t)$, i.e. $X(t_1)$, $X(t_2)$, is completely specified by the set of means $\{\mu_X(t_1), \mu_X(t_2)\}$ and the set of autocovariance functions $\{C_X(t_1, t_1),$ $C_X(t_1, t_2), C_X(t_2, t_1), C_X(t_2, t_2)\}$. A Gaussian process has some interesting properties, which are not true in general for other processes. The important properties of a Gaussian process can be summarized as follows:

– The Gaussian process makes mathematical analysis simple and analytic results possible, and in system analysis, the Gaussian process is often the only one for which a complete statistical analysis can be carried out.
– The Gaussian process, due to central limit theorem, provides a good mathematical model for numerous physically observed random time-varying phenomena.
– The Gaussian process from which Gaussian random variables are derived can be completely specified, in a statistical sense, from only first and second moments.
– If a Gaussian process is the input to a linear system, then its output is also a Gaussian process, as the linear filter only modifies the first two moments.
– The input and output of a linear system are jointly Gaussian processes.
– The linear operation, such as a sum, derivative, and integral, on a Gaussian process results in another Gaussian process.

- The Gaussian process is one of the few random processes for which the moments at the output of a nonlinear transformation can be found.
- If a Gaussian process satisfies the conditions for wide-sense stationary, then the random process is also strictly stationary.
- If the set of random variables obtained by sampling a Gaussian process are all uncorrelated, the set of random variables are then statistically independent.
- If the joint pdf of the random variables resulting from sampling a Gaussian process is Gaussian, then the resulting marginal pdfs and conditional pdfs are all individually Gaussian.
- Linear transformation of a set of Gaussian random variables, obtained by sampling a Gaussian process, produces another set of Gaussian random variables.
- The linear combinations of jointly Gaussian variables, resulting from sampling a Gaussian process, is also jointly Gaussian.
- For Gaussian random variables resulting from sampling a Gaussian process, nonlinear and linear mean-square estimates are identical.
- The Gaussian independent, identically distributed random process has the property that the value at every time instant is independent of the value at all other time instants.
- If the autocorrelation function of a stationary Gaussian random process is absolutely integrable, it is then ergodic.

Example 11.16

Suppose $X(t)$ and $Y(t)$ are independent Gaussian processes with zero means and the same covariance function $C(t_1, t_2)$, and we also have $Z(t) = X(t)\cos t + Y(t)\sin t$. Determine the pdf of the random process $Z(t)$.

Solution

In order to determine the pdf of $Z(t)$, we first find its mean and variance at time t as follows:

$$E[Z(t)] = E[X(t)\cos t + Y(t)\sin t] = (\cos t)E[X(t)] + (\sin t)E[Y(t)] = 0$$

Since $X(t)$ and $Y(t)$ are independent, we have

$$E[X(t)Y(t)] = E[X(t)]\,E[Y(t)]$$

Therefore, we have

$$
\begin{aligned}
C_Z(t_1, t_2) &= E[(X(t_1)\cos t_1 + Y(t_1)\sin t_1)(X(t_2)\cos t_2 + Y(t_2)\sin t_2)] \\
&= E[(X(t_1)\cos t_1)(X(t_2)\cos t_2)] \\
&\quad + E[(X(t_1)\cos t_1)(Y(t_2)\sin t_2)] \\
&\quad + E[(Y(t_1)\sin t_1)(X(t_2)\cos t_2)]
\end{aligned}
$$

$$+ E[(Y(t_1) \sin t_1)(Y(t_2) \sin t_2)]$$
$$= \cos t_1 \cos t_2 C(t_1, t_2) + \sin t_1 \sin t_2 C(t_1, t_2)$$
$$= C(t_1, t_2) \cos(t_1 - t_2)$$
$$\rightarrow C_Z(t, t) = C(t, t)$$

Since $Z(t)$ is a linear combination of two Gaussian processes $X(t)$ and $Y(t)$, the pdf of $Z(t)$ is a Gaussian random process whose mean is then zero and covariance function is $C(t, t)$. The pdf of $Z(t)$ is thus as follows:

$$f_{Z(t)}(z) = \left(\frac{1}{\sqrt{2\pi C(t, t)}} \right) \exp\left(-\frac{z^2}{2C(t, t)} \right) \qquad \blacksquare$$

Example 11.17

Assuming X and Y are independent identically distributed Gaussian random variables with zero mean and unit variance, we have $Z(t) = X \cos t + Y \sin t$. Determine the joint pdf of the random processes $Z(t)$ and $Z(t + \tau)$.

Solution

Since X and Y are jointly Gaussian random variables, $Z(t)$ and $Z(t + \tau)$ are also jointly Gaussian. In order to determine the joint pdf of $Z(t)$ and $Z(t + \tau)$, we find the mean vector and covariance matrix at time t and at $t + \tau$ as follows:

$$E[Z(t)] = E[X \cos t + Y \sin t] = (\cos t)E[X] + (\sin t)E[Y] = 0$$
$$E[Z(t + \tau)] = E[X \cos(t + \tau) + Y \sin(t + \tau)]$$
$$= (\cos(t + \tau))E[X] + (\sin(t + \tau))E[Y] = 0$$

and

$$C_Z(t_1, t_2) = E[(X \cos t_1 + Y \sin t_1)(X \cos t_2 + Y \sin t_2)]$$
$$= E[(X \cos t_1)(X \cos t_2)] + E[(X \cos t_1)(Y \sin t_2)]$$
$$+ E[(Y \sin t_1)(X \cos t_2)] + E[(Y \sin t_1)(Y \sin t_2)]$$
$$= \cos t_1 \cos t_2 + \sin t_1 \sin t_2 = \cos(t_1 - t_2)$$

We thus have

$$\mu_Z = \begin{pmatrix} 0 \\ 0 \end{pmatrix} \quad \text{and} \quad Z = \begin{pmatrix} z_1 \\ z_2 \end{pmatrix}$$

and

$$K_Z = \begin{pmatrix} 1 & \cos \tau \\ \cos \tau & 1 \end{pmatrix} \rightarrow |K_Z| = \sin^2 \tau$$
$$\text{and} \quad K_Z^{-1} = \frac{1}{\sin^2 \tau} \begin{pmatrix} 1 & -\cos \tau \\ -\cos \tau & 1 \end{pmatrix}$$

The joint pdf of $Z(t)$ and $Z(t+\tau)$ is thus as follows:

$$f_{Z(t),Z(t+\tau)}(z_1,z_2) = \frac{\exp\left(-\frac{z_1^2-2(\cos\tau)z_1z_2+z_2^2}{2\sin^2\tau}\right)}{2\pi|\sin\tau|}$$

∎

The Wiener process, also known as the Brownian motion process, is an example of a continuous-time, continuous-value random process. A zero-mean random process $X(t)$, where $t \geq 0$ and $X(0) = 0$, is called a **Weiner process** if it has stationary independent increments, and the increment $X(t) - X(s)$ has a Gaussian distribution with mean zero and variance $E[X^2(1)](t-s)$ for any $0 \leq s < t$, noting that $\sigma_X^2 = E[X^2(1)]$ is a parameter of the Wiener process that must be determined from observations.

The Wiener process $X(t)$ is a Gaussian process with mean zero and variance $\sigma_X^2 t$, we thus have

$$f_{X(t)}(x) = \left(\frac{1}{\sqrt{2\pi\sigma_X^2 t}}\right)\exp\left(-\frac{x^2}{2\sigma_X^2 t}\right) \quad -\infty < x < \infty \quad (11.16)$$

where its autocorrelation function is as follows: $R_X(t,s) = \sigma_X^2 \min(t,s)$. The Wiener process is a nonstationary correlated random process. The joint pdf of the samples of a Wiener process is a multivariate Gaussian where its mean vector is equal to zero and the elements of its covariance matrix are the same.

11.7 Poisson Processes

The Poisson process, which is the simplest counting process, has many applications from statistical characterization of noise in electronic devices to statistical analysis of queueing systems in telecommunications and transportation networks. The Poisson process is an example of a continuous-time, discrete-value random process. A counting process $X(t)$ represents the total number of occurrences of random events in the interval $[0, t]$. In a counting process, events occur at random instants of time, such that the average rate of events per second is equal to the constant λ. A sample function is shown in Figure 11.3. Note that $X(t)$ is a nondecreasing, integer-valued, continuous-time process, where we assume $X(0) = 0$.

A random counting process is said to be a **Poisson process** with average rate $\lambda > 0$, if it has the stationary independent increments property and the time between events is exponentially distributed, i.e. the number of event occurrences in the interval $[0, t]$ has a Poisson distribution with mean λt, as defined by

$$P(X(t) = k) = \frac{(\lambda t)^k}{k!}\exp(-\lambda t) \quad k = 0, 1, 2, \ldots \quad (11.17)$$

$X(t)$

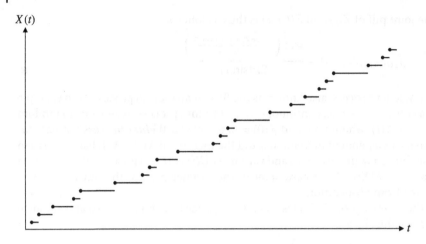

Figure 11.3 A sample function of the Poisson process.

Note that the inter-arrival times is an exponential random variable with mean $\frac{1}{\lambda}$. The Poisson process is thus a memoryless counting process in which an arrival at a particular instant is independent of an arrival at any other instant. The number of arrivals in any interval $(t_0, t_1]$, that is $X(t_0) - X(t_1)$, is a Poisson random variable with expected value $\lambda (t_1 - t_0)$. Moreover, for any pair of nonoverlapping intervals, the number of arrivals in each interval are independent random variables. The mean, variance, and autocorrelation functions of a Poisson process $X(t)$ with rate λ are as follows:

$$E[X(t)] = \lambda t$$
$$\text{Var}[X(t)] = \lambda t$$
$$R_X(t, s) = \lambda \min(t, s) + \lambda^2 ts \qquad (11.18)$$

It is thus evident that the Poisson process is not strict-sense stationary, nor is it even wide-sense stationary.

Example 11.18
Patients arrive at the doctor's office according to a Poisson process with the average rate of six patients per hour. The doctor will not see a patient until at least two patients are in the waiting room. Determine the expected waiting time until the first patient is admitted to see the doctor, and the probability that there are three patients in the first 20 minutes and four patients in the last 40 minutes of the day.

Solution
The expected waiting time is as follows:

$$E[X(t)] = \lambda t \rightarrow 2 = 6t \rightarrow t = \frac{1}{3} \text{ hours}$$

We have two nonoverlapping intervals $I_1 = \frac{1}{3}$ hours and $I_2 = \frac{2}{3}$ hours, the probability of the two events, which is the product of individual probabilities, is as follows:

$$P(3 \text{ in } I_1 \text{ and } 4 \text{ in } I_2) = P(3 \text{ in } I_1) \times P(4 \text{ in } I_2)$$

$$= \left(\frac{\left(\frac{6}{3} \right)^3}{3!} \exp\left(-\frac{6}{3} \right) \right) \times \left(\frac{\left(\frac{12}{3} \right)^4}{4!} \exp\left(-\frac{12}{3} \right) \right)$$

$$= \left(\frac{4}{3} \exp(-2) \right) \times \left(\frac{32}{3} \exp(-4) \right) \cong 0.035\,25 \qquad \blacksquare$$

The sum of two independent Poisson processes with rates λ_1 and λ_2 is a Poisson random processes with the rate $\lambda_1 + \lambda_2$. The Poisson process can be generalized by considering an arrival rate that is a nonnegative function of time or modeling two-dimensional or spatial Poisson processes.

11.8 Summary

In this chapter, the random processes were first classified and then characterized. All major aspects of joint moments of random processes were introduced. The important concepts of stationarity and ergodicity in most general form were briefly discussed, and the widely known Gaussian process, along with its important properties, as well as the Poisson process were briefly discussed.

Problems

11.1 Consider $Y(t) = A\cos(2\pi f_c t + \Theta) + A\sin(2\pi f_c t + \Theta)$, where the amplitude A and the frequency f_c are both nonrandom constants and the initial phase Θ is a uniform random variable in the interval $[0, 2\pi]$. Determine the expected value and variance of the random process $Y(t)$.

11.2 Consider $X(t) = A\cos(2\pi f_c t + \Theta)$, where the amplitude A and the frequency f_c are both nonrandom constants and the initial phase Θ is a uniform random variable in the interval $[0, \pi]$. Is $X(t)$ a wide-sense stationary process?

11.3 Show that that autocorrelation function $R(\tau)$ of a wide-sense stationary random process $X(t)$ is an even function, i.e. $R(\tau) = R(-\tau)$, and has its maximum value at $\tau = 0$.

11.4 Consider a random process $X(t)$ defined by $X(t) = A\cos(2\pi t)$, where $t > 0$ and A is a uniform random variable over the interval $[0, 1]$. Determine the pdf's of $X(t)$ at $t = 0$, 0.125, 0.25, and 0.375.

11.5 Let $X(t) = A|\cos t|$ be a rectified cosine signal having a random amplitude A whose pdf is as follows:

$$f_A(a) = \begin{cases} \exp(-a) & a \geq 0 \\ 0 & \text{otherwise} \end{cases}$$

Determine the pdf of $X(t)$.

11.6 Consider a random process $X(t)$ defined by $X(t) = A\cos(2\pi t)$, where $t > 0$ and A is a uniform random variable over the interval $[0, 1]$. Determine the mean, autocorrelation, and covariance functions of $X(t)$.

11.7 Consider a random process $X(t)$ defined as follows:

$$X(t) = A\cos(t) + B\sin(t)$$

where A and B are independent uniform random variables over the interval $[0, 1]$. Is $X(t)$ strict-sense stationary?

11.8 Suppose A, B, and C are independent zero-mean unit-variance Gaussian random variables, and we have the two random processes $X(t) = A + Bt$ and $Y(t) = A + Ct$. Determine the cross-correlation of the random processes $X(t)$ and $Y(t)$.

11.9 Consider the random process $X(t) = \sin(2\pi F t)$, where the frequency F is a uniform random variable over the interval $[0, W]$ and W is a known positive real number. Show that $X(t)$ is nonstationary.

11.10 Consider two zero-mean stationary random processes $X(t)$ and $Y(t)$ that are related by

$$Y(t) = X(t) - \rho X(t + \tau)$$

where ρ and τ are both nonrandom constants. Determine the value of ρ that minimizes the mean-square value of $Y(t)$ for a given value of τ.

11.11 The random processes $X(t)$ and $Y(t)$ are jointly wide-sense stationary and their autocorrelation functions are $R_X(\tau)$ and $R_Y(\tau)$, respectively. Show that their cross-correlation function is bounded as follows:

$$|R_{XY}(\tau)| \leq \frac{1}{2}[R_X(0) + R_X(0)]$$

11.12 Give an example of two random processes $X(t)$ and $Y(t)$ for which $R_{XY}(t+\tau, t)$ is a function of τ, but $X(t)$ and $Y(t)$ are not stationary processes.

11.13 Can $R_X(\tau) = \sin(2\pi\tau)$ be the autocorrelation function of a wide-sense stationary process $X(t)$? Explain your answer.

11.14 Consider the random processes $Y(t) = X(t)\cos(t+\Theta)$ and $Z(t) = X(t)\sin(t+\Theta)$, where $X(t)$ is a wide-sense stationary random process and is independent of the random variable Θ that is uniformly distributed over the interval $[0, 2\pi]$. Show that $Y(t)$ and $Z(t)$ at some fixed value of time t are orthogonal to each other.

11.15 Suppose the random process $X(t)$ is defined as follows:

$$X(t) = N \quad \forall t$$

where N is a zero-mean unit-variance random variable. Is this random process stationary? Is it ergodic?

12

Analysis and Processing of Random Processes

When a random process is applied to a linear system, the output is also a random process. Even if the probability distribution of the input signal is known, it is, in general, difficult to determine the probability distribution of the output signal. It is, however, possible to determine some major characteristics of the output random process. The focus of this chapter is on the analysis and processing of random signals in linear systems.

12.1   Stochastic Continuity, Differentiation, and Integration

In this section, the continuous-time continuous-valued random processes are exclusively discussed. In system analysis, as a sample function of a random process is a deterministic signal, the system output is a sample function of another random process. However, in order to determine the output random process when the input is a random process, probabilistic methods, to address continuity, differentiation, and integration of an ensemble of sample functions as a whole, are required. Our focus is on the mean-square sense, as it provides simple tractability along with useful applications in the study of linear systems with random inputs.

12.1.1   Mean-Square Continuity

A random process $X(t)$ is said to be ***continuous in the mean-square*** or ***mean-square continuous***, if we have

$$\lim_{\epsilon \to 0} E[(X(t + \epsilon) - X(t))^2] = 0 \tag{12.1a}$$

If all sample functions of a random process are continuous at a time t, then the random process will be mean-square continuous at the time t. However, the mean-square continuity of a random process does not imply that all the sample

Probability, Random Variables, Statistics, and Random Processes: Fundamentals & Applications,
First Edition. Ali Grami.
© 2020 John Wiley & Sons, Inc. Published 2020 by John Wiley & Sons, Inc.
Companion Website: www.wiley.com/go/grami/PRVSRP

functions are continuous. For instance, the Poisson process is mean-square continuous, but its sample functions have discontinuities. The random process $X(t)$ is mean-square continuous at the time t, if its autocorrelation function $R_X(t_1, t_2)$ is continuous at the time (t, t). If a random process is mean-square continuous, then its mean is continuous, i.e. we have

$$\lim_{\epsilon \to 0} \mu_X(t + \epsilon) = \mu_X(t) \to \lim_{\epsilon \to 0} E[X(t + \epsilon)] = E[\lim_{\epsilon \to 0} X(t + \epsilon)] \tag{12.1b}$$

Hence, for a mean-square continuous random process, the order of expectation and limiting are interchangeable. Assuming the random process $X(t)$ is wide-sense stationary, then it is mean-square continuous if its autocorrelation function $R_X(\tau)$ is continuous at $\tau = 0$.

12.1.2 Mean-Square Derivatives

A random process $X(t)$ is said to have a **mean-square derivative** $X'(t)$, if we have

$$\lim_{\epsilon \to 0} E\left[\left(\frac{X(t + \epsilon) - X(t)}{\epsilon} - X'(t) \right)^2 \right] = 0 \tag{12.2a}$$

The mean-square derivative of $X(t)$, denoted by $X'(t)$, exists, if the following derivative of its autocorrelation function exists:

$$\frac{\partial^2}{\partial t_1 \partial t_2} R_X(t_1, t_2) < \infty \tag{12.2b}$$

If $X'(t)$ exists, then its mean is as follows:

$$E[X'(t)] = \frac{d}{dt} E[X(t)] \tag{12.2c}$$

This indicates that the operations of differentiation and expectation may be interchanged. Moreover, the autocorrelation function of the derivative is as follows:

$$R_{X'}(t_1, t_2) = \frac{\partial^2}{\partial t_1 \partial t_2} R_X(t_1, t_2) \tag{12.2d}$$

Note that if all sample functions are differentiable, then the mean-square derivative exists. However, the existence of the mean-square derivative does not imply the existence of the derivative for all sample functions. If the mean-square derivative of a Gaussian process exists, then its mean-square derivative is also a Gaussian process. The mean-square derivative of a wide-sense stationary random process $X(t)$ exists if $R_X(\tau)$ has the first and second derivatives at $\tau = 0$. Moreover, the autocorrelation function of the mean-square derivative is then as follows:

$$R_{X'}(\tau) = -\frac{d^2}{d\tau^2} R_X(\tau) \tag{12.2e}$$

Note that since the mean of a wide-sense stationary process is a constant, then the mean of its derivative is 0.

12.1.3 Mean-Square Integrals

The **mean-square integral** of the random process $X(t)$ is defined as follows:

$$Y(t) = \int_{t_0}^{t} X(u)\,du \tag{12.3a}$$

The random process $Y(t)$ exists if the following double integral exists:

$$\int_{t_0}^{t} \int_{t_0}^{t} R_X(u, v)\,du\,dv < \infty \tag{12.3b}$$

If $X(t)$ is a mean-square continuous random process, then its mean-square integral $Y(t)$ exists. If $Y(t)$ exists, then its mean is as follows:

$$E[Y(t)] = E\left[\int_{t_0}^{t} X(u)\,du \right] = \int_{t_0}^{t} E[X(u)]\,du \tag{12.3c}$$

This indicates that the operations of integration and expectation may be interchanged. Moreover, the autocorrelation function of the integral is then as follows:

$$R_Y(t_1, t_2) = \int_{t_0}^{t_1} \int_{t_0}^{t_2} R_X(u, v)\,du\,dv \tag{12.3d}$$

Note that if $X(t)$ is wide-sense stationary, then the integrand $R_X(u, v)$ is replaced by $R_X(u - v)$. Moreover, the integral of a Gaussian random process is also a Gaussian process.

12.2 Power Spectral Density

To analyze deterministic time-domain signals in the frequency domain, we use the Fourier transform. In case of random processes, as a sample function is deterministic, we could use the Fourier transform of a sample function; however, a single sample function almost always falls short of being a representative of all sample functions in a random process. A random process is an ensemble of sample functions, and the spectral characteristics of these signals determine the spectral characteristics of the random process. A statistical average of the sample functions is thus a more meaningful measure to reflect the spectral components of the random process.

For a wide-sense stationary process, the autocorrelation function is an appropriate measure for the average rate of change of a random process. Figure 12.1 shows two sets of sample functions for slowly varying and rapidly

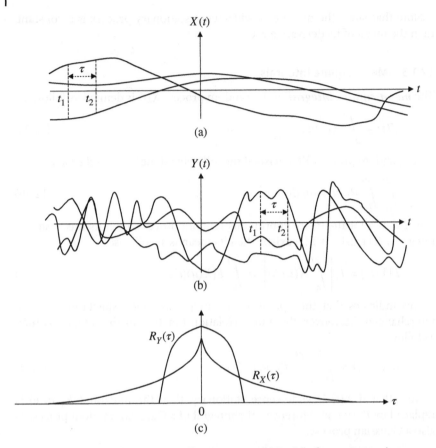

Figure 12.1 (a) Slowly-fluctuating random processes. (b) Rapidly-fluctuating random processes. (c) Autocorrelation functions of slowly and rapidly fluctuating random processes.

varying wide-sense stationary random processes, and their representative autocorrelation functions. If a random signal is slowly time-varying, then the Fourier representation of the random process will mostly possess low frequencies and its power will be mostly concentrated at lower frequencies. For such signals, the autocorrelation function is long in duration, and decreases slowly as the time difference between two samples is increased. On the other hand, if a random signal changes very fast, then most power in the random process will be at high frequencies. For such processes, the autocorrelation function is short in duration and decreases rapidly as the time difference between two samples is increased.

We now define the power spectral density or power spectrum of a wide-sense stationary random process, which determines the distribution of the power of the random process at different frequencies. The power spectral density of a random process $X(t)$, denoted by $S_X(f)$, is measured by watts per hertz (W/Hz). The power spectral density of a wide-sense stationary random process, when integrated over a band of frequencies, gives rise to the average power within that band.

The **Wiener–Khintchine theorem** states that the power spectral density of a wide-sense stationary process and its autocorrelation function form a Fourier transform pair, as given by

$$S_X(f) = F\{R_X(\tau)\} = \int_{-\infty}^{\infty} R_X(\tau)e^{-j2\pi f\tau}\, d\tau$$

$$R_X(\tau) = F^{-1}\{S_X(f)\} = \int_{-\infty}^{\infty} S_X(f)e^{j2\pi f\tau}\, df \tag{12.4}$$

where $j = \sqrt{-1}$. The power spectral density of a wide-sense stationary random process is always real, nonnegative, and for a real-valued random process, it is an even function of frequency.

The **mean-square value** of a wide-sense stationary process equals the total area under the graph of the power spectral density, i.e. to determine the total power, integrate the power spectral density over all frequencies, which in turn results in $R_X(0)$, we thus have

$$E[X^2(t)] = \int_{-\infty}^{\infty} S_X(f)df = R_X(0) \tag{12.5}$$

Example 12.1
Suppose $X(t)$ and $Y(t)$ are uncorrelated, zero-mean, wide-sense stationary random processes. If $Z(t) = X(t) + Y(t)$, then determine the power spectral density of $Z(t)$.

Solution:
We have

$$\begin{aligned} R_Z(\tau) &= E[Z(t)Z(t + \tau)] = E[(X(t) + Y(t))(X(t + \tau) + Y(t + \tau))] \\ &= R_X(\tau) + R_Y(\tau) + R_{XY}(\tau) + R_{YX}(\tau) \end{aligned}$$

Since $X(t)$ and $Y(t)$ are uncorrelated, we have

$$R_{XY}(\tau) = E[X(t)]\, E[Y(t + \tau)] = R_{YX}(-\tau)$$

Since $X(t)$ and $Y(t)$ are also zero-mean, their cross-correlation functions are 0, i.e. $R_{XY}(\tau) = R_{YX}(-\tau) = 0$, we thus have

$$R_Z(\tau) = R_X(\tau) + R_Y(\tau)$$

Noting that the Fourier transform is a linear operation, we can thus have the following:

$$S_Z(f) = S_X(f) + S_Y(f) \qquad\qquad \blacksquare$$

Example 12.2

Let $X(t) = A \cos(2\pi t + \Theta)$, where the random variable Θ is uniformly distributed in the interval $[0, 2\pi]$. Find its power spectral density and the total power.

Solution:

Using a product-to-sum trigonometric identity, we have

$$R_X(t_1, t_2) = E[X(t_1)X(t_2)] = E[A \cos(2\pi t_1 + \Theta)A \cos(2\pi t_2 + \Theta)]$$

$$= \frac{A^2}{2} E[\cos(2\pi t_1 + 2\pi t_2 + 2\Theta) + \cos(2\pi t_1 - 2\pi t_2)]$$

$$= \frac{A^2}{2} \cos(2\pi \tau)$$

Taking the Fourier transform, we get

$$S_X(f) = \frac{A^2}{4}(\delta(f + 1) + \delta(f - 1))$$

Noting that $X(t)$ has a zero mean, we can obtain the total power P as follows:

$$P = R_X(0) = \int_{-\infty}^{\infty} S_X(f)df = \frac{A^2}{2}$$ ∎

Suppose a wide-sense stationary random process $X(t)$ is applied to a linear time-invariant (LTI) system whose impulse response is $h(t)$ and frequency response is $H(f)$, the system output is then a wide-sense stationary random process $Y(t)$. The **mean of the output random process** $Y(t)$ is as follows:

$$E[Y(t)] = E[X(t)] \int_{-\infty}^{\infty} h(s) \, ds = E[X(t)] \, H(0) \tag{12.6}$$

Therefore, if the system is not a low-pass filter or the input is a zero-mean random process, the output is then a zero-mean random process. The **power spectral density of the output random process** $Y(t)$ is equal to the power spectral density of the input $X(t)$ multiplied by the squared magnitude of the transfer function $H(f)$ of the system, hence we have

$$S_Y(f) = |H(f)|^2 S_X(f) \tag{12.7}$$

We can conclude that the system phase response has no bearing on the output power spectral density. Note that the autocorrelation function of the output can be easily found through the inverse Fourier transform of the power spectral density of the output.

Example 12.3

The power spectral density of a zero-mean wide-sense stationary random process is constant $\frac{N_0}{2}$. This random process is passed through an ideal low-pass filter whose bandwidth is $B \neq 0$ Hz. Determine the autocorrelation function of the output, and the instants of time for which the samples of the output signal are uncorrelated.

Solution:
We have

$$S_Y(f) = |H(f)|^2 S_X(f) = \begin{cases} |H(0)|^2 \left(\dfrac{N_0}{2}\right) & \text{if } 0 \leq |f| \leq B \\ 0 & \text{otherwise} \end{cases}$$

The inverse Fourier transform of the output power spectral density gives rise to the output autocorrelation function, we have

$$R_Y(\tau) = |H(0)|^2 BN_0 \operatorname{sinc}(2B\tau)$$

Noting the input mean is 0, the output mean is also 0. We have uncorrelated samples at the output, if we have the following:

$$R_Y(\tau) = |H(0)|^2 BN_0 \left(\frac{\sin(2\pi B\tau)}{(2\pi B\tau)}\right) = 0$$

which means $\frac{\sin(2\pi B\tau)}{(2\pi B\tau)} = 0$, or equivalently, $2\pi B\tau = \pi k$, or $\tau = \frac{k}{2B}$, where $k = \pm 1, \pm 2, \ldots$ ∎

Example 12.4
Suppose we have $Y(t) = A\,X(t)\cos(2\pi t + \Theta)$, where the amplitude A is a non-random constant, $X(t)$ is a wide-sense stationary random process, and Θ is a uniform random variable in the interval $[0, 2\pi]$. Moreover $X(t)$ and Θ are assumed to be independent. Determine the power spectral density of $Y(t)$ in terms of the power spectral density of $X(t)$.

Solution:
Using a product-to-sum trigonometric identity, and noting $X(t)$ and Θ are independent, we can obtain the following:

$$\begin{aligned} R_Y(\tau) &= E[Y(t+\tau)Y(t)] \\ &= E[AX(t+\tau)\cos(2\pi t + 2\pi\tau + \Theta)AX(t)\cos(2\pi t + \Theta)] \\ &= 0.5A^2 R_X(\tau)E[\cos(2\pi\tau) + \cos(4\pi t + 2\pi\tau + 2\Theta)] \\ &= \frac{A^2}{2}R_X(\tau)\cos(2\pi\tau) \end{aligned}$$

We now find the Fourier transform of $R_Y(\tau)$, noting that multiplication in one domain translates into convolution in the other, the power spectral density of the output is thus as follows:

$$\begin{aligned} S_Y(f) &= \frac{A^2}{2}F\{R_X(\tau)\} * F\{\cos(2\pi\tau)\} \\ &= \frac{A^2}{2}S_X(f) * \left(\frac{1}{2}(\delta(f-1) + \delta(f+1))\right) \end{aligned}$$

where the symbol * denotes the convolution operation. After carrying out the convolution operation, we then obtain the following:

$$S_Y(f) = \frac{A^2}{4}(S_X(f-1) + S_X(f+1))$$

∎

As the power spectral density provides a measure of the frequency distribution of a single random process, cross-spectral densities provide a measure of the frequency inter-relationship between two random processes. Assuming the input and output processes $X(t)$ and $Y(t)$ are two jointly wide-sense stationary random processes with their cross-correlation functions denoted by $R_{XY}(\tau)$ and $R_{YX}(\tau)$ and their power spectral densities by $S_{XY}(f)$ and $S_{YX}(f)$, which are the Fourier transforms of the respective cross-correlation functions, we have

$$S_{XY}(f) = S_X(f)\,H^*(f)$$
$$S_{YX}(f) = S_X(f)\,H(f) \tag{12.8}$$

Note that $H^*(f)$ is the complex conjugate of $H(f)$. Even though $S_X(f)$ and $S_Y(f)$ are both real nonnegative functions, $S_{XY}(f)$ and $S_{YX}(f)$ may be complex functions. Figure 12.2 shows relations for various spectral densities.

When a random process is transmitted through a linear system, the first two moments – mean and variance functions as well as autocorrelation and covariance functions – can generally provide an adequate statistical characterization of the system. However, when the system is nonlinear, valuable information may be contained in higher-order moments of the resulting random process.

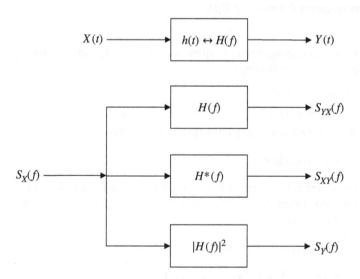

Figure 12.2 Input–output relations for the power spectral and cross-spectral densities.

12.3   Noise

Noise refers to the unwanted, uncontrollable, and ever-present time-varying waveforms that can degrade the transmission of signals in communication systems. A major communication design objective is to always minimize the effect of noise. There are two distinct categories of sources of noise, one is external to the system, such as atmospheric noise and man-made noise, and the other is internal to the system, such as the transmitting and receiving equipment causing shot and thermal noise.

12.3.1   White Noise

White noise is often used to model the thermal noise in electronic devices. *White noise* is ideally defined as a wide-sense stationary random process $N(t)$ whose frequency components all appear with equal power. The adjective white is used in the sense that white light contains equal amounts of all frequencies within the visible frequency band. White noise is a mathematical idealization and cannot exist as a physically realizable process. Since the power spectral density of white noise is a constant for all frequencies, its autocorrelation

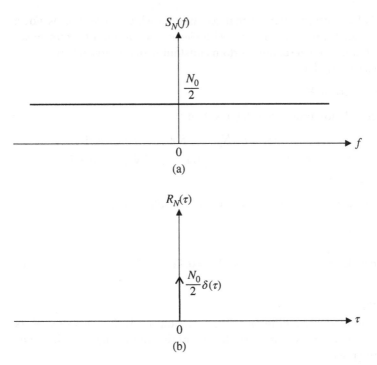

Figure 12.3 White noise: (a) power spectral density; (b) autocorrelation function.

function consists of a delta (impulse) function. The power spectral density and autocorrelation functions of white noise, as shown in Figure 12.3, are as follows:

$$S_N(f) = \frac{N_0}{2}$$

$$R_N(\tau) = \frac{N_0}{2}\delta(\tau) \tag{12.9}$$

where $\frac{N_0}{2}$ is a positive constant, and $\delta(\tau)$ represents the unit impulse function, i.e., $\delta(\tau) = 1$ for $\tau = 0$ and $\delta(\tau) = 0$ for $\tau \neq 0$. It is thus important to highlight that any two different samples of white noise – no matter how closely together in time they are taken, i.e. $\tau \neq 0$ – are uncorrelated. Strictly speaking, white noise has infinite average power as such it is not physically realizable. White noise is in fact an approximation to the noise that is present in real systems. As long as the bandwidth of a noise process at the input of a system is significantly larger than that of the system itself, the noise process can be modeled as white noise.

Example 12.5
Show that white noise is a zero-mean wide-sense stationary random process.

Solution:
The proof is by contradiction, in that we assume white noise $N(t)$ is not a zero-mean random process. It can thus be viewed as the sum of a zero-mean process $N_Z(t)$ and a nonzero, nonrandom constant K that represents the mean of white noise, i.e. we have

$$N(t) = N_Z(t) + K$$

The autocorrelation function of $N(t)$ is then as follows:

$$R_N(\tau) = E[N(t)\,N(t+\tau)] = E[(N_Z(t)+K)(N_Z(t+\tau)+K)]$$
$$= E[N_Z(t)N_Z(t+\tau)] + K\,E[N_Z(t)] + K\,E[N_Z(t+\tau)] + K^2$$
$$= R_Z(\tau) + K^2$$

Since the noise is white, we must then have the following:

$$R_N(\tau) = R_Z(\tau) + K^2 = \frac{N_0}{2}\delta(\tau)$$

We thus conclude that we must have $K^2 = 0$, i.e. $K = 0$. ∎

Example 12.6
An LTI system works as a differentiator, i.e. its output is the derivative of its input. Determine the power spectral density of the output signal, if the input signal is white noise.

Solution:
The frequency response of a differentiator is the following linear function of frequency:

$$H(f) = j2\pi f$$

We thus have

$$S_Y(f) = |H(f)|^2 S_X(f) = (2\pi f)^2 \frac{N_0}{2} = 2\pi^2 N_0 f^2$$ ∎

Note that a stationary Gaussian noise with zero mean and flat power spectral density is called a **white Gaussian noise**. White Gaussian noise can also be described as the derivative of Wiener process. Any two samples of white Gaussian noise, no matter how closely together in time they are taken, are statistically independent. In a sense, white Gaussian noise represents the ultimate in randomness. It is important to note that noise in a communication system is generally assumed to be **additive white Gaussian noise (AWGN)**. Note that by analogy to colored light that has only a portion of the visible frequencies in its spectrum, **colored noise** is defined as any noise that is not white.

12.4   Sampling of Random Signals

Sampling is an indispensable signal processing tool to bridge between continuous-time signals and discrete-time signals and, in turn, paves the way for digital signal processing and transmission. The samples of a random signal are usually taken at periodic points (i.e. spaced uniformly) in time, through which the original continuous signal can be reconstructed in the mean-square sense. This can be achieved if the signal is band-limited. However, since real signals are always time-limited, they cannot be truly band-limited. The spectrum of any practical signal decreases as frequency gets farther away from center band. To this effect, there is usually some bandwidth of important frequencies in any signal such that outside this bandwidth the spectral components can be considered negligible, and thus approximated as zero.

A wide-sense stationary process $X(t)$ is called band-limited if its power spectral density vanishes for frequencies beyond a specific frequency W Hz and it has finite power, as reflected below:

$$S_X(f) = 0 \quad |f| > W$$

$$R_X(0) < \infty \tag{12.10}$$

When a continuous wide-sense stationary random process is sampled, a wide-sense stationary random sequence results. The random process $X(t)$ can be reconstructed from the sequence of its samples taken at a minimum rate of twice the highest frequency component, i.e. $2W$ samples per second. The sampled version of the process equals the original continuous process in the mean-square sense for all time t, and it is as follows:

$$\hat{X}(t) = \sum_{n=-\infty}^{\infty} X\left(\frac{n}{2W}\right) \frac{\sin(\pi(2Wt - n))}{(\pi(2Wt - n))} \quad -\infty < t < \infty \quad (12.11)$$

where $X\left(\frac{n}{2W}\right)$ is the random variable obtained by sampling the process $X(t)$ at $t = \frac{n}{2W}$, where n is an integer. Note that the equality in the mean-square sense does not imply equality for all sample functions. It can be shown that if the sampling rate is at least twice the highest frequency component of the power spectral density of the continuous random process, then the **mean-square error** between the continuous random process and its sampled version is 0, i.e. we have the following:

$$\overline{e^2} = E[(X(t) - \hat{X}(t))^2] = 0 \quad (12.12)$$

Suppose we have $S_X(f) \neq 0$, for $|f| > W$, i.e. $X(t)$ is not band-limited to W Hz. If the random process $X(t)$ is applied to an ideal low-pass filter whose bandwidth is W Hz and the resulting output process $Y(t)$ is sampled at a rate equal to $2W$ samples per second to produce $\hat{Y}(t)$, then the resulting mean-square error between the original (unfiltered) signal $X(t)$ and the sampled version of the filtered signal $\hat{Y}(t)$ is as follows:

$$\overline{e^2} = E[(X(t) - \hat{Y}(t))^2] = 2 \int_{W}^{\infty} S_X(f) df \neq 0 \quad (12.13)$$

Example 12.7
The power spectral density of a wide-sense stationary random process is as follows:

$$S_X(f) = \begin{cases} -10^{-14}|f| + 10^{-7} & |f| \leq 10^7 \\ 0 & \text{otherwise} \end{cases}$$

The signal is sampled at $2W$ Hz. Determine the mean-square value of the sampling error for $W = 10$ MHz and $W = 5$ MHz.

Solution:
This problem highlights the relationship between the bandwidth of a random signal on the one hand, and the sampling rate and the resulting mean-square error on the other hand. The bandwidth of this wide-sense stationary random signal is 10 MHz. In case of $W = 10$ MHz, we have $\overline{e^2} = 0$, as the sampling rate

is twice the highest frequency component. In case of $W = 5$ MHz, since the sampling rate is not high enough, we must find the nonzero mean-square error:

$$e^2 = 2 \int_{\frac{10^7}{2}}^{10^7} (-10^{-14}f + 10^{-7})df = 0.25$$

■

12.5   Optimum Linear Systems

A system is optimum when it maximizes or minimizes a certain criterion while meeting a set of specific constraints. In designing any optimum system, input specification, system constraints, and a criterion of optimality are required.

Input specification implies that some knowledge about the input to the system is available. We assume that the continuous-time input signal arrives along with white additive noise, though the same general procedure to design an optimum system can also be used for nonwhite additive noise. In addition, the input signal and the additive noise are assumed to be uncorrelated.

In terms of system constraints, we assume the system is LTI. With Gaussian noise, there is no nonlinear system that will do a better job than the optimum linear system. However, for non-Gaussian noise, the linear system may not be the best, while noting that determining the optimum nonlinear system is usually a very difficult task. We also assume the system need not be causal, as causality can introduce undue mathematical complexity, while noting that it is possible to select a realizable (causal) filter to approximate the optimum filter.

The choice of a criterion of optimality is often influenced by the nature of the input signal, that is, whether it is deterministic or random. In either case, there are a number of criteria that might make sense. However, only the most common criteria will be discussed in this section.

12.5.1   Systems Maximizing Signal-to-Noise Ratio

As shown in Figure 12.4a, we assume the input to an LTI system consists of the sum of the deterministic signal $s(t)$ and white noise $N(t)$, and the system output consists of the sum of the deterministic signal $r(t)$, which is the response to $s(t)$, and the colored noise $M(t)$, which is the response to $N(t)$. The purpose of the system is generally to detect the presence of a signal of known shape or to measure the time at which such a signal occurs.

The goal is to design a system (or filter) that would enhance its output power at some instant in time while reducing its output average noise power. A filter that maximizes the output signal-to-noise ratio (SNR) is called a **matched**

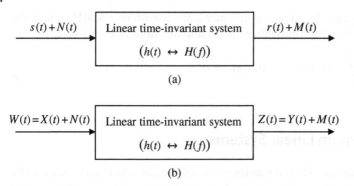

Figure 12.4 Notation for optimum filter: (a) maximum signal-to-noise-ratio criterion; (b) minimum mean-square-error criterion.

filter, and the objective is to determine its impulse response $h(t)$ or equivalently transfer function $H(f)$. The output SNR to be maximized at time t_0 is thus as follows:

$$\text{SNR}\big|_{\text{Output}} = \frac{|r(t_0)|^2}{E[M^2(t_0)]} = \frac{\left|\int_{-\infty}^{\infty} S(f)H(f)\exp(j2\pi f t_0)df\right|^2}{\left(\frac{N_0}{2}\right)\int_{-\infty}^{\infty}|H(f)|^2 df} \tag{12.14}$$

where $r(t)$ is the output signal, $S(f)$ is the Fourier transform of the input signal $s(t)$ and $\frac{N_0}{2}$ is the power spectral density of white noise $N(t)$ at the input. In order to maximize the above ratio, we need to apply the Schwarz inequality, which is as follows:

$$\left|\int_{-\infty}^{\infty} V(f)W(f)df\right|^2 \le \int_{-\infty}^{\infty}|V(f)|^2\,df \int_{-\infty}^{\infty}|W(f)|^2\,df \tag{12.15}$$

where $V(f)$ and $W(f)$ are two complex functions of the real variable f, and the equality holds only when $W(f)$ is proportional to the complex conjugate of $V(f)$, i.e. we have $W(f) = C\,V^*(f)$ and C is any arbitrary real constant. With $V(f) = \sqrt{\frac{N_0}{2}}H(f)$ and $W(f) = \sqrt{\frac{2}{N_0}}S(f)\exp(j2\pi f t_0)$, the output SNR then becomes as follows:

$$\text{SNR}\big|_{\text{output}} \le \int_{-\infty}^{\infty}\left|\sqrt{\left(\frac{2}{N_0}\right)}S(f)\exp(j2\pi f t_0)\right|^2 df = \frac{2}{N_0}\int_{-\infty}^{\infty}|S(f)|^2 df \tag{12.16}$$

The maximum of this ratio occurs when we have the following:

$$\sqrt{\frac{2}{N_0}}S(f)\exp(j2\pi f t_0) = C\sqrt{\frac{N_0}{2}}H^*(f) \rightarrow H_{\text{opt}}(f) = k\,S^*(f)\exp(-j2\pi f t_0) \tag{12.17}$$

where $k = \frac{2}{N_0 C}$, an arbitrary constant. The optimum filter is thus related to the spectrum of the input signal and the time that the ratio is maximum. Using the inverse Fourier transform and assuming the input signal is real, i.e. $s(t) = s^*(t)$, the impulse response of the optimum filter is as follows:

$$h_{opt}(t) = ks(t_0 - t) \tag{12.18}$$

k is simply a gain constant that does not affect the SNR, it can be set equal to any value, oftentimes $k = 1$ is assumed. The **impulse response of the matched filter** is a time-reversed and delayed version of the input $s(t)$, that is, matched to the input signal, and hence the name matched filter. Since t_0 is a parameter that a designer may have some latitude in choosing, its value may be selected in some cases to make the optimum filter causal.

Example 12.8
Suppose the input signal to a matched filter is as follows:

$$s(t) = \begin{cases} \sqrt{2}\cos(2\pi t) & 0 \le t \le 1 \\ 0 & \text{elsewhere} \end{cases}$$

Determine the impulse response of a filter matched to $s(t)$ as well as the output signal $r(t)$. Show that the filter output attains its maximum value at $t = 1$.

Solution:
The impulse response of a filter matched to $s(t)$ is thus as follows:

$$h_{opt}(t) = s(1 - t) = \begin{cases} \sqrt{2}\cos(2\pi t) & 0 \le t \le 1 \\ 0 & \text{elsewhere} \end{cases}$$

The impulse response of the matched filter happens to be the same as the input signal $s(t)$. The filter output is the inverse Fourier transform of the product of the Fourier transform of the input signal and the Fourier transform of the impulse response. It can be thus shown the filter output $r(t)$ is as follows:

$$r(t) = F^{-1}\{S(f)H(f)\} = \begin{cases} t\cos(2\pi t) & 0 \le t \le 1 \\ (2 - t)\cos(2\pi t) & 1 \le t \le 2 \\ 0 & \text{elsewhere} \end{cases}$$

The output signal $r(t)$ thus attains its maximum at $t = 1$, and the maximum value is $r(1) = 1$. ∎

12.5.2 Systems Minimizing Mean-Square Error

As shown in Figure 12.4b, we assume the input $W(t)$ to an LTI system consists of the sum of the random signal $X(t)$ and white noise $N(t)$, and the system output $Z(t)$ consists of the sum of the random signal $Y(t)$, which is the response to

$X(t)$, and the colored noise $M(t)$, which is the response to $N(t)$. The purpose of the system is to observe the unknown signal for purposes of measurement and control. We assume that the input random process $X(t)$, which has a power spectral density $S_X(f)$, and the input noise $N(t)$, which is white with power spectral density $\frac{N_0}{2}$, are jointly wide-sense stationary. Since the input signal $X(t)$ and the input noise $N(t)$ are assumed to be uncorrelated, we have the following:

$$W(t) = X(t) + N(t) \rightarrow \begin{cases} S_W(f) = S_X(f) + \dfrac{N_0}{2} \\ S_{WX}(f) = S_X(f) \end{cases} \qquad (12.19)$$

The goal is to design a system (or filter) that would minimize the mean-square error between the system output $Z(t)$ and the input signal $X(t)$. The difference between the system output $Z(t)$ and the true value of the signal $X(t)$ consists of two components. One component is the signal error and represents the difference between the input and output when there is no input noise. The second component is the noise error, which also represents an error in the output. The total error is the sum of these two, and the quantity to be minimized is the mean-square value of this error. The optimum filter, which minimizes total mean-square error, is referred to as the **Wiener filter**. The objective is to determine a real filter (i.e. $H(f) = H^*(-f)$) that minimizes the following:

$$E[e^2(t)] = E[(X(t) - Z(t))^2] = R_X(0) - 2R_{ZX}(0) + R_Z(0) \qquad (12.20)$$

We also have the following functions in terms of their power spectral densities:

$$R_X(0) = \int_{-\infty}^{\infty} S_X(f)df \qquad (12.21a)$$

$$R_Z(0) = \int_{-\infty}^{\infty} S_W(f)|H(f)|^2 df \qquad (12.21b)$$

$$R_{ZX}(0) = E[X(t)Z(t)] = E\left[X(t)\int_{-\infty}^{\infty} h(\alpha)W(t-\alpha)d\alpha\right]$$

$$= \int_{-\infty}^{\infty} R_{WX}(\alpha)h(\alpha)d\alpha$$

$$= \int_{-\infty}^{\infty} \left(\int_{-\infty}^{\infty} S_{WX}(f)\exp(jf\alpha)df\right) h(\alpha)d\alpha$$

$$= \int_{-\infty}^{\infty} S_{WX}(f) \left(\int_{-\infty}^{\infty} h(\alpha)\exp(jf\alpha)d\alpha\right) df$$

$$= \int_{-\infty}^{\infty} S_X(f) \left(\int_{-\infty}^{\infty} h(\alpha) \exp(jf\alpha) d\alpha \right) df$$

$$= \int_{-\infty}^{\infty} S_X(f) H(-f) df$$

$$= \int_{-\infty}^{\infty} S_X(f) A(f) \exp(-jB(f)) df \qquad (12.21c)$$

where $A(f)$ is an even function representing the magnitude response of the real filter $H(f)$ and $B(f)$ is an odd function representing its phase response. We can then write the mean-square error as follows:

$$E[e^2(t)] = \int_{-\infty}^{\infty} S_X(f) df + \int_{-\infty}^{\infty} \left(S_X(f) + \frac{N_0}{2} \right) A^2(f) df$$

$$- 2 \int_{-\infty}^{\infty} S_X(f) A(f) \exp(-jB(f)) df \qquad (12.22)$$

In order to minimize $E[e^2(t)]$, we first find the optimum phase response $B(f)$ and then the optimum magnitude response $A(f)$. It can be shown that we must then have the following $A(f)$ and $B(f)$ to have the optimum filter:

$$\begin{cases} B(f) = 0 \\ A(f) = \dfrac{S_X(f)}{S_X(f) + \dfrac{N_0}{2}} \end{cases} \rightarrow H_{opt}(f) = \dfrac{S_X(f)}{S_X(f) + \dfrac{N_0}{2}} \qquad (12.23)$$

Note that the optimum filter, referred to as the **Wiener filter**, can be considered as a cascade of two parts, the first part is the prewhitening filter and the second part does the actual filtering. The optimum filter attenuates heavily the band where noise is relatively stronger. Although this causes some signal distortion, the noise is attenuated more heavily. It can be shown that the resulting minimum mean-square error is as follows:

$$E[e^2(t)]_{min} = \frac{N_0}{2} \int_{-\infty}^{\infty} \frac{S_X(f)}{S_X(f) + \dfrac{N_0}{2}} df \qquad (12.24)$$

Note that if the noise is 0, then there is no need to filter the signal. In other words, there must be a direct connection from input to output, and i.e. we have $H_{opt}(f) = 1$, a result that is intuitively agreeable. Moreover, if the noise is very large and quite dominant, then the filter is turned off, that is we have $H_{opt}(f) = 0$. However, the Wiener filter is practical when the noise level is rather large.

Example 12.9

Suppose the autocorrelation function of the input $R_X(\tau)$ signal is as follows:

$$R_X(\tau) = \exp(-|\tau|)$$

Determine the impulse response of the Wiener filter.

Solution:

Noting that the autocorrelation function of white noise is $\delta(\tau)$ and using the Fourier transform, we have the following:

$$\begin{cases} R_X(\tau) = \exp(-|\tau|) \\ R_N(\tau) = \delta(\tau) \end{cases} \rightarrow \begin{cases} S_X(f) = \dfrac{2}{1 + 4\pi^2 f^2} \\ S_N(f) = 1 \end{cases}$$

This yields the following:

$$H_{\text{opt}}(f) = \frac{S_X(f)}{S_X(f) + S_N(f)} = \frac{\dfrac{2}{1 + 4\pi^2 f^2}}{\dfrac{2}{1 + 4\pi^2 f^2} + 1} = \frac{2}{3 + 4\pi^2 f^2}$$

Using the inverse Fourier transform, the Wiener filter impulse response is as follows:

$$h_{\text{opt}}(t) = \left(\frac{1}{\sqrt{3}}\right) \exp(-\sqrt{3}|t|)$$

∎

12.6 Summary

In this chapter, our focus was on applying wide-sense stationary random signals to systems to assess the output random signals. The systems, which were briefly discussed, were LTI systems, samplers, and optimum linear systems to meet certain criteria.

Problems

12.1 Consider $X(t) = A\cos(2\pi f_c t + \Theta)$, where A and f_c denote the nonrandom amplitude and frequency, respectively, and Θ denotes the random phase uniformly distributed in the interval $[0, 2\pi]$. If $X(t)$ is the input to a simple differentiator, where $H(f) = j2\pi f$, then determine the mean as well as the power spectral density of the output process $Y(t)$.

12.2 Consider a random process $X(t)$ consisting of a sinusoidal component $\cos(2\pi t + \Theta)$, where Θ is a uniformly distributed random variable in the interval $[0, 2\pi]$ and a zero-mean white noise component $W(t)$ whose power spectral density is 1 W/Hz. Noting that the sinusoidal and noise components are independent, determine the autocorrelation function of the random process $X(t)$ and identify the values for which $X(t)$ samples are uncorrelated.

12.3 Suppose the wide-sense stationary processes $X(t)$ and $Y(t)$ are related as follows:

$$Y(t) = X(t)\cos(2\pi f_c t + \Theta) - X(t)\sin(2\pi f_c t + \Theta)$$

where the frequency f_c is a nonrandom constant and the initial phase Θ, which is independent of $X(t)$, is a continuous uniform random variable over the interval $[0, 2\pi]$. Suppose $S_X(f)$, the power spectral density of $X(t)$, is band-limited to W Hz. Determine $S_Y(f)$, the power spectral density of $Y(t)$.

12.4 A stationary zero-mean white Gaussian process $X(t)$, whose power spectral density is 1 W/Hz, is applied to a linear time-invariant filter whose impulse response is as follows:

$$h(t) = u(t) - u(t - 1)$$

where $u(t)$ is the unit step function. A sample is taken at the filter output at time 1; determine the mean, variance, and probability density function (pdf) of this sample.

12.5 Suppose the magnitude square of a linear time-invariant system is as follows:

$$|H(f)|^2 = \frac{1}{1 + 4\pi^2 f^2}$$

If the input is white noise with power spectral density $S_N(f) = 2$ W/Hz, determine the average output power.

Bibliography

Books

Adam, C. (2010). *Essential Mathematics and Statistics for Forensic Science*. Wiley-Blackwell. ISBN: 978-0470742525.

Beyer, W.H. (1991). *CRC Standard Probability and Statistics*. CRC. ISBN: 0-8493-0680-9.

Boncelet, C. (2016). *Probability, Statistics, and Random Signals*. Oxford. ISBN: 978-0-19-020051-0.

Cooper, G.R. and McGillem, C.D. (1998). *Probabilistic Methods of Signal and System Analysis*, 3e. Oxford University. ISBN: 0195123549.

Cox, D.R. and Miller, H.D. (1977). *The Theory of Stochastic Processes*. Chapman and Hall. ISBN: 978-0412151705.

Daniel, W.W. and Cross, C.L. (2013). *Biostatistics*, 3e. Wiley. ISBN: 978-1-118-30279-8.

Deep, R. (2006). *Probability and Statistics*. Academic Press. ISBN: 978-0-12-369463-8.

De Veaux, R.D., Velleman, P.D., and Bock, D.E. (2010). *Stats: Data and Models*, 3e. Pearson. ISBN: 978-0321692559.

Fine, T.L. (2006). *Probability and Probabilistic Reasoning for Electrical Engineering*. Pearson. ISBN: 0-13-020591-5.

Forbes, C., Evans, M., Hastings, N., and Peacock, B. (2011). *Statistical Distributions*, 4e. Wiley. ISBN: 978-0-470-39063-4.

Goodman, W.M. (2009). *Modern Statistics*. Nelson. ISBN: 978-0-17-625179-6.

Gordis, L. (2014). *Epidemiology*, 5e. Elsevier. ISBN: 978-1-4557-3733-8.

Gossett, E. (2009). *Discrete Mathematics with Proof*, 2e. Wiley. ISBN: 978-0-470-45793-1.

Grami, A. (2016). *Introduction to Digital Communications*. Academic Press. ISBN: 978-0-12-407682-2.

Probability, Random Variables, Statistics, and Random Processes: Fundamentals & Applications,
First Edition. Ali Grami.
© 2020 John Wiley & Sons, Inc. Published 2020 by John Wiley & Sons, Inc.
Companion Website: www.wiley.com/go/grami/PRVSRP

Gubner, J.A. (2006). *Probability and Random Processes for Electrical and Computer Engineers*. Cambridge. ISBN: 978-0-521-86470-1.

Haddad, A.H. (2006). *Probabilistic Systems and Random Signals*. Pearson. ISBN: 0-13-009455-2.

Hand, D.J. (2014). *The Improbability Principle: Why Coincidences, Miracles, and Rare Events Happen Everyday*. Scientific American. ISBN: 978-0-374-17534-4.

Haykin, S. (1978). *Communication Systems*. Wiley. ISBN: 0-471-02977-7.

Hayter, A. (2007). *Probability and Statistics for Engineers and Scientists*, 3e. Thomson. ISBN: 0-495-10757-3.

Hogg, R.V. and Tanis, E.A. (2010). *Probability and Statistical Interference*, 8e. Pearson. ISBN: 978-0-321-58475-5.

Hsu, H.P. (1996). *Probability, Random Variables, and Random Processes*. McGraw-Hill. ISBN: 0-07-030644-3.

Ibe, O.C. (2014). *Fundamentals of Applied Probability and Random Processes*. Academic Press. ISBN: 978-0-12-800852-2.

Johnson, R.A. (2011). *Probability and Statistics for Engineers*, 8e. Prentice Hall. ISBN: 978-0-321-64077-2.

Karli, S. and Taylor, H.M. (1975). *A First-Course in Stochastic Processes*, 2e. Academic Press. ISBN: 0-12-398552-8.

Kay, S. (1993). *Fundamentals of Statistical Signal Processing: Estimation Theory*. Prentice-Hall. ISBN: 0-13-345711-7.

Kay, S. (2006). *Intuitive Probability and Random Processes Using MATLAB*. Springer. ISBN: 978-0-387-24157-9.

Kleinrock, L. (1975). *Queueing Systems Volumes I: Theory*. Wiley. ISBN: 0-471-49110-1.

Lathi, B.P. and Ding, Z. (2009). *Modern Digital and Analog Communication Systems*, 4e. Oxford. ISBN: 978-0-19-533145-5.

Leon-Garcia, A. (2008). *Probability, Statistics, and Random Processes for Electrical Engineering*, 3e. Pearson. ISBN: 97-0-13-147122-1.

Lindley, D.V. (2014). *Understanding Uncertainty, Revised Edition*. Wiley. ISBN: 978-1-118-65012-7.

Lipschutz, S. and Lipson, M. (2007). *Discrete Mathematics*, 3e. McGaw-Hill. ISBN: 978-0-07-161586-0.

Lipschutz, S. and Lipson, M. (2011). *Probability*, 2e. McGaw-Hill. ISBN: 978-0-07-175561-0.

Maisel, L. (1971). *Probability, Statistics and Random Processes*. Simon and Schuster.

Olofsson, P. (2007). *Probabilities: The Little Numbers That Rule Our Lives*. Wiley. ISBN: 978-0-470-04001-0.

Oppenheim, A.V. and Verghese, G.C. (2017). *Signals, Systems and Inference*. Pearson. ISBN: 9781292156200.

Peebles, P.Z. Jr. (2001). *Probability, Random Variables and Random Signal Principles*, 4e. McGraw-Hill. ISBN: 978-0-07-366007-3.

Pishro-Nik, H. (2014). *Introduction to Probability, Statistics, and Random Processes.* Kappa Research. ISBN: 978-0990637202.

Proakis, J.G. and Salehi, M. (2014). *Fundamentals of Communication Systems*, 2e. Pearson. ISBN: 0-13-147135-X.

Root, W.L. and Davenport, W.B. (1987). *An Introduction to the Theory of Random Signals and Noise.* Wiley-IEEE Press. ISBN: 978-0-87942-235-6.

Rosen, K.H. (2012). *Discrete Mathematics and Its Applications*, 7e. McGraw-Hill. ISBN: 978-0-07-338309-5.

Rosner, B. (2016). *Fundamentals of Biostatistics*, 8e. Nelson. ISBN: 9781305268920.

Ross, S. (2012). *A First Course in Probability*, 9e. Prentice Hall. ISBN: 978-0321794772.

Roussas, G. (2007). *Introduction to Probability.* Academic Press. ISBN: 978-0-12-088595-4.

Santos, D.A. (2011). *Probability: An Introduction.* Jones and Bartlett. ISBN: 978-0-7637-8411-9.

Scheaffer, R.L. and Young, L.J. (2010). *Introduction to Probability and its Applications*, 3e. Cengage Learning. ISBN: 978-0-534-38671-9.

Scheaffer, R.L., Mulekar, M.S., and McClave, J.T. (2011). *Probability and Statistics for Engineers*, 5e. Thomson. ISBN: 978-0-534-40302-7.

Shanmugam, K.S. and Breipohl, A.M. (1988). *Random Signals: Detection, Estimation and Data Analysis.* Wiley. ISBN: 0-471-81555-1.

Spiegel, M.R. (1975). *Probability and Statistics.* McGraw-Hill. ISBN: 0-07-060220-4.

Spiegel, M.R. and Stephens, L.J. (2011). *Statistics.* McGraw-Hill. ISBN: 978-0-07175549-8.

Stark, H. and Woods, J.W. (2012). *Probability, Statistics, and Random Processes for Engineers*, 4e. Pearson. ISBN: 978-013-231123-6.

Upton, G. and Cook, I. (2014). *Dictionary of Statistics.* Oxford. ISBN: 978-0-19-967918-8.

Van, H. (1968). *Trees, Detection, Estimation, and Modulation Theory.* Wiley. ISBN: 471-89955-0.

Walpole, R.E., Myers, R.H., Myers, S.L., and Ye, K. (2012). *Probability and Statistics for Engineers and Scientists*, 9e. Prentice Hall. ISBN: 978-0-321-62911-1.

Woolfson, M.M. (2012). *Everyday Probability and Statistics*, 2e. Imperial College. ISBN: 978-1-84816-761-2.

Wozencraft, J.M. and Jacobs, I.M. (1965). *Principles of Communication Engineering.* Wiley. ISBN: 0-471-96240-6.

Yates, R.D. and Goodman, D.J. (2014). *Probability and Stochastic Processes*, 3e. Wiley. ISBN: 978-1-118-32456-1.

Internet Websites

www.britannica.com
www.khanacademy.org
https://ocw.mit.edu/index.htm
www.nist.gov
www.randomservices.org
www.statsref.com
www.wikipedia.org
www.wolframalpha.com

Answers

Chapter 1

1.1 0

1.2 (a) 0.8, (b) 0.6, and (c)0.2

1.3 (a) 0.56, (b) 0.06, and (c) 0.94

1.4 $\dfrac{1}{36}$

1.5 0.25

1.6 Event A

1.7 $\dfrac{16}{45}$

1.8 $\dfrac{1}{3}$

1.9 $1 - \left(\dfrac{7}{10}\right)^n$

1.10 (i) $\dfrac{5}{36}$, (ii) $\dfrac{1}{24}$, and (iii) $\dfrac{5}{648}$,

It is more likely to get the sum of 6 with two fair dice.

1.11 $\dfrac{1}{3}$

1.12 0.2

Probability, Random Variables, Statistics, and Random Processes: Fundamentals & Applications,
First Edition. Ali Grami.
© 2020 John Wiley & Sons, Inc. Published 2020 by John Wiley & Sons, Inc.
Companion Website: www.wiley.com/go/grami/PRVSRP

1.13 $\dfrac{1}{3}$

1.14 p

1.16 $\dfrac{1}{26}$

1.17 (a) $\dfrac{8}{225}$ and (b) $\dfrac{4}{91}$

1.18 $x = 6$

1.19 A wins as he is the first to toss.

1.20 $\dfrac{29}{45}$

1.21 No, they are not independent.

1.22 They are not mutually exclusive, but they are statistically independent.

1.23 Events A and B are independent. Events A and C are not independent. Events B and C are not independent.

1.24 Yes, they are statistically independent.

1.25 $\dfrac{7}{51}$

1.26 In case 1, P(both tosses are heads) $= 0.255025$. In case 2, P(both tosses are heads) $= 0.25505$, and the unfair coin is more likely to have been picked.

1.27 $\dfrac{2}{7}$

1.28 (a) $P(R) = 0.3642$ and (b) $P(A|R) = \dfrac{1}{3642}, P(B|R) = \dfrac{16}{3642},$
$P(C|R) = \dfrac{225}{3642}, P(D|R) = \dfrac{900}{3642}, P(E|R) = \dfrac{2500}{3642}$

1.29 $\dfrac{10}{11}$

1.30 62%

Chapter 2

2.1 $\dfrac{mp}{1+(m-1)p}$

2.2 $\dfrac{p}{r+(1-r)p}$

2.3 There are two orders of leaving that are the most likely one: FFFMFMFM or FFMFMFMF. The most likely order of leaving has thus the probability $\dfrac{1}{56}$.

2.4 $x \leq 27$

2.5 $\dfrac{3}{8}$. If there are n doors, the switching can increase the probability of winning by $\dfrac{1}{n(n-2)}$.

2.6 $(1-p^{n_1})(1-p^{n_2})\dots(1-p^{n_n})$

2.7 $1-(1-(1-p)^{m_1})(1-(1-p)^{m_2})\dots(1-(1-p)^{m_m})$

2.8 $k \geq 1375$

2.9 The candidate which is as good as or better than the first candidate is chosen.
$P(A) = \dfrac{11}{24}, P(B) = \dfrac{7}{24}, P(C) = \dfrac{4}{24}, P(D) = \dfrac{2}{24}$

2.10 $\dfrac{1}{362} \cong 0.27\%$. Obviously, this is a very low probability. Nevertheless, it amounts to the incredible loss of precious lives of 27 innocent individuals in that country. Hence, it is an awful miscarriage of justice!

Chapter 3

3.1 156

3.2 $2^{\frac{n}{2}}$

3.3 (a) 66 and (b) 22

3.4 10

3.5 126

3.6 2880

3.7 32

3.8 50 400

3.9 75

3.10 47

3.11 96

3.12 (a) 5040 and (b) 576

3.13 (a) 5005, (b) 715, (c) 1716, (d) 4770, and (e) 235

3.14 (a) 3 628 800 and (b) 2 540 160

3.15 3

3.16 1260

3.17 (a) 224224 and (b) 0.083

3.18 151 200

3.19 $\dfrac{52!}{13!13!13!13!} \cong 5.364 \times 10^{28}$

3.20 120

3.21 80

3.22 24

3.23 120

3.24 132 300

3.25 (a) 5040 and (b) 720

3.26 12

3.27 (a) 207 360 and (b) 8 709 120

3.28 2

3.29 $r = 3, n \geq 3$

3.31 (a) $30! \cong 2.65 \times 10^{32}$ and (b) $18! \times 12! \cong 3.06 \times 10^{24}$

3.32 (a) 1771 and (b) 10 626

3.33 (a) 495 and (b) 360

3.34 $\dfrac{4}{11}$

3.35 $\dfrac{r}{n}$

3.36 $\dfrac{((l + m + q) - (i + j + p))!}{(l - i)!(m - j)!(q - p)!}$

Chapter 4

4.2 $\dfrac{3}{5}$

4.3 6

4.4 (a) $f_X(x) = \begin{cases} 3x^2 & 0 \leq x < 1 \\ 0 & \text{otherwise} \end{cases}$ and (b) $\dfrac{13}{32}$

4.5 0.25

4.6 $E[X] = 1.4,$
$\sigma_X^2 = 0.84$

4.7 $b = 4,$
$P(1 < X \leq 2) \cong 0.009$

4.8 $p = \dfrac{1}{4}$

$$f_X(x) = \begin{cases} 0 & x < 0 \\ 0.25 & 0 \le x < 1 \\ 0.25 & x = 1 \\ 0.5 & 1 < x \le 2 \\ 0 & x > 2 \end{cases},$$

$$F_X(x) = \begin{cases} 0 & x < 0 \\ 0.25x & 0 \le x < 1 \\ 0.5x & 1 \le x \le 2 \\ 1 & x > 2 \end{cases}$$

4.9 $c = 1$,

$$f_X(x) = \begin{cases} 0 & x \le 0 \\ 2x & 0 < x \le 1, \\ 0 & x > 1 \end{cases}$$

$P(0.1 < x \le 0.4) = 0.15$

4.10 Mean $= 1.60$,
Median $= 9 - 4.5\sqrt{2} \cong 1.623$,
Mode $= \sqrt{3} \cong 1.732$

4.11 $a = 2, b = 0$

4.12 $2m = t(1 - p)$

4.13 $P(X \le 20) \cong 0.632$,
$P(X \le 20 | X \le 40) \cong 0.644$

4.14 $p_X(2|B) = p_X(3|B) = p_X(5|B) = p_X(7|B) = p_X(11|B) = \dfrac{1}{5}$

4.15 $f_X(x|2 < X \le 3) = \begin{cases} \dfrac{2x}{5} & 2 < x \le 3 \\ \\ 0, & \text{otherwise} \end{cases}$

$E[X|(2 < X \le 3)] = \dfrac{38}{15}$

4.16 $E[X|X > 0.5] = \dfrac{2}{3}$

4.17 $f_X(x|1 < X \leq 2) = \begin{cases} \dfrac{\frac{2x}{9}}{\frac{1}{3}} = \dfrac{2x}{3} & 1 < x \leq 2 \\ \\ 0 & \text{otherwise} \end{cases}$

$E[X|(1 < X \leq 2)] = \dfrac{14}{9}$

4.18 $P(Y = 2) = P(Y = 12) = \dfrac{1}{36}$,

$P(Y = 3) = P(Y = 11) = \dfrac{2}{36}$,

$P(Y = 4) = P(Y = 10) = \dfrac{3}{36}$,

$P(Y = 5) = P(Y = 9) = \dfrac{4}{36}$,

$P(Y = 6) = P(Y = 8) = \dfrac{5}{36}$,

$P(Y = 7) = \dfrac{6}{36}$,

$E[Y] = 7$,

$P(Y = 5|5 \leq Y \leq 9) = \dfrac{1}{6}$,

$P(Y = 6|5 \leq Y \leq 9) = \dfrac{5}{24}$,

$P(Y = 7|5 \leq Y \leq 9) = \dfrac{1}{4}$,

$P(Y = 8|5 \leq Y \leq 9) = \dfrac{5}{24}$,

$P(Y = 9|5 \leq Y \leq 9) = \dfrac{1}{6}$,

$E[Y|5 \leq Y \leq 9] = 7$

4.19 $y_1 = 2 \rightarrow p_Y(y_1) = 0.3$, $y_2 = \dfrac{2}{3} \rightarrow p_Y(y_2) = 0.3$,

$y_3 = \dfrac{4}{3} \rightarrow p_Y(y_3) = 0.4$

4.20 $f_Y(y) = \begin{cases} 0, & -1 \leq Y < 0 \\ 0.5\delta(Y), & Y = 0 \\ |Y|, & 0 < Y \leq 1 \end{cases}$

4.21 All six different values of Y are all equally likely.
$$P(10 < Y \le 24) = P(2 < X < 5) = \frac{1}{3}$$

4.22 $y = -\dfrac{\ln(1-x)}{\lambda}$ $\qquad 0 < x < 1$

4.23 $f_Y(y) = \dfrac{4}{y^5}$

4.24 $f_X(x) = \dfrac{1}{\pi x^2}(1 - \cos x)$

4.25 $f_Y(y) = \begin{cases} \left(\dfrac{1}{2\sqrt{2\pi y}}\right) \exp\left(-\dfrac{y}{2}\right) & y > 0 \\[2ex] 0.5\delta(y) & y = 0 \\[2ex] 0 & y < 0 \end{cases}$

4.26 $P[|X - 15| \ge 5] \le 0.36$

4.29 $F_X(x) = \begin{cases} 0 & x \le 0 \\[1ex] \displaystyle\int_0^x x\, dx = \frac{1}{2}x^2 & 0 < x \le 1 \\[2ex] \dfrac{1}{2} + \displaystyle\int_1^x (-x + 2)\, dx = -\frac{1}{2}x^2 + 2x - 1 & 1 < x \le 2 \\[2ex] 1 & x > 2 \end{cases}$

4.30 $E[X] = 4.2,$
$\sigma_X^2 = 2.76$

4.31 $P(8 < X < 18) \ge \dfrac{15}{16} \cong 93.7\%$

4.32 Mean $= \dfrac{13}{6}$,
Median $= \sqrt{5}$,
Mode $= 3$

4.33 $E[Y] = 1.1,$

$\sigma_X^2 = \dfrac{0.97}{3}$

$$f_Y\left(y|\frac{4}{5} < Y \le \frac{5}{4}\right) = \begin{cases} 0 & y \le 0.8 \\ \dfrac{0.4}{0.23} & 0.8 < y \le 1 \\ \dfrac{0.6}{0.23} & 1 < y \le 1.25 \\ 0 & y > 1.25 \end{cases}$$

4.34 $P(X \ge 11) \le \dfrac{1}{2},$

$P(X \ge 11) = \dfrac{1}{30.25}$

4.35 $\Psi_X(\omega) = \cos\omega$

4.36 $M_X(v) = e^{\lambda(e^v - 1)}$

Chapter 5

5.1 $E[X] = p, \sigma^2 = p(1-p)$

5.2 $E[X] = np, \sigma_X^2 = np(1-p)$

5.3 $p_X(0) = (1-p)^n, \dfrac{p_X(k)}{p_X(k-1)} = \dfrac{(n-k+1)p}{k(1-p)}$

5.4 $P(X \le 2) \cong 0.919\,71$

5.5 $P(X \le k) = 1 - (1-p)^k$

5.6 $E[X] = \dfrac{1}{p},$

$\sigma_X^2 = \dfrac{1-p}{p^2}$

5.7 $P(X = 100\,00) \cong 0.999\,99 \times 10^{-9}$

5.8 0.0037,
0.634

5.11 $E[X] = \lambda$,
$\sigma_X^2 = \lambda$

5.12 $E[X] = \dfrac{1}{\lambda}$,
$\sigma_X^2 = \dfrac{1}{\lambda^2}$

5.13 $f_Y(y) = \dfrac{1}{\pi\sqrt{1 - y^2}} \quad -1 < y < 1$

5.14 $H = -\log_2\left(\dfrac{p(1-p)^{\frac{1}{p}}}{1 - p}\right)$

5.15 $P\left(\pi \leq x \leq \dfrac{3\pi}{2}\right) = \dfrac{1}{4}$

5.17 $f_Y(y) = \begin{cases} \dfrac{1}{y} & 1 < y < e \\[2ex] 0 & \text{otherwise} \end{cases}$

5.18 $E[X] = \dfrac{a + b}{2}$,
$\sigma_X^2 = \dfrac{(b - a)^2}{12}$

5.19 $r(t) = a\lambda^a t^{a-1}$,
$r(t) = 0.02t$

5.20 $E[X] = \dfrac{a}{a + b}$,
$\sigma_x^2 = \dfrac{ab}{(a + b)^2(a + b + 1)}$

5.21 $E[X] = a^{-\frac{1}{b}}\Gamma\left(1 + \dfrac{1}{b}\right)$,
$\sigma_X^2 = a^{-\frac{2}{b}}\Gamma(1 + 2b^{-1}) - (\Gamma(1 + b^{-1}))^2$

5.22 $E[X] = \dfrac{\alpha}{\lambda}$,

$\sigma_X^2 = \dfrac{\alpha}{\lambda^2}$

5.23 $E[X] = a, \sigma_X^2 = \dfrac{2}{b^2}$

5.24 0.203 24

5.25 $j + 4$, where j is an integer.

5.26 $P(998 \leq X \leq 1000) = \sum_{x=998}^{1000} \binom{1000}{x} (0.001)^x (1 - 0.001)^{1000-x}$

5.27 Bernoulli $\rightarrow M_X(v) = 1 - p + pe^v$,
Binomial $\rightarrow M_X(v) = (1 - p + pe^v)^n$,
Poisson $\rightarrow M_X(v) = e^{\lambda(e^v - 1)}$

5.28 $P(X = 4) \cong 0.075$,
$P(X = 0 \mid X \leq 2) \cong 0.028$

5.29 $E[T] = 0.5$,
$f_T(1) \cong 13.5\%$,
$f_T(0.5) \cong 36.8\%$,
$t \cong 2.3$

Chapter 6

6.1 X and Y are dependent, orthogonal, and uncorrelated.

6.2 $1 - (1 - z)^2$

6.3 $\dfrac{\pi}{4}$

6.4 Yes, X and Y are independent.

6.5 $c = 0.5 \tan^{-1} \left(\dfrac{2\rho_{X,Y}\sigma_X\sigma_Y}{(\sigma_Y^2 - \sigma_X^2)} \right)$

6.6 X and Y are uncorrelated, but completely dependent, as we have $Y = X^4$.

6.8 $F_X(x) = (1 - \exp(-x))u(x)$,
$F_Y(y) = (1 - \exp(-y))u(y)$,
$P(X > x, Y > y) = \exp(-x - y)$

6.9 $f_{X|Y}(x|y) = \frac{x+y}{2(y+1)}$, $\quad 0 < x < 2$, $\quad 0 < y < 2$
$f_{Y|X}(y|x) = \frac{x+y}{2(x+1)}$, $\quad 0 < x < 2$, $\quad 0 < y < 2$
$P(0 < Y \le 0.5 | X = 1) = \frac{5}{32}$

6.10 X and Y are not statistically independent.

6.11 X and Y are statistically independent.

6.12 $f_Z(z) = \begin{cases} 0 & z \le 0 \\ \int_0^z (1)\exp((x-z))dx = 1 - e^{-z} & 0 < z \le 1 \\ \int_0^1 (1)\exp((x-z))dx = (e-1)e^{-z} & z > 1 \end{cases}$

6.13 X and Y are not uncorrelated.

6.14 X and Y are not independent.

6.15 X and Y are uncorrelated, but dependent.

6.16 $f_{Y|X}(y|x) = \begin{cases} \dfrac{1}{-y \ln x} & x \le y \le 1 \\ 0 & \text{otherwise} \end{cases}$

6.17 $P(X > 1 | Y = 2) = \exp(-0.5)$

6.18 X, Y, and Z are independent.

6.20 $f_{X,Y}(x, y) = \begin{cases} 1 & |y| \le x \le 1 \\ 0 & \text{elsewhere} \end{cases}$
$f_Y(y) = 1 - |y|$

Chapter 7

7.1 $\sigma = \sqrt{-\dfrac{x_0{}^2}{2\ln p_0}}$,

For $p_0 = 0.001$ and $x_0 = 2$, we obtain $\sigma \cong 0.538$.

7.2 $E[X] = \sigma\sqrt{\dfrac{\pi}{2}}$,

$\sigma_X^2 = \sigma^2\left(\dfrac{4-\pi}{2}\right)$

7.3 $f_Z(z) = \begin{cases} 0 & z \le 0 \\ 0.5z & 0 < z \le 1 \\ 0.5 & 1 < z \le 2 \\ 0.5(3-z) & 2 < z \le 3 \\ 0 & z > 3 \end{cases}$

7.4 $x = \sigma\sqrt{2\ln 2}$

7.5 $Q(0.4) \cong 0.345$

7.7 $P(0 < X \le 9) = 1 - Q\left(\dfrac{2}{3}\right) - Q\left(\dfrac{5}{3}\right)$

7.8 $K = \dfrac{1}{3\sqrt{2\pi}}$,

$P(X \le 0) = Q\left(\dfrac{4}{3}\right)$

7.9 $P(e) = 0.5\displaystyle\int_{-\infty}^{0} \dfrac{1}{\sqrt{2\pi}} \exp\left(-\dfrac{(x-1)^2}{2}\right) dx +$

$0.5\displaystyle\int_{0}^{\infty} \dfrac{1}{\sqrt{2\pi}} \exp\left(-\dfrac{(x+1)^2}{2}\right) dx = Q(1)$

7.10 $Q(\sqrt{2})$

7.11 $P(X \ge a) \le \exp\left(-\dfrac{(a-E[X])^2}{2\sigma_X^2}\right)$

7.12 $f_Y(y) = \dfrac{1}{y} f_X(\ln x) = \left(\dfrac{1}{y\sqrt{2\pi\sigma_X^2}} \right) \exp\left(-\dfrac{(\ln y - \mu_X)^2}{2\sigma_X^2} \right) \quad 0 < y < \infty$

7.14 $\text{Var}(X|X > 0) = 1 - \dfrac{2}{\pi}$

7.15 $f_Y(y) = \left(\dfrac{1}{y\sqrt{2\pi n}} \right) \exp\left(-\dfrac{(\ln y + n)^2}{2n} \right) \quad y > 0$

7.16 It is a Gaussian random variable whose mean is 0 and variance is 25.

7.18 $f_{R,\Theta}(r, \theta) = \dfrac{r}{2\pi\sigma^2} \exp\left(\dfrac{-r^2}{2\,\sigma^2} \right),$

$f_R(r) = \dfrac{r}{\sigma^2} \exp\left(\dfrac{-r^2}{2\sigma^2} \right),$

$f_\Theta(\theta) = \dfrac{1}{2\pi}$

7.19 $P(X > 1) = 0.159,$
$P(X > 1|X > 0) = 0.318$

7.20 $P(X > 1 \mid Y > 0) = 1 - Q(1) \cong 0.841$

7.21 $E[X|X > 0] = \sqrt{\dfrac{2\sigma_X^2}{\pi}},$

$\text{Variance } (X|X > 0) = \dfrac{\sigma_X^2(\pi - 2)}{\pi}$

7.22 $\text{COV}(Z, W) = 2 \neq 0$, hence they are correlated.

7.23 $E[X|(X < b)] = \dfrac{1}{1 - Q\left(\frac{b - \mu_X}{\sigma_X}\right)} \displaystyle\int_{-\infty}^{b} x f_X(x)dx$

7.24 $P(Y > m) = \displaystyle\int_m^\infty \dfrac{1}{\sqrt{2\pi a^2\sigma_X^2}} \exp\left(-\dfrac{(Y - a\,\mu_X - b)^2}{2a^2\sigma_X^2} \right) dy$

7.25 $n \geq 1746$

7.26 $P(X \leq 0.6) = \Phi(-0.8) \cong 0.212$,
0.51 seconds

7.27 $E[S_N] = 100$,
$\sigma_{S_N}^2 = 10\,100$

7.29 $\dfrac{1}{2}$,

$$1 - Q\left(\frac{2}{\sqrt{3}}\right) - Q\left(\frac{1}{\sqrt{3}}\right)$$

7.30 $P(780 \leq S_{100} \leq 820) \cong 0.682$

7.31 $f_{Y_1 Y_2 Y_3}(y_1, y_2, y_3) = \dfrac{1}{3(2\pi)^{\frac{3}{2}}} \exp\left(0.5\left(\left(\dfrac{y_1 + 2y_2 + y_3}{3}\right)^2 \right.\right.$
$$\left.\left. + \left(\frac{y_1 - y_2 + y_3}{3}\right)^2 + \left(\frac{y_1 - y_2 - 2y_3}{3}\right)^2 \right)\right)$$

Chapter 8

8.1 1, 2, 5, 7, 9, 9, 10, 11, and 27,
Mean = Median = Mode = 9

8.2 $\sigma_6^2 = \dfrac{13}{6}$

8.3 In scheme (i), the mean of the new price is increased to $5500, but the standard deviation remains $1250. In scheme (ii), the mean of the new price is increased by 10% to $5500, and the standard deviation of the new price is also increased by 10% to $1375.

8.4 16th percentile
95

8.5 Sample mean = 7.45,
Sample median = 8.5,
Unbiased sample variance $\cong$ 6.55,
Sample standard deviation $\cong$ 2.56,
Upper and lower sample quartiles are 9.35 and 5.3,
Coefficient of variation $\cong$ 0.34

Chapter 9

9.2 $E[\widehat{\Omega}] = \theta + \pi$

9.3 $\widehat{P}_{ML} = \dfrac{\overline{X}_n}{m}$, where $\overline{X}_n$ is the sample mean.

9.4 $\widehat{A}_{ML} = n\left(\sum_{i=1}^{n} \ln\left(\dfrac{X_i}{k}\right)\right)^{-1}$

9.5 $\widehat{A^2}_{ML} = \dfrac{1}{2n}\sum_{i=1}^{n} X_i^2$

9.6 $\widehat{\Theta} = \max_{1 \le i \le n}(X_i)$

9.7 $\widehat{\Lambda}_B = \dfrac{n+1}{\alpha + n\overline{X}_n}$

9.8 $n = 62$

9.9 $\widehat{Y} = E[Y|X]$

9.10 $\widehat{Y} = \dfrac{X}{2}$

9.11 $k = E[Y]$

9.12 $a = \rho_{X,Y}\dfrac{\sigma_Y}{\sigma_X}, b = \mu_Y - a\,\mu_X$

9.13 Best mean square estimator: $\widehat{Y} = X^2$.
Linear mean square estimator: $a = 0$ and $b = \dfrac{1}{3}$

9.14 $\widehat{Y}_{MAP} = \dfrac{X}{2}$

Chapter 10

10.1 Two-sided test → Acceptance region: $\{37, \dots, 63\}$.
One-sided test → Acceptance region: $\{0, \dots, 62\}$

10.2 p-value $\cong 0.153 \to$ Accept H_0

10.3 p-value $\cong 0.0005 \to$ Reject H_0

10.4 $p - \text{value} \cong 0.000\ 0317 \rightarrow$ Contradicts the claim

10.5 $n \cong 22$

10.6 $n \underset{H_0}{\overset{H_1}{\gtrless}} \dfrac{\ln \eta + (\lambda_1 - \lambda_0)}{\ln \left(\frac{\lambda_1}{\lambda_0} \right)}$

10.7 $n \underset{H_0}{\overset{H_1}{\gtrless}} \dfrac{\sigma^2}{2} \ln \left(\dfrac{1-p}{p} \right) = \gamma$

$\text{BER} = pQ \left(\dfrac{\gamma + 1}{\sigma_X} \right) + (1-p)Q \left(\dfrac{\gamma - 1}{\sigma_X} \right)$

10.8 $\sum_{i=1}^{n}(s_{1i} - s_{0i})r_i \underset{H_0}{\overset{H_1}{\gtrless}} \sum_{i=1}^{n} 0.5(s_{1i}^2 - s_{0i}^2)$

10.9 $\overline{C} = C_{00}P(D_0|H_0)P(H_0) + C_{10}P(D_1|H_0)P(H_0) +$
$C_{01}P(D_0|H_1)P(H_1) + C_{11}P(D_1|H_1)P(H_1) = C_{00}(0.5)P(H_0) +$
$C_{10}(0.5)P(H_0) + C_{01}(0.25)P(H_1) + C_{11}(0.75)P(H_1)$

Chapter 11

11.1 $E[Y(t)] = 0, \sigma_{Y(t)}^2 = A^2$

11.2 $X(t)$ is not wide-sense stationary.

11.4 $f_{X(0)}(x) = \begin{cases} 1 & 0 \leq x \leq 1 \\ 0 & \text{otherwise} \end{cases}$

$f_{X(0.125)}(x) = \begin{cases} \sqrt{2} & 0 \leq x \leq \dfrac{1}{\sqrt{2}} \\ 0 & \text{otherwise} \end{cases}$

$f_{X(0.25)}(x) = \delta(0)$

$f_{X(0.375)}(x) = \begin{cases} \sqrt{2} - \dfrac{1}{\sqrt{2}} \leq x \leq 0 \\ 0 \qquad\qquad \text{otherwise} \end{cases}$

11.5 $f_{X(t)}(x) = \left(\dfrac{1}{\cos t} \right) \exp \left(-\dfrac{x}{\cos t} \right) \qquad x \geq 0$

11.6 $E[A] = 0.5$,
$E[X(t)] = 0.5 \cos(2\pi t)$,

$$R_X(t_1, t_2) = \frac{1}{3} \cos(2\pi t_1) \cos(2\pi t_2),$$

$$C_X(t_1, t_2) = \frac{1}{12} \cos(2\pi t_1) \cos(2\pi t_2),$$

11.7 $X(t)$ is not strict-sense stationary.

11.8 $C_{XY}(t_1, t_2) = 1$.

11.9 It is not even wide-sense stationary, let alone strict-sense stationary.

11.10 $\rho = \dfrac{R_X(\tau)}{R_X(0)}$

11.12 $X(t) = X \exp(t)$ and $Y(t) = Y \exp(-t)$, where X and Y are different random variables.

11.13 No

11.15 It is stationary, but not ergodic.

Chapter 12

12.1 Mean $= 0$,
$S_Y(f) = A^2 \pi^2 f_c^2 (\delta(f + f_c) + \delta(f - f_c))$

12.2 $R_X(\tau) = 0.5 \cos(2\pi\tau) + \delta(\tau)$,
$\tau = \dfrac{1}{4} \pm \dfrac{k}{2}, \quad k = 0, 1, 2, \ldots$

12.3 $S_Y(f) = \dfrac{1}{2}(S_X(f - f_c) + S_X(f + f_c))$

12.4 $E[Y(1)] = 0$,
$\sigma_Y^2(1) = 1$
$$f_{Y(1)}(y) = \frac{1}{\sqrt{2\pi}} \exp\left(-\frac{y^2}{2}\right)$$

12.5 1 W

Index

Probability, Random Variables, Statistics, and Random Processes: Fundamentals & Applications,
First Edition. Ali Grami.
© 2020 John Wiley & Sons, Inc. Published 2020 by John Wiley & Sons, Inc.
Companion Website: www.wiley.com/go/grami/PRVSRP